# Dams, Fish and Fisheries: Opportunities, Challenges and Conflict Resolution

# Dams, Fish and Fisheries: Opportunities, Challenges and Conflict Resolution

Contributors

**Carlos Edwar de Carvalho Freitas, Alexandre A. F. Rivas et al.**

AURIS
Reference

www.aurisreference.com

# Dams, Fish and Fisheries: Opportunities, Challenges and Conflict Resolution

Contributors: Carlos Edwar de Carvalho Freitas, Alexandre A. F. Rivas et al.

**Published by Auris Reference Limited**

**www.aurisreference.com**

United Kingdom

**Copyright 2016**
**Printed in 2017 for Sale in the Indian Subcontinent**

**Dams, Fish and Fisheries: Opportunities, Challenges and Conflict Resolution**

ISBN: 978-1-78154-978-0

British Library Cataloguing in Publication Data
A CIP record for this book is available from the British Library

Printed in the United Kingdom
Exclusively distributed by CBS Publishers & Distributors Pvt. Ltd.

Sales & Distribution Rights only for India, Pakistan, Bangladesh, Sri Lanka, Nepal and Bhutan.This book is not to be sold outside these territories.

# Contents

# List of Abbreviations

| | |
|---|---|
| AEI | Adverse Environmental Impact |
| AFMA | Australian Fisheries Management Authority |
| ANPP | Aquatic net primary production |
| BTA | Best technology available |
| CRP | C-reactive protein |
| CV | Coefficient of variation |
| CVMP | Committee for Medicinal Products for Veterinary Use |
| CWIS | Cooling water intake structures |
| DO | Dissolved oxygen |
| EBFM | Ecosystem based fisheries management |
| EBM | Ecosystem based management |
| EEZ | Exclusive economic zone |
| EMEA | European Medicines Agency |
| EPA | Environmental Protection Agency |
| ER | Ecosystem respiration |
| ESS | Evolutionary stable strategy |
| EU | European Union |
| FARAD | Food Animal Drug Avoidance Databank |
| FAT | Fluorescent antibody assays |
| FDA | Drug Administration |
| FWS | Fish and Wildlife Service |
| GABT | Great Australian Bight Trawl |
| GAM | Generalized additive model |
| GCMS | Global Climate Models |
| GHG | anthropogenic greenhouse gas |
| GHGS | Greenhouse gases |
| GPP | Gross primary productivity |
| HSR | Heat stress response |
| IM | Intramuscular route |
| IP | Intraperitoneal |
| IPCC | International Panel on Climate Change |
| ITQ | Individually transferable quotas |
| ITQS | Individual transferable quotas |
| IUCN | Union for Conservation of Nature |
| MAF | Macrophage activating factor |
| MHC | Major Histocomapatibility Complex |
| MHC | Major histocompatibility complex |
| MMSY | Multispecies maximum sustainable yield |
| MRLS | Maximum residue limits |
| MSE | Management strategy evaluation |
| MSE | strategy evaluation |
| NBT | Nitroblue tetrazolium reaction |

| | |
|---|---|
| NRC | National Research Council |
| PAF | Plateled activating factor |
| PCR | Polymerase chain reaction |
| PD | Pharmacodynamics |
| PK | Pharmacokinetics |
| QMS | Quota management system |
| SESSF | Southern and Eastern Scalefish and Shark Fishery |
| SSBPR | Spawning Stock Biomass per Recruit |
| SST | Surface temperature |
| TAC | Total allowable catch |
| TNF | Tumour-necrosis factor |
| UNFCCC | United Nations Framework Convention on Climate Change |
| UV | Ultra- violet |
| VMPS | Veterinary Medicinal Products |

# List of Contributors

**Carlos Edwar de Carvalho Freitas**
Federal University of Amazonas, Manaus, Amazonas, Brazil
Washington and Lee University, Manaus, Amazonas, Brazil

**Alexandre A. F. Rivas**
Federal University of Amazonas, Manaus, Amazonas, Brazil
Washington and Lee University, Manaus, Amazonas, Brazil

**Caroline Pereira Campos**
National Institute for Amazonian Research, Manaus, Amazonas, Brazil

**Igor Sant'Ana**
Federal University of Amazonas, Manaus, Amazonas, Brazil

**James Randall Kahn**
Federal University of Amazonas, Manaus, Amazonas, Brazil
Washington and Lee University, Manaus, Amazonas, Brazil

**Maria Angélica de Almeida Correa**
Federal University of Amazonas, Manaus, Amazonas, Brazil

**Michel Fabiano Catarino**
National Institute for Amazonian Research, Manaus, Amazonas, Brazil

**Masami Fujiwara**
Department of Wildlife and Fisheries Sciences, Texas A&M University, College Station, Texas, United States of America

**Elizabeth A. Fulton**
Wealth from Oceans Flagship, Commonwealth Scientific and Industrial Research Organisation, Hobart, Tasmania, Australia,

**Anthony D. M. Smith**
Wealth from Oceans Flagship, Commonwealth Scientific and Industrial Research Organisation, Hobart, Tasmania, Australia,

**David C. Smith**
Wealth from Oceans Flagship, Commonwealth Scientific and Industrial Research Organisation, Hobart, Tasmania, Australia

**Penelope Johnson**
Rural Environment and Agriculture Statistics Branch, Australian Bureau of Statistics, Hobart, Tasmania, Australia

**Selim Sekkin**
Adnan Menderes University Turkey

**Cavit Kum**
Adnan Menderes University Turkey

**Gordon W. Holtgrieve**
School of Aquatic and Fishery Sciences, University of Washington, Seattle, Washington, United States of America

**Mauricio E. Arias**
Department of Civil and Natural Resources Engineering, University of Canterbury, Christchurch, New Zealand

**Kim N. Irvine**
Department of Geography/Planning and Center for Southeast Asia Environment and Sustainable Development, Buffalo State, State University of New York, Buffalo, New York, United States of America

**Dirk Lamberts**
Laboratory of Aquatic Ecology, Evolution and Conservation, University of Leuven, Leuven, Belgium

**Eric J. Ward**
Conservation Biology Division, Northwest Fisheries Science Center, National Marine Fisheries Service, National Oceanic and Atmospheric Administration, Seattle, Washington, United States of America

**Matti Kummu**
Water and Development Research Group, Aalto University, Espoo, Finland

**Jorma Koponen**
Environmental Impact Assessment Centre of Finland Ltd, Espoo, Finland

**Juha Sarkkula**
Finnish Environment Institute, Helsinki, Finland

**Jeffrey E. Richey**
School of Oceanography, University of Washington, Seattle, Washington, United States of America

**Bimal P Mohanty**
Central Inland Fisheries Research Institute, Barrackpore, Kolkata 700120; India.

**Sasmita Mohanty**
School of Biotechnology,KIIT University, Bhubaneswar 751024; India.

**Jyanendra K Sahoo**
Orissa University of Agriculture & Technology, College of Fisheries, Berhampur760007; India.

**Anil P Sharma**
Central Inland Fisheries Research Institute, Barrackpore, Kolkata 700120; India.

**David E. Bailey**
Mirant Corporation, 8711 Westphalia Road, Upper Marlboro, MD 20772, USA

**Kristy A.N. Bulleit**
Hunton & Williams, 1900 K Street, N.W., Washington, D.C. 20006-1109, USA

**Tomislav Vladić**
Department of Zoology, Stockholm University, Stockholm, Sweden

**Y. Hamed**
Civil Engineering Department, Faculty of Engineering, Port Said University, Egypt

**Sh. Salem**
Ministry of Water Resources and Irrigation, Egypt

**A. Ali**
Irrigation and Hydraulics Department, Faculty of Engineering, Ain Shams University, Egypt

**A. Sheshtawi**
Civil Engineering Department, Faculty of Engineering, Port Said University, Egypt

**Jeffrey J. Polovina**
Pacific Islands Fisheries Science Center, NOAA Fisheries, Honolulu, Hawaii, United States of America

**Phoebe A. Woodworth-Jefcoats**
Pacific Islands Fisheries Science Center, NOAA Fisheries, Honolulu, Hawaii, United States of America

**Cavit Kum**
University of Adnan Menderes, Turkey

**Selim Sekkin**
University of Adnan Menderes, Turkey

**Hans-Joachim Rätz**
European Commission, Joint Research Centre, Institute for Protection and Security of the Citizen, Ispra (VA), Italy
Johann Heinrich von Thünen-Institut, Federal Research Institute for Rural Areas, Forestry and Fisheries, Institute for Sea Fisheries, Hamburg, Germany

# Preface

The extent to which fisheries can be developed, sustained or protected along riverine ecosystems modified by dams reflects basin topography, geological features, watershed hydrology, and climate, as well as engineering features of the dam itself, and operational programs for retention and release of water from the reservoir , through the dam and into the tailwaters. The text *Dams, Fish and Fisheries: Opportunities, Challenges and Conflict Resolution* determines the impact of dams on river ecosystems and their associated fisheries depends on spatial and temporal scale of interest. The potential impacts of global climatic changes and dams on Amazonian fish and their fisheries have been investigated in first chapter. Second chapter discusses the causes and general concerns of global climate change and deals, specifically, on the impacts of climate change on fisheries and aquaculture, possible mitigation options and development of suitable monitoring tools. Application and effects of antibacterial drugs in fish farms have been focused in third chapter. The objective of fourth chapter is to review the proximate mechanisms of sperm competition and its evolutionary implications in the two sympatrically occurring salmonid species exhibiting alternative male maturation tactics. Fifth chapter focuses on environmental effect of using polluted water in new/old fish farms. Sixth chapter deals with fishery-induced changes in the subtropical pacific pelagic ecosystem size structure. Seventh chapter discusses on immune system drugs in fish. The purpose of eighth chapter is to contribute to the development of § 316(b) regulations that will both protect living aquatic resources and reflect sound social policy. Last chapter deals with demographic diversity and sustainable fisheries.

# Chapter 1

# THE POTENTIAL IMPACTS OF GLOBAL CLIMATIC CHANGES AND DAMS ON AMAZONIAN FISH AND THEIR FISHERIES

Carlos Edwar de Carvalho Freitas[1, 3], Alexandre A. F. Rivas[1, 3], Caroline Pereira Campos[2], Igor Sant'Ana[1], James Randall Kahn[1, 3], Maria Angélica de Almeida Correa[1] and Michel Fabiano Catarino[2]

[1] Federal University of Amazonas, Manaus, Amazonas, Brazil
[2] National Institute for Amazonian Research, Manaus, Amazonas, Brazil
[3] Washington and Lee University, Manaus, Amazonas, Brazil

## INTRODUCTION

The Amazon River Basin, which encompasses the world's largest remaining tropical rainforest, has the highest diversity of fish species of any region in the world [1]. Some of these species represent highly abundant fish stocks that have supported an important fishery for many decades, or many centuries if the history prior to European colonization is included. The importance of fishing in the Amazon River Basin can easily be observed from the high fish consumption, which is mainly attributed to people who live in rural areas near rivers and lakes (Table 1). Regardless of this importance for food, there is no integrative strategy for fishery management, and the activity in this basin as a whole is highly vulnerable to externalities, including those resulting from environmental changes and man-made interventions.

**Table 1:** Fish consumption in the Amazon River Basin.

| Reference | Sub-basin | Social group | g/per capita.day | Kg/per capita. year |
|---|---|---|---|---|
| [2] | Rio Negro | urban | 53.95 | 19.69 |
| [3] | Rio Negro | urban | 121.70 | 44.42 |
| [4] | Rio Amazonas | rural | 369.00 | 135.00 |

| [5] | Rio Solimões | rural | 510.00 to 600.00 | 186.00 to 219.00 |
| [6] | Rio Solimões and Rio Japurá | rural | 509.00 to 805.00 | 186.00 to 294.00 |
| [7] | Rio Madeira | rural | 243.00 | 88.00 |
| [8] | Rio Amazonas | rural | 511.00 to 643.00 | 187.00 to 235.00 |

There is a consensus that fishery production is directly related to biological productivity, which is a function of a set of environmental characteristics in the aquatic system. The majority of the Rio Amazonas and its tributaries are accompanied by large floodplains, which is where most of the biological production occurs. A key factor for biological production in the floodplains of large Amazonian rivers is the flood pulse [9], which generates tremendous variation in the input of nutrients over the course of the year, primarily at river headwaters located in the pre-Andean areas, such as Madeira, Purus, Juruá, and Solimões. The flood pulse is the driving element that structures the landscape of the floodplains adjacent to the river channels, forming a mobile ecotone that is referred to as ATTZ or the aquatic-terrestrial transition zone [9].

The Amazonian hydrological cycle is annual and quite predictable. The flood intensity and timing is controlled by several factors, including those that act on a global scale. The cyclical phenomenon of warming in the Pacific Ocean near the cost of Peru, termed *El Niño*, is related to severe drought in the Amazon Basin. Alternatively, *La Niña* is associated with strong floods. The simultaneous occurrence of other climatic phenomena, such as the warming of the Tropical North Atlantic Ocean, has been used to explain extreme climatic events [10].

Despite the lack of models describing the relationship between flood intensity and timing and fishing success, the life strategy of many species of Amazonian fish is synchronized with the hydrological cycle. For example, several species of Characiforms, including *Colossoma macropomum*, *Brycon amazonicus*, *Prochilodus nigricans*, *Semaprochilodus insignis*, *S. taenirus*, *Piaractus brachypomum* and others, begin their reproductive migration at the beginning of the rainy season when the waters begin flooding [11, 12, 13]. This life strategy was most likely developed to ensure the colonization of the floodplain with newly hatched larvae. The availability of food and places of refuge in the colonized floodplain may determine the strength of the annual recruitment of these species.

There is ample evidence that the Earth's climate is changing more rapidly now than it has in the past [14], with potential effects on the Amazon basin [15, 16]. These effects include physical alterations and changes in nutrient flow [14]. Although uncertainties associated with how local climates change in response to global climate change exist, global circulation models employed by the Intergovernmental Panel on Climate Change (IPCC) have found an increased likelihood of a significant increase in the mean global temperature [14]. A rise in sea level is also predicted, with estimates varying between 0.75 m to 1.90 meters by the end of twenty-first century [17]. Other environmental changes in freshwater systems, such as stratification, productivity reduction and acidification, have no consistent patterns. These changes are very difficult to generalize based on the available evidence; however, there is a consensus that the impact on fish will be species-specific, that is directly related to the biological characteristics of each species.

Nevertheless, the effects on fisheries should be a result of a series of effects that start at the organism level. At an individual level, all fish have an optimal thermal interval that is limited on the upper and lower boundaries by their critical thermal maxima and minima, respectively [18], thus reducing the analysis of the warming effects. Therefore, fish exposed to temperatures within the sub-lethal interval, excluding the optimal thermal interval, may be affected by warming, and the consequences of this temperature effect should be evident by physiological responses. The high energetic cost necessary to compensate for these unfavorable environmental conditions may affect the growth rates or reproduction success of the fish (Figure 1). Realistically, general effects from water warming can be expected. Because biochemical reaction rates are a function of body temperature, all aspects of an individual fish's physiology, including growth, reproduction and activity, are directly influenced by changes in temperature [19].

When the environment changes, these temperature effects should continue to increase and may be perceptible at the population and community levels. Although different species are affected by environmental change in different ways, the abundance patterns of the entire community, and thus the fishing production, should be influenced (Figure 1).

Dams cause local changes at the sub-basin level but also have the potential to exhibit regional effects. In essence, the impact of dams on the hydrological river cycle, which primarily involves flood timing, is the most important effect of change in freshwater fisheries because the flood regime is the most important determining force in Neotropical rivers [20]. Dam construction can affect environments and fisheries by changing the timing and quantity of river flows; altering the water temperature, nutrient and sediment transport; reducing adjacent floodplains and other wetlands; and blocking fish migrations.

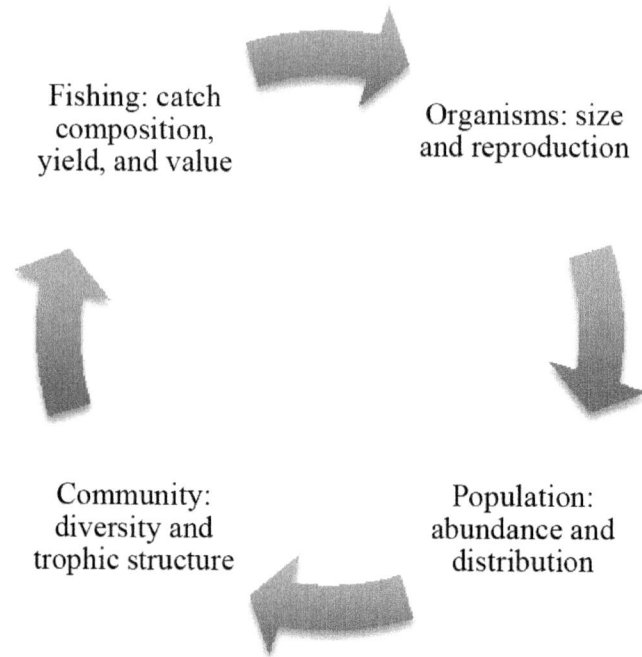

Fishing: catch composition, yield, and value

Organisms: size and reproduction

Community: diversity and trophic structure

Population: abundance and distribution

**Figure 1:** The effects of global climatic changes on different levels.

Currently, there is a large proliferation of hydroelectric dams within the Amazon region. At the western boundary near the Andean and Pre-Andean areas, there are plans for 151 new hydroelectric dams with greater than 2 MW of power over the next 20 years, which is more than a 300% increase [21]. Similarly, in the Brazilian region near the south and southeast boundaries of the Amazon, there are several hydroelectric dam projects that have the potential to completely fragment the river basins with headwaters on the Brazilian Plateau. Similar to climate change, the impact of dams should be associated with the life strategies of different fish species.

The most important species captured by small-scale fisheries in the Amazon basin belong to three groups: Characiforms, which are primarily from the Prochilodontidae, Characidae, and Serrasalmidae families; Siluriforms, which are primarily from the Pimelodidae family and include piramutaba (*Brachyplatystoma vailantii*), dourada (*B. rouseauxii*) and piraíba (*B. filamentosum* and *B. capapretum*); and Perciforms, which are primarily from the genus *Cichla*. Over evolutionary time, members of these groups have developed specific life strategies designed to optimize the survival of Amazonian environmental conditions. Alterations induced by global changes or man-made interventions may directly influence these strategies, with negative effects on both the recruitment and stock abundance of these species, as well as on the socio-economic conditions of the Amazonian people that exploit these fish stocks for food and income.

Therefore, the goals of this chapter are as follows:

- Review the main scenarios for environmental alterations in the Amazon Basin, which is predicted to be a function of global climatic changes and dams.

- Identify the potential impacts of different scenarios of environmental alterations in the Amazon Basin on Amazonian freshwater fish populations.

- Identify the consequences of the predicted impacts on the Amazonian freshwater fish populations, taking into account the main characteristics of the population dynamics.

- Illustrate the potential social and economic consequences for the local and regional fisheries and the people who depend on these fisheries.

## GLOBAL CLIMATIC CHANGES AND DAMS: WHAT WE CAN EXPECT AND WHAT ARE THE POTENTIAL EFFECTS

Despite the uncertainties at the local level, it is highly likely that there will be more frequent occurrences of extreme climatic events. Most of the global climate models (GCMs) proposed by the IPCC [14] project significant Amazonian drying during the 21st century. Pacific sea surface temperature (SST) variation, which is dominated by the El Niño–Southern Oscillation (ENSO), is the main driving force for wet-season rainfall. However, dry-season rainfall is strongly influenced by the Tropical Atlantic north-south SST gradient. Therefore, an intensification of this gradient from the warming of northern SSTs relative to those of the south would move the Inter-Tropical Convergence Zone north and strengthen the Hadley Cell circulation. This change would enhance the duration and intensity of the dry season in much of southern and eastern Amazonia, which already has occurred in 2005. Studies indicated that the most extreme droughts in Amazonia were a result of the strong events of the El Niño-Southern Oscillation (ENSO), the large temperature increase of the sea surface in the Tropical North Atlantic or a combination of these events [22]. Changes in precipitation during the dry season are likely the most critical determinant of the climatic fate of the Amazon [14].

These extreme droughts, even if short in duration, can be catastrophic for aquatic organisms because of the strong reduction in the area of the aquatic environments. Floodplain lakes are the most impacted, and the areas of these lakes can be reduced by several orders of magnitude (Figure 2). Although several species of fish are able to relocate to the river channel during the dry season, some lake resident species remain in the lakes and are unable to survive if the drought is severe. Some studies have observed that fish assemblages seemed to recover rapidly from normal drought seasons [23], but there are indications that extreme droughts occasionally alter fish assemblages. Some species that are vulnerable to these catastrophic events may disappear at a local level [23].

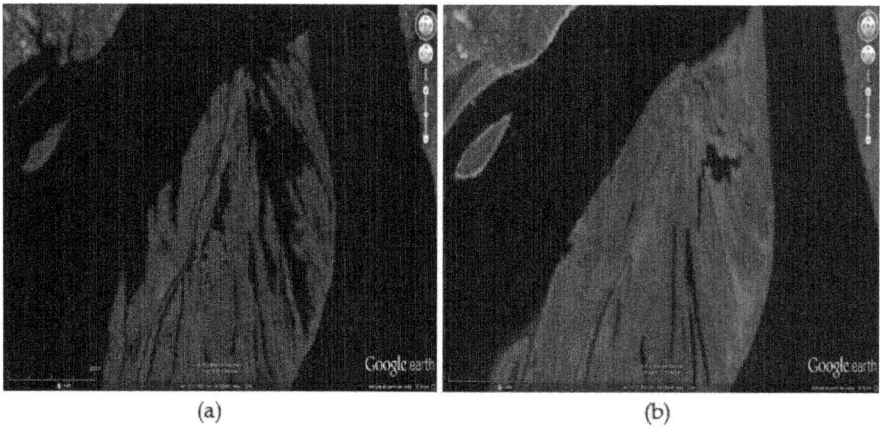

(a)         (b)

**FIGURE 2:** A floodplain area of the Rio Amazonas during the flood season (A) and dry season (B) when an extreme drought occurred.

Another likely climatic change is global warming [14]. Over the next two decades, a warming of approximately 0.2°C per decade is projected for a range of SRES emission scenarios. Even if the concentrations of all greenhouse gases and aerosols had been maintained at the levels present in the year 2000, a further warming of approximately 0.1°C per decade would be expected [14]. This warming represents an increase of 1-7°C in the mean global temperature within the next one hundred years.

Freshwater fish may explore habitats within an optimal thermal interval and thermo-regulate behaviorally and physiologically. Temperature tolerance ranges are species-specific and range from stenothermal species that support only a narrow thermal range to eurythermal species that are able to live in a wide thermal range. Fish populations subjected to changing thermal regimes may increase or decrease in abundance, experience range expansions or contractions or face extinction [19]. There is also an inverse relationship between the temperature and concentration of dissolved oxygen in water. Thus, an increase in the temperature can exacerbate the hypoxia or anoxia conditions naturally observed in some lentic habitats of freshwater fish.

A direct consequence of global warming is a rise in sea level. Despite the uncertainties related to the dynamics of ice sheets and glaciers, there are models that predict a rise in sea level, which were summarized previously [14]. One model proposed a relationship between global sea level variations and the global mean temperature and predicts a rise in sea level ranging from 75 to 190 cm for the period 1990-2100 [17]. However, some scenarios [14] predict a rise of 4.0 meters (Table 2).

What are the potential effects of a rise in sea level for the Amazon Basin? With regard to its physical characteristics, we can anticipate that the sea will be a hydraulic barrier and will flood areas that are not currently flooded but which are primarily within the floodplain adjacent to the river channel. Other environmental consequences of this barrier can also be expected: a reduction in water flow, an increase in the sedimentation rate and an increase of the flooded area. It is possible that the hydrological cycle will also be affected.

These changes to the hydrological cycle can be magnified by the fragmentation of the environment that will occur as a result of the introduction of hydroelectric dams. We can identify at least four phenomena associated with the introduction of dams:

- Blockage of the sediment flow in whitewater systems (e.g., Rio Madeira).
- Change of the flood pulse.
- Blockage of fish migration.
- Reduction in oxygen levels both above and below the dams.

Blockage of the sediments might have a large effect on fish communities. Whitewater rivers originate in Pre-Andean areas and are heavily loaded with volcanic soil sediment. The dams act as a barrier and result in a reduction of water speed, thus improving the rate of decantation. The end result is an impoverishment of the river below the dam. Thus, the species that have evolved in the presence of high levels of nutrients will not be able to adapt to the rapid loss in primary productivity that is associated with the reduction of nutrient content. This result will favor a change in the composition of local species and will have serious impacts on fishing activity.

**TABLE 2:** Projected global average surface warming and associated sea level rise at the end of the 21$^{st}$century. Source: [14]

|  | Temperature change ((C at 2090-2099 relative to 1980-1999) | | Sea level rise (m at 2090-2099 relative to 1980-1999) |
|---|---|---|---|
|  | Best estimate | Likely range | Model-based range excluding future rapid dynamic changes in ice flow |
| Constant Year 2000 concentration | 0.6 | 0.3 - 0.9 | NA |
| B1 scenario | 1.8 | 1.1 - 2.9 | 0.18 - 0.38 |
| A1T scenario | 2.4 | 1.4 - 3.8 | 0.20 - 0.45 |
| B2 scenario | 2.4 | 1.4 - 3.8 | 0.20 - 0.43 |
| A1B scenario | 2.8 | 1.7 - 4.4 | 0.21 - 0.48 |
| A2 scenario | 3.4 | 2.0 - 5.4 | 0.23 - 0.51 |
| A1F1 scenario | 4.0 | 2.4 - 6.4 | 0.26 - 0.59 |

Similarly, Amazonian fish species evolved in a system regulated by an annual and predictable flood pulse, developing life strategies to explore the several habitats available during the hydrological cycle. The elimination or change in the timing or duration of this pulse can destroy signals that trigger reproduction and other life cycle events, which will potentially influence fish recruitment.

The blockage of the fish migration can be critical, with significant impacts for some species that participate in long-distance migrations from the estuary to the headwaters of whitewater rivers to spawn. As a result, some populations may be locally extinct.

The fourth phenomenon concerning the fall in oxygen levels is relatively self-explanatory. The large amount of organic material in the reservoirs will remove a great deal of oxygen from a system that is already low in oxygen content due to the water temperature. The synergy between this phenomenon and global warming is quite evident. The results may include the loss of species with less tolerance for low oxygen conditions.

Clearly, these phenomena can be completely integrated and synergistic, and their effects on fish communities can be magnified and strongly disruptive. As is most often the case with multiple sources of environmental stress, the combined stress resulting from several sources is greater than the sum of the individual stresses. This point is emphasized by [21], who stated that the impact of hydroelectric dams in the Amazon Basin should be considered in a broad perspective, including the planned projects of other Amazonian countries, such as Bolivia, Colombia, Ecuador and Peru. The fragmentation of Amazonian rivers originating in Pre-Andean areas may result in severe nutrient depletion of the rivers because the mountains and associated uplands are the main source of sediments that form the basis for the high primary productivity observed in the Amazonian floodplains.

An unavoidable effect of dams is the shift in species composition and abundance. This shift includes the extreme proliferation of some populations and a reduction, or even elimination, of others [25]. The obstacles in the migratory routes, the loss of natural nursery areas placed upstream of dams and the modification of the hydrological regime downstream of dams, in addition to the rheophilic behavior of the community, are factors directly linked to failures in recruitment and the limited distribution of adults in reservoirs [25], which strongly affect fisheries [26].

## LIFE STRATEGIES OF FRESHWATER AMAZONIAN FISH

In the Amazon Basin, the life strategies associated with migratory and reproductive processes can be employed to distinguish three fish groups. First, groups can be distinguished by their migration length. The fish species that participate in long-distance migrations are from the family Pimelodidae and belong to a unique genus: *Brachyplatystoma rousseauxii, B. vailantii, B. filamentosum, B. capapretum* and *B. platynemum*[13]. These species migrate up to 3,000 km to complete their life cycle. They migrate from the Amazonian estuary to the border of the Andean mountains in Bolivia, Colombia and Peru [27]. The estuary is the nursery area, and the fish remain there approximately one year prior to beginning their migration. The floodplain areas of the Central Amazon Basin are feeding habitats where the immature fish grow up and store fats prior to their reproductive migration toward the Pre-Andean areas [13, 27,28, 29]. This process is synchronized with the hydrological cycle. The gonads of *B. rousseauxii* are in an advanced stage of development starting at the beginning of the flood season, while *B. flamentosum,B. platynemum* and *B. vailantii* show the highest reproductive activity at the end of the flood season [28].

The short-distance migratory species belong to several groups, including Siluriforms such as*Pinirampus pirinampu, Calophysus macropterus, Hypophthalmus marginatus, H. edentates, H. fimbriatus, Phractocephalus hemiliopterus, Pseudoplatystoma punticfer, P. fasciatum* and *P. tigrinum*[13, 30, 31]. These species are also called floodplain migratory fish because they participate in short-distance migrations between the main stem of the Amazon

River and its tributaries and floodplain lakes. Despite the absence of published studies, evidence from field research indicates that the migrations of this group do not appear to exhibit a pattern associated with reproductive events. Because these fish are predator species, these short-distance migrations have trophic causes and are developed to find prey in general small and medium size characins.

Another short-distance migratory species is a highly diverse group of Characiformes, which are extensively exploited by the small-scale fishing fleet from the Amazon Basin. Species such as *Colossoma macropomum*, *Prochilodus nigricans*, *Semaprochilodus insignis*, *S. taenirus* and several Myleinae and Curimatidae evolved for a life strategy strongly associated with the hydrological cycle of the Amazon Basin [13, 32, 33]. These species build large schools at the beginning of the rainy season and participate short-distance migrations from their feeding habitat, which is generally within black water tributaries, to white water rivers where spawning occurs [13, 32, 33, 34, 35]. The parental schools are very large, containing hundreds of thousands of individuals, and there is no parental care after spawning [13, 36]. Adult fish move toward flooded areas, which are rich in food, aiming to store energy for a new reproductive cycle. There are some differences between the timing of migration for these species; however, the schematic in Figure 3 shows the synchronism with the rainy season when the water starts to rise and the importance of the newly inundated floodplains as a place of refuge and feeding for the young fish [37].

Lastly, there are species that do not need to migrate to complete their life cycle. These fish are a diversified group with species from several orders; however, some cichlids from the genus *Cichla* are highly important for regional fisheries. These cichlids are called peacock bass and are the main target of recreational fisheries that are located primarily in black water rivers. The peacock bass is also a to predator that moves in several environments for trophic reasons [37]. Ornamental fish compose another group of non-migratory species that are exploited by fishing. This group is highly diverse, with species belonging to several orders, including Characiforms, Siluriforms, Perciforms, Osteoglossiforms and Gymnotiforms. In general, these are small sized fish with high levels of endemism.

## POTENTIAL EFFECTS OF GLOBAL CLIMATIC CHANGES AND DAMS ON FRESHWATER AMAZONIAN FISH

The intensity and direction (positive or negative) of the potential effects of environmental changes will vary among populations and species in the

Amazonian fish fauna. Some global scenarios are catastrophic [38], proposing that 75% of global freshwater fish will become extinct before the end of the 21st century due to a reduction in river discharge. Nevertheless, the possible effect is local extinction, which would be a critical event for endemic species. Two species of the small fish*Paracheirodon*, which are exploited as ornamental species, exist in the middle to upper Rio Negro in Brazil and in the upper Rio Orinoco in Colombia and Venezuela. A study conducted at an inter-fluvial palm camp of the Middle Rio Negro found that these two species are rarely observed in the same habitat. The *P. simulans* habitat water temperature ranged from a low of 24.6 to a high of 35.2 °C, while the *P. axelroldi* habitat temperature varied between 25.1 and 29.9 °C [39]. The authors propose that because inter-fluvial areas flood as a function of rainfall, a decrease in regional precipitation could alter the hydrologic balance of these wetlands, especially during dry periods, which would lower water levels and increase the water temperature. This scenario would be extremely adverse for *P. simulans*, which exists only in very shallow inter-fluvial areas. A decrease in precipitation could dry out these areas completely, ultimately leading to the local extinction of this species.

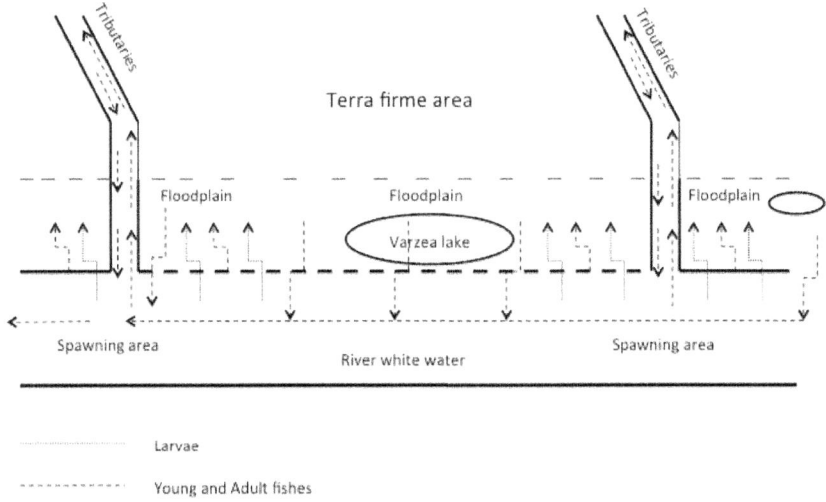

**Figure 3:** A general description of the Characiforms migrations.

At the beginning of rising waters season, the adults move down river from tributaries of black and clear waters to spawning in the turbid and rich environment of white water rivers. After breeding event, these fish move toward the flooded forest for feeding. The larvae are carried by drift toward flooded areas of the floodplain. After six months, when the water starts to recede, large schools of adults and young fish move toward tributaries.

The survivorship or abundance of fish species in a dynamic environment is dependent on three factors: the intensity of change, the velocity at which change will occur and the ability of organisms to adapt in the midst of these changes. Plasticity is a characteristic inherent to each species of fish. However, some common characteristics are useful for classifying the fish into groups and for discussing the most probable effects of environmental changes. For example, the impact of changes in the water temperature should be related to the lethal, sub-lethal and optimal thermal limits of each species. Despite of a scarcity of data on the physiology, life history and behavior of the Amazonian species, some information about population dynamics is available and can be used to hypothesize the effects of the environmental changes resulting from climate change and new dams.

The inverse relationship between temperature and dissolved oxygen in the water may result in an expansion of hypoxia zones. Although the Amazonian fish exhibit a variety of strategies related to oxygen intake in response to hypoxia [40], there are limits to these adaptive strategies that are a result of a long evolutionary time.

Another example of the types of analyses that can be conducted involves the use of the match-mismatch hypothesis (originally proposed by Cushing [41, 42] to describe the relationship between starvation and recruitment), which has clear connections to climate variability. This hypothesis recognizes that early-stage fish need food to survive and grow. It also recognizes that periods of strong food production in the ocean can be variable and are often controlled by climate, which depends on the strength of the wind, the frequency of storms and the amount of heating or fresh water supplied to the surface layers in the ocean. The hypothesis examines the timing match or mismatch between when and where food is available and when and where early-stage fish are able to encounter and consume this food. Assuming that there is a synchrony between the flood pulse and the spawning season of the Amazonian Characiformes, this cycle can be modified by climatic changes with substantial consequences on species that are the most important sources of protein consumption in the Amazon Basin.

In fact, some species or groups of species may be positively affected. A study of the reproduction of fish from the Rio Cuiabá in Upper Pantanal indicated that both the reproductive dynamics and the hydrological regime were closely related. The authors of the study showed that the intense events of floods were positively related with gonadal development of species that participate in long-distance migration and parental care [43].

A potentially useful approach to develop strategies to study the effects of global climate change could be classifying their effects by the type of relationship between the phenomenon and the impact. [44] classified the range of effects that climate change will have on freshwater, estuarine and marine fish into primary, secondary and tertiary categories. These authors found that the primary impacts are climate-related changes that directly affect the behavior, physiology, fitness and survivorship of fish without intermediary causal drivers. Secondary effects are primarily related to changes in the quality or quantity of habitats. Lastly, the tertiary impacts are related to the interactions between several causal factors.

In contrast, impacts from dams are generally related to the fragmentation of the area in which the individual members of a species live, creating obstacles for migration. The decline in the abundance of long-distance migratory species is the most distinct consequence of such filters. The community of these species undergoes seasonal migrations toward spawning habitats located upstream and consequently requires free-flowing stretches of river. Therefore, recruitment success depends on the presence of and accessibility to spawning areas, which are located in the upstream stretches of the main channel and its tributaries, as well as nursery habitats, which are located in flooded areas downstream [25, 45]. The loss of nursery habitat can critically impact several species, including several species of Characins and Perciforms. Dams also alter water temperature and quality [26, 46], which affects the community structure as a whole.

In addition, there are predictable changes in the species composition of the fish assemblages above the dam in the altered environment of the reservoir, and pre-adapted species could become abundant in this location. However, an impoverishment in the fish diversity as a whole would be expected [25, 26]. A study developed to analyze the alterations of the fish communities due to pollution and the damming of highly impacted rivers from Southeast Brazil, which were fragmented and polluted in their upper stretches, and also detected a synergic effect due to these two impact sources [47]. These authors observed a noticeable decrease in species richness in the polluted stretches of the river, with one or two species dominating. However, the artificial control of floods and discharge levels should have direct impacts on recruitment success. An analysis on the influence of the mean annual water level (m), the amplitude (maximum water level of the river in a given year; m) and the flood duration (number of days above 3.5 m; yearly total and for each season; summer and autumn were considered together) on the recruitment of *Prochilodus scrofa* for the fishery conducted at the Itaipu Reservoir and observed that flood duration is more important than flooding amplitude [48].

Table 3 summarizes the effects of global warming, sea level rise and dams on freshwater Amazonian fish, taking into account our level of knowledge. Fish faced with a changing environment must adapt, migrate or perish [19]. In addition to the high level of uncertainty at the species level, some evidence is available to predict that the resulting stress of a temperature increase will affect fauna as a whole, including fish. The effects of the higher energy demand to compensate the stress would start at the physiological level and would include size reduction and reproductive failure. This evolution affects the community structure when the dominant species has more adaptive capacity. Therefore, another possible effect of climate change is the loss of biodiversity through the extinction of specialized or endemic fish species [48]. This pattern of environmental change inducing effects will initiate from a rise in sea level and the introduction of dams.

## SOCIAL AND ECONOMIC EFFECTS OF CLIMATE CHANGES AND DAMS ON AMAZONIAN FISHERIES

Fisheries are very important activity worldwide. Gross revenues from marine capture fisheries worldwide are estimated between US$ 80 billion and 85 billion annualy [49]. However, some authors stated that the global marine fisheries are underperforming economically due to overfishing, pollution and habitat degradation [50]. As is the case in many regions of the world, fish are a key source of animal protein, essential amino acids and minerals, mainly for low-income population who live in the Amazon basin [3, 5, 6]. A recent paper examines if marine fisheries and aquaculture can supply fish demand for a growing human population, taking into account climate change [51]. The authors claim that an effective management of fisheries is necessary to assure sustainability for world fish stocks. The authors also called for a reduction in the amount of wild fish employed to produce animal feed.

In general, Brazilian fish production followed the world tendency, with mean rates of growth of 2.48% and 10.82%, for fishing and aquaculture, respectively [52]. Analyzing just Amazonas State, the main producer of fish exclusively from freshwater, we can see that the state follows the same trend of the region, for the last ten years. On average, the Amazonas State contributed 29% of the region's fisheries production.

A closer analysis of the data shed some light on the impacts of environmental changes, as a result of climate changes or dams, on fisheries and its consequences for well being. Figure 5 shows the Amazonas state Gross Domestic Product (GDP) per capita and an index of fish production growth, for the period between 1992 and 2010, taking 1992 as base year.

**Table 3:** The potential impacts of global climate change and dams and their effects on freshwater Amazonian fish

| Impacts | Effects |
|---|---|
| Global Climate Change<br>– Global warming | Siluriformes–Pimelodidae: long-distance migrations<br>– Alterations in physiological functions to survive in environmental conditions out of an optimal specific interval;<br>– Medium size reduction;<br>– Size of adult stock reduced due to reductions in prey abundance.<br>Characiformes– Short-distance migrations<br>– Alterations in physiological functions to survive in environmental conditions out of an optimal specific interval;<br>– Failure in recruitment due to the mismatch between young fish and food;<br>Perciformes–Cichlidae<br>– Alterations in physiological functions to survive in environmental conditions out of an optimal specific interval;<br>– Failure in recruitment due to impacts on reproductive functions.<br>– Failure in recruitment due to loss of habitat; |
| – Sea level rise | Siluriformes–Pimelodidae: long-distance migrations<br>– Strong recruitment due to expansion of nursery area at estuary.<br>Characiformes– Short-distance migrations<br>– Reduction of the abundance of less-adapted species for the altered environment;<br>– Changes in the community structure as a response to the alterations in the environment.<br>Perciformes–Cichlidae<br>– Alterations in physiological functions to survive in environmental conditions out of an optimal specific interval;<br>– Failure in recruitment due to impacts on reproductive functions. |
| Dams | Siluriformes–Pimelodidae: long-distance migrations<br>– Stock abundance reduced due to reductions in the livable area;<br>– Blockage of fish migrations, which create obstacles for freshwater species to complete their life cycles.<br>Characiformes– Short-distance migrations<br>– Change in species abundance because of alterations in the habitat;<br>– Loss of important habitats for young fish (e.g., areas that are seasonally flooded);<br>Perciformes–Cichlidae<br>– Failure in recruitment due to the loss of spawning habitat. |

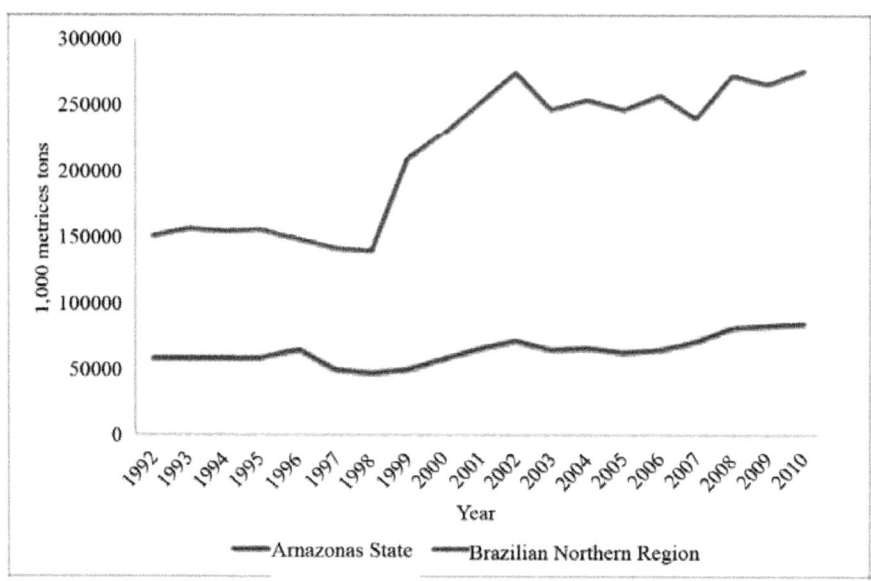

**Figure 4:** Amazonas State and Brazilian Northern region fisheries production between 1992-2010. (Source: 53, 54).

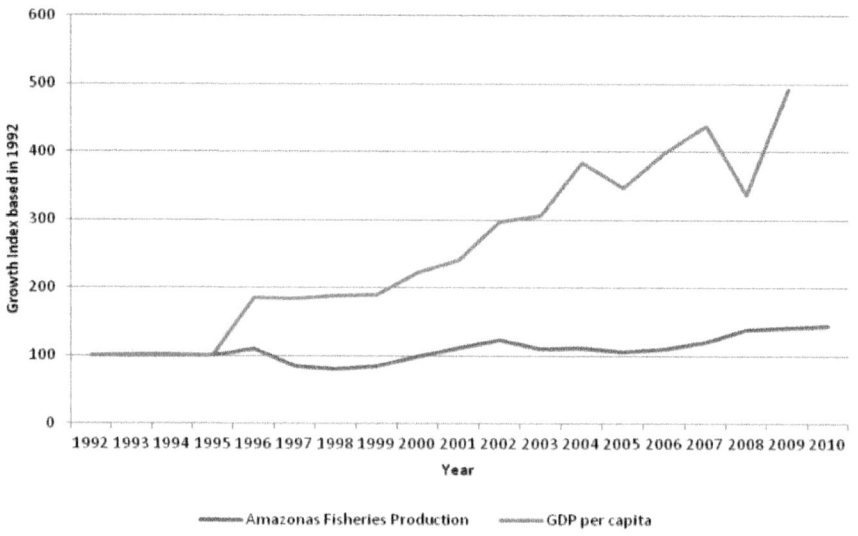

**Figure 5:** Amazonas State GDP per capita and an index of fish production growth [53, 54].

Taking 1992 as the base year both for a fish production index and GDP, it is possible to see that starting in 1995, GDP per capita grew continuously with fluctuations around the trend due to business cycles. At the same time, the index of fish production also grew but with much less intensity. It is clear from the graph that while GDP per capita presented a strong growth trend, increments in fish production were very slow.

What relationships do these trends have with global warming? It is clear that fish production does not drive GDP growth, so global climate change is not likely to have much impact on the GDP of Amazonas through the impact of global change on fisheries. This is not to imply that there would not be critically important impacts through other sectors of the economy. Of course, impacts on fisheries would have a large impact on the income of those participating in the commercial fishing industry, and it could have a significant impact on the GDP of small cities in Amazonia (5 to 10 thousand inhabitants) that do not have alternative sources of income. This is particular true for small cities that are sufficiently close to Manaus to sell their catch in this large urban market.

It should be noted that GDP is a measure of market output, not economic benefits and there are a number of ways in which fisheries impact social welfare, both within the urban/industrial area and within the rural communities. For example, the output of subsistence fisheries is explicitly excluded from the measure of GDP because the output is not traded in formal markets. Clearly, small communities will suffer immensely if fisheries are highly impacted by climate change. Moreover, Amazonas has a fish culture, as opposed to the beef culture of the rest of Brazil. As Table I indicates, fish consumption is extraordinarily high, even in urban centers. If fish become scarcer and more expensive, the welfare of the urban centers will be diminished as they are forced to substitute meat and poultry for their traditional fish dishes.

The social welfare impacts of global climate change induced impacts on fisheries are difficult to calculate. The reason for this is that the direct impacts on fisheries may change economic behavior, which could then lead to a series of indirect impacts that could compound the impacts of climate change. These reactions could occur between urban and rural communities, within the fishery sector, or within the subsistence communities.

If the impacts of global climate change on fisheries reduce the quality of life of small communities, it could spur additional migration from the small communities to the urban centers. This would increase the urban externalities associated with population increases as a whole, and those associated with immigration of a group of people without training to participate in the service or industrial sectors of the urban center. Moreover, the introduction of more people with a high preference for fish consumption into the urban center will increase the urban demand for fish, putting more pressure on the fisheries near the urban centers, which are already stressed and showing evidence of decline.

People that remain in small communities will continue to be negatively impacted by the decreased fish populations because of the impacts of global climate change. This could lead to several negative impacts. First, they may react to the change by fishing more intensively to try to compensate for the decline in populations. This will further stress the populations that are already stressed by global climate change. Second, they may switch species, trying to capture species that previously were not high priority, but remain more abundant. In general, these species will be smaller, and the trophic cascading associated with the decline in forage fish is difficult to predict. Third, they may turn to more hunting to supply their protein needs, leading to other negative impacts on the ecosystem. In particular, the hunting of caiman could lead to impacts on biodiversity as the controlling predators are eliminated from the ecosystem. This would be in addition to the impacts of reductions in intermediate level aquatic predators such as peacock bass (*Cichla* spp.) which would suffer from the negative impacts of global climate change.

Worse impacts could potentially occur if the rural populations increased their participation in other extractive activities, including agriculture, timbering and non-timber forest products. Although the collection of non-timber forest products, such as fruit and fibers is likely to be have a relatively benign direct impact, areas of the forest that were previously not the subject of economic activity could become the subject of economic activity. The heavy presence of people in these previously unharvested areas could lead to impacts on fisheries and wildlife, interfering with the ability of these areas to serve as a reserve for repopulating depleted areas.

Increased participation in timbering and agriculture will lead to deforestation, which has a negative impact on the biodiversity of the forest. Moreover, it will have a negative impact on the aquatic systems. If communities are successful in developing markets for these extractive products, it could lead to a reverse migration of people from urban areas back to the forest, leading to an increasing cycle of degradation.

Both the direct effects of global climate change and the indirect effects associated with the reaction to global climate change will have negative impacts on the social welfare of both urban and rural populations. It is likely that a feedback cycle could develop where the reaction to degradation is more degradation, dramatically reducing both social welfare and ecosystem function.

## CONCLUSIONS

Climate changes and dams are likely to represent the most important threats to freshwater fish around the world. The effects of climate change on the ecosystem will include alterations of the timing, distribution and form of precipitation, as well as the timing of the flood pulse, and the intensity and frequency of floods and droughts. The impacts of these changes on the fish fauna and fisheries are, at that moment, unpredictable at both the species and ecosystem level. Actually, the degree of uncertainty and the low level of knowledge about the biology of the most of Amazonia fish species, make it hard to determine the current impacts of climate changes for each species and for the ecosystem as a whole, and even harder to predict future impacts. In addition to regional impacts, it will be very difficult to predict the impacts of dams for each fish species from the Amazon basin. As we discussed earlier, dams block fish migration, which could be very critical for many freshwater fish species that need to do migrations to complete their life cycles.

One pertinent question is what we can do to minimize the impacts of global climate changes and dams. Actually, we need to identify clearly the possible strategies to avoid human contribution to the magnitude of both sources of impacts. We realize that any intervention on the crescendo of climate change needs action of a very large scale or a very large package of small scale actions. Both of these necessitate development strategies arising from a coordinated source, such as a global agreement. Unhappily, the global negotiations on this issue have made little progress, and we remain distant from an agreement. On the other hand, the effect associated with potential new dams are in the sphere of national decision-making and these impacts could be avoided if the proposed construction does not take place. Thus, it is not too late to find alternatives to Amazonian hydropower for power supply. A global goal to minimize the impacts of both climate change and dams on freshwater fisheries is needed in order to avoid these severe impacts

## REFERENCES

1.  Freitas CEC, Siqueira-Souza FK, Prado KLL, Yamamoto KC, Hurd LE. Fish diversity in Amazonian floodplain lakes. International Journal of Medical and Biological Frontiers 2010;16:128-142.

2.  Honda EMS, Correa CM, Castelo FP, Zapellini EA. Aspectos gerais do pescado no Amazonas. Acta Amazonica 1975;5(1):87-94.

3.  Shrimpton R, Giugliano R. Consumo de alimentos e alguns nutriente em Manaus, Amazonas. Acta Amazonica 1979;9(1):117-141.

4.  Cerdeira RGP, Ruffino ML, Isaac VJ. Consumo de pescado e outros alimentos pela população ribeirinha do Lago Grande de Monte Alegre. Acta Amazonica 1997;27(3):213-227.

5.  Batista VS, Inhamuns AJ, Freitas CEC, Freire-Brasil D. Characterization of the fishery in river communities in the low-Solimões/high-Amazon

region. Fisheries Management and Ecology 1998;5(5):419-435.

6.  Fabré N, Alonso J. Recursos ícticos no Alto Amazonas. Sua importância para as populações ribeirinhas. Boletim do Museu Paraense Emilio Goeldi, Serie Zoologia 1998;14(1):19-55.

7.  Boischio AAP, Henshel D. Fish consumption, fish lore, and mercury pollution - risk contamination for the Madeira River people. Environmental Research 2000;84:108-126.

8.  Murrieta RSS, Dufour DL. Fish and farinha: protein and energy consumption in Amazonia rural communities on Ituqui Island, Brazil. Ecology of Food and Nutrition 2004;43(3):231-255.

9.  Junk WJ, Bayley PB, Sparks RE. The flood pulse concept in river floodplain systems. Special publication of the Canadian Journal of Fisheries and Aquatic Sciences 1989;106:110-127.

10. Oliver LP et al. Drought sensitivity of the Amazon rainforest. Science 2009;323:1344-1347.

11. Vieira EF, Isaac VJ, Fabré NN. Biologia reprodutiva do tambaqui, Colossoma macropomum Cuvier, 1818 (Teleostei, Serrasalmidae), no baixo Amazonas. Acta Amazonica 1999;29(4):625-638.

12. Araujo-Lima CARM, Oliveira EC. Transport of larval fish in the Amazon. Journal of Fish Biology 1998;29:1-11.

13. Araujo-Lima CARM, Ruffino ML. Migratory Fishes of the Brazilian Amazon. In: Carosfeld J, Harvey B, Ross C, Baer A (eds.) Migratory Fishes of South America: Biology, Fisheries and Conservation Status. Ottawa. The World Bank; 2003. p233-302.

14. IPCC. Summary for Policymakers. Climate Change 2007: The Physical Science Basis. Contribution of Working Group I to the Fourth Assessment Report of the Intergovernmental Panel on Climate Change (Solomon S, Qin D, Manning M, Chen Z, Marquis M, Avryt KB, Tignor M, Miller HL (eds.). Cambridge University Press, Cambridge, UK and New York, USA, 2007, 113p.

15. Shukla J, Nobre CA, Sellers P. Amazon deforestation and climate change. Science 1990;247:1322-1325.

16. Malhi Y, Roberts T, Betts RA, Killen TJ, Li W, Nobre CA. Climate Change, Deforestation, and the Fate of the Amazon. Science 2008;319:169-172.

17. Vermeer M, Rahmstorf S. Global sea level linked to global temperature. PNAS 2009; www.pnas.prg/cgi/doi/10.1073/pnas.0907765106 (acessed 5 July 2012).

18. Becker CD, Genoway RG. Evaluation of the critical thermal maximum for determining thermal tolerance of freshwater fish. Environmental Biology of Fishes 1979;4:245-256.

19. Ficke AD, Myrick CA, Hansen LJ. Potential impacts of climate changes on freshwater fisheries. Review in Fish Biology and Fisheries 2007;17:581-613.

20. Lowe-McConnell RH. Ecological Studies in Tropical Fish Communities. Cambridge University Press, Cambridge, England 1987;382pp.

21. Finer M, Jenkins CN. Amazon and its implications for Andes-Amazon connectivity. PLoS ONE 2012;7(4): e35126. doi:10.1371/journal. pone.0035126.

22. Marengo JA, Nobre CA, Tomasella J, Oyama MD, Oliveira GS, Oliveira R, Camargo H, Alves LM, Brown IF. The drought of Amazonia in 2005. Journal of Climatology 2008;21:495-516.

23. Humphries P, Baldwin DS. Drought and aquatic systems: an introduction. Freshwater Biology 2003;48:1141-1146.

24. Matthews WJ, Marsh-Matthews E. Effects of drought on fish across axes of space, time and ecological complexity. Freshwater Biology 2003;48:1232-1253.

25. Agostinho AA, Pelicice FM, Gomes LC. Dams and the fish fauna of the Neotropical region: impacts and management related to diversity and fisheries. Brazilian Journal of Biology 2008;68(4,suppl.):1119-1132.

26. Agostinho AA, Gomes LC, Veríssimo S, Okada EK. Flood regime, dam regulation and fish in the Upper Paraná River: effects on assemblages attributes, reproduction and recruitment. Reviews in Fish Biology and Fisheries 2004;14(1):11-19.

27. Barthem RB, Goulding M. The catfish connection: ecology, migration, and conservation of Amazon predators. Biology and Resource Management in the Tropics Series. Columbia Press, New York 1997;144 p.

28. Agudelo E, Salinas Y, Sánches CL, Muñoz-Sosa DL, Alonso JC, Arteaga ME, Rodrígues OJ, Anzola NR, Acosta LE, Núñez M, Valdés H. Bagres da la Amazonia Colombiana: Uno recurso sin fronteras. Fabré NN, Donato JC, Alonso JC (Eds). Instituto Amazónico de Investigaciones Científicas SINCHI. Programa de Ecosistemas Acuáticos. Editorial Scripto, Bogotá 2000;252p.

29. Barthem RB, Ribeiro MCLB, Petrere M. Life strategies of some long-distance migratory catfish in relation to hydroelectric dams in the Amazon Basin. Biological Conservation, 1991;55:339–345.

30. Ruffino ML, Isaac VJ. Dinamica populacional do Surubim-Tigre, Pseudoplatystoma tigrinum (Valenciennes, 1840) no médio Amazonas (Siluriformes, Pimelodidae). Acta Amazonica 1999;29(3):463-476.

31. Pérez A, Fabré NN. Seasonal growth and life history of the catfish Calophysus macropterus (Lichtenstein, 1819) (Siluriformes: Pimelodidae) from the Amazon floodplain. Journal of Applied Ichthyology 2009;25:343–349.

32. Loubens G, Panfili J. Biologie de Colossoma macropomum (Teleostei: Serrasalmidae) dans le bassin du Mamoré (Amazonie bolivienne). Ichthyological Exploration of Freshwaters 1997;8:1–22.

33. Ribeiro MCLB, Petrere Jr. M. Fisheries ecology and management of the Jaraqui (Semaprochilodus taeniurus, S. insignis) in Central Amazonia. Regulated Rivers: Research and Management 1990;5:195–215.

34. Fernandes CC. Lateral migrations of fishes in Amazon floodplain. Ecology of Freshwater Fish 1997;6:36-44.

35. Mota SQ, Ruffino ML. Biologia e pesca do curimatá (Prochilodus nigricans Agassiz, 1829) (Prochilodontidae) no Médio Amazonas. Revista UNIMAR 1997;19:493-508.

36. Ruffino ML, Isaac VJ. Life cycle and biological parameters of several Brazilian Amazon fish species. The ICLARM Quaterly 1995;18:41-45.

37. Jepsen DB, Winemiller KO, Taphorn DC. Age structure and growth of peacock cichlids from rivers and reservoirs of Venezuela. Journal of Fish Biology 1999;55:433-450.

38. Xenopoulos MA, Lodge DM, Alcamo J, Märker M, Shulze K, Van Vuuren DP. Scenarios of freshwater fish extinctions from climate change and water withdrawl. Global Change Biology 2005;11:1557-1564.

39. Marshall BG, Forsberg BR, Hess LL, Freitas CEC. Water temperature differences in interfluvial palm swamp habitats of Paracheirodon axelroldi and P. simulans (Osteichthyes: Characidae) in the middle Rio Negro, Brazil 2011;22(4):377-383.

40. Almeida-Val V, Val AL, Duncan WP, Souza FCA, Paula-Silva MN, Land S. Scaling effects on hypoxia tolerance in the Amazon fish Astronotus ocellatus (Perciformes: Cichlidae): contribution of tissue enzyme levels. Comparative Biochemistry and Physiology;125B:219-226.

41. Cushing DH. Climate and fisheries. Academic Press, London 1982;373p.

42. Cushing DH. Plankton production and year-class strength in fish populations: an update of the match/mismatch hypothesis. Advances in Marine Biology 1990;26:249-293.

43. Bailly D, Agostinho AA, Suzuki HI. Influence of the flood regime on the reproduction of fish species with different reproductive strategies in the Cuiabá River, Upper Pantanal, Brazil. River Research and Applications 2008;24:1218-1229.

44. Koehn JD, Hobday AJ, Pratchett MS, Gillanders BM. Climate change and Australian marine and freshwater environments, fishes and fisheries: synthesis and options for adaptation. Marine An Freshwater Research 2011;62:1148-1164.

45. Agostinho AA, Vazzoler AEAM, Gomes LC, Okada EK Estratificacíon especial y comportamiento de Prochilodus scrofa en distintas fases del ciclo de vida, en la planicie de inundacíon del alto río Paraná y embalse de Itaipu, Paraná, Brazil. Revue due Hydrobiologie Tropicale 1993;26:79-90.

46. Barrela W, Petrere Jr M. Fish community alterations due to pollution and damming in Tietê and Paranapanema Rivers (Brazil). River Research and Applications 2003;19:59-76.

47. Porto L, McLaughlin R, Noakes D. Low-head barrier dams restrict the movements of fishes in two lake Ontario streams. North American Journal of Fisheries and Management 1999;4:1028-1036.

48. Gomes LC, Agostinho AA. Influence of the flooding regime on the nutritional state and juvenile recruitment of the curimba, Prochilodus scrofa, Steindachner, in Upper Paraná river, Brazil. Fisheries Management and Ecology 1997;4:263-274.

49. Angermeier PL. Ecological attributes of extinction-prone species: loss of freshwater fishes of Virginia. Conservation Biology 1995;9:143-158.

50. Food and Agriculture Organization. The State of World Fisheries and Aquaculture 2010. Rome 2011.

51. Sumaila UR, Cheung WWL, Lam VWY, Pauly D, Herrick S. Climate change impacts on the biophysics and economics of world fisheries. Nature Climate Change 2011; doi: 10.1038/NCLIMATE1301.

52. Merino G, Barange M, Blanchard JL, Harle J, Holmes R, Allen I, Allison EH, Badjeck MC, Dulvy NK, Holt J, Jennings S, Mullon C, Rodwell LD. Can marine fisheries and aquaculture meet fish demand from a growing human population in a changing climate? Global Environment Change 2012;22:795-806. Doi: 10.1016/j.gloenvcha.2012.03.003.

53. Food and Agriculture Organization. Statistical databases. Available in: <http://www.fao.org>. Acessed on September 2012.

54. Instituto Brasileiro do Meio Ambiente e dos Recursos Naturais

Renováveis – IBAMA. Estatística da Pesca 2007. Brasil – Grandes Regiões e Unidades da Federação 2007.

55. Ministério da Pesca e Aquicultura. Produção Pesqueira e Aquícola – Estatística 2008 e 2009 – 2010.

# Chapter 2

## CLIMATE CHANGE: IMPACTS ON FISHERIES AND AQUACULTURE

Bimal P Mohanty[1], Sasmita Mohanty[2], Jyanendra K Sahoo[3] and Anil P Sharma[1]

[1]Central Inland Fisheries Research Institute, Barrackpore, Kolkata 700120; India.
[2]School of Biotechnology,KIIT University, Bhubaneswar 751024; India.
[3]Orissa University of Agriculture & Technology, College of Fisheries, Berhampur760007; India.

Climate change has been recognized as the foremost environmental problem of the twenty-first century and has been a subject of considerable debate and controversy. It is predicted to lead to adverse, irreversible impacts on the earth and the ecosystem as a whole. Although it is difficult to connect specific weather events to climate change, increases in global temperature has been predicted to cause broader changes, including glacial retreat, arctic shrinkage and worldwide sea level rise. Climate change has been implicated in mass mortalities of several aquatic species including plants, fish, corals and mammals. The present chapter has been divided in to two parts; the first part discusses the causes and general concerns of global climate change and the second part deals, specifically, on the impacts of climate change on fisheries and aquaculture, possible mitigation options and development of suitable monitoring tools.

## GLOBAL CLIMATE CHANGE: CAUSES AND CONCERNS

Climate change is the variation in the earth's global climate or in regional climates over time and it involves changes in the variability or average state of the atmosphere over durations ranging from decades to millions of years. The United Nations Framework Convention on Climate Change (UNFCCC) uses the term 'climate change' for human-caused change and 'climate variability' for other changes. In last 100 years, ending in 2005, the average global air temperature near the earth's surface has been estimated to increase at the rate of 0.74 +/- 0.18 °C (1.33 +/- 0.32 °F) (IPCC 2007). In recent usage, especially in the context of environmental policy, the term 'climate change' often refers to changes in the modern climate.

# CAUSES OF CLIMATE CHANGE

There are both natural processes and anthropogenic activities affecting the earth's temperature and the resultant climate change. The steep increases in the global anthropogenic greenhouse gas (GHG) emissions over the decades are major contributors to the global warming.

## Natural processes affecting the earth's temperature

Sun is the primary source of energy on earth. Though the sun's output is nearly constant, small changes over an extended period of time can lead to climate change. The earth's climate changes are in response to many natural processes like orbital forcing (variations in its orbit around the Sun), volcanic eruptions, and atmospheric greenhouse gas concentrations. Changes in atmospheric concentrations of greenhouse gases and aerosols, land-cover and solar radiation alter the energy balance of the climate system and causes warming or cooling of the earth's atmosphere. Volcanic eruptions emit many gases and one of the most important of these is sulfur dioxide ($SO_2$) which forms sulfate aerosol ($SO_4$) in the atmosphere.

## Greenhouse gases

Greenhouse gases (GHGs) are those gaseous constituents of the atmosphere, both natural and anthropogenic, that are responsible for the greenhouse effect, leading to an increase in the amount of infrared or thermal radiation near the surface. While water vapor ($H_2O$), carbon dioxide ($CO_2$), nitrous oxide ($N_2O$), methane ($CH_4$), and ozone ($O_3$) are the primary greenhouse gases in the Earth's atmosphere, there are a number of entirely human-made greenhouse gases in the atmosphere, such as the halocarbons and other chlorine- and bromine-containing substances. Halocarbons such as CFCs (chlorofluorocarbons) are completely artificial (man-made), and are produced from the chemical industry in which they are used as coolants and in foam blowing.

Increases in $CO_2$ are the single largest factor contributing more than 60% of humanenhanced increases and more than 90% of rapid increase in past decade. Most $CO_2$ emissions are from the burning of fossil fuels such as coal, oil, and gas. Rising $CO_2$ is also related to deforestation, which eliminates an important carbon sink of the terrestrial biosphere (www.ncdc.noaa.gov/oa/climate/

globalwarming.html; Shea et al., 2007). Currently, the atmosphere contains about 370 ppm of $CO_2$, which is the highest concentration in 420000 years and perhaps as long as 2 million years. Estimates of $CO_2$ concentrations at the end of the 21st century range from 490 to 1260 ppm, or a 75% to 350% increase above preindustrial concentrations (WMO World Data Centre for Greenhouse Gases. Greenhouse gas bulletin, 2006; Shea KM and the Committee on Environmental Health, 2007).

## IMPACTS OF CLIMATE CHANGE

Although it is difficult to connect specific weather events to global warming, an increase in global temperatures may in turn cause broader changes, including glacial retreat, arctic shrinkage, and worldwide sea level rise. Changes in the amount and pattern of precipitation may result in flooding and drought. Other effects may include changes in agricultural yields, addition of new trade routes, reduced summer stream flows, species extinctions, and increases in the range of disease vectors (Understanding and responding to Climate Change. 2008: http://www.national-academies.org).

Most models on Global climate change indicate that snow pack is likely to decline on many mountain ranges in the west, which would bring adverse impact on fish populations, hydropower, water recreation and water availability for agricultural, industrial and residential use. Partial loss of ice sheets on polar land could imply meters of sea level rise, major changes in coastlines and inundation of low-lying areas, with greatest effects in river deltas and low-lying islands. Such changes are projected to occur over millennial time scales, but more rapid sea level rise on century time scales cannot be excluded. Current models of climate change predict a rise in sea surface temperatures of between 2 °C and 5 °C by the year 2100 (IPCC Third Assessment Report, 2001: Done et al., 2003). Climate change will affect ecosystems and human systems like agricultural, transportation and health infrastructure. The regions that will be most severely affected are often the regions that are the least able to adept. Bangladesh is projected to lose 17.5 % of its land if sea level rises about 1 meter (39 inches), displacing millions of people. Several islands in the South Pacific and Indian oceans may disappear. Many other coastal regions will be at increased risk of flooding, especially during storm surges, threatening animals, plants and human infrastructure such as roads, bridges and water supplies. There are many ways in which climate change might affect human health, including heat stress, heat (sun) stroke, increased air pollution, and food scarcities due to drought and other agricultural stresses. Because many disease pathogens and carriers are strongly influenced by temperature, humidity and other climate variables, climate change may also influence the spread of infectious diseases or the intensity of disease outbreaks. During the last 100 years, anthropogenic activities related to burning fossil fuel, deforestation and agriculture has led to a 35% increase in the CO2 levels in the temperature and this has resulted in increased trapping of heat and the resultant increase in the earth's atmosphere. Most of the observed increase in globally-averaged temperatures has been attributed to the greenhouse gas concentrations. The globally averaged surface temperature rise has been projected to be 1.1-6.4 °C by end of the 21st century (2090-2099) which is mainly due to thermal expansion of the ocean (www.searo.who.int/en/Section260/Section2468_14335.htm, 2008). The global average sea level rose at an average rate of 1.8 mm per year from 1961 to 2003 and the total rise during the 20th century was estimated to be 0.17 m (The Fourth Assessment Report of IPCC, 2007). Due to such surface warming it is predicted that heat waves and heavy precipitations will continue to become more frequent with more intense and devastating tropical cyclones (typhoons and hurricanes). Due to the resultant disruption in ecosystem's services to support human health and livelihood, there will be strong negative impact on the health system. IPCC has projected an increase in malnutrition and consequent disorders, with implications for child growth and development. Increased burden of diarrheal diseases and infectious disease vectors are expected due to the erratic rainfall patterns.

Climate change is likely to lead to some irreversible impacts. Approximately 20- 30 % of species assessed so far are likely to be at increased risk of extinction if increases in global average warming exceed 1.5-2.5 °C (relative to 1980-1999). As global average temperature increase exceeds about 3.5 °C, model projections suggest significant extinctions (40-70 % of species assessed) around the globe. Some projected regional impacts of Climate change have been systematically listed in the IPCC Fourth Assessment Report, 2007.

## IMPACTS OF CLIMATE CHANGE ON FISHERIES AND AQUACULTURE

Fish has been an important part of the human diet in almost all countries of the world. It is highly nutritious; it can provide vital nutrients absent in typical starchy staples which dominate poor people's diets (FAO, 2005a; FAO, 2007a). Fish provides about 20 % of animal protein intake (Thorpe et al., 2006) and is one of the cheapest sources of animal proteins as far as availability and affordability is concerned. While it serves as a health food for the affluent world owing to the fish oils rich in polyunsaturated fatty acids (PUFAs), for the people in the other extreme of the nutrition scale, fish is a health food owing to its proteins, oils, vitamins and minerals and the benefits associated with the consumption of small indigenous fishes (Mohanty et al.,

2010a). Although aquaculture has been contributing an increasingly significant proportion of fish over recent decades, approximately two-thirds of fish are still caught in capture fisheries. The number of people directly employed in fisheries and aquaculture is estimated at 43.5 million, of which over 90 % are small –scale fishers (FAO, 2005a). In addition to those directly employed in fishing, over 200 million people are thought to be dependent on smallscale fishing in developing countries, in terms of other economic activities generated by the supply of fish (trade, processing, transport, retail, etc.) and supporting activities (boat building, net making, engine manufacture and repair, supply of services to fisherman and fuel to fishing boats etc.) in addition to millions for whom fisheries provide a supplemental income (FAO, 2005a). Fisheries are often available in remote and rural areas where other economic activities are limited and can thus be important sources for economic growth and livelihoods in rural areas with few other economic activities (FAO, 2005a)

## Potential impacts of climate change on fisheries

Climate change is projected to impact broadly across ecosystems, societies and economics, increasing pressure on all livelihoods and food supplies. The major chunk of earth is encompassed by water that harbors vast majority of marine and freshwater fishery resources and thus likely to be affected to a greater extent by vagaries of climate change. Capture fisheries has unique features of natural resource harvesting linked with global ecosystem processes and thus is more prone to such problems. Aquaculture complements and increasingly adds to the supply chain and has important links with capture fisheries and is likely to be affected when the capture fisheries is affected. The ecological systems which support fisheries are already known to be sensitive to climate variability. For example, in 2007, the International Panel on Climate Change (IPCC) highlighted various risks to aquatic systems from climate change, including loss of coastal wetlands, coral bleaching and changes in the distribution and timing of fresh water flows, and acknowledged the uncertain effect of acidification of oceanic water which is predicted to have profound impacts on marine ecosystems (Orr et al., 2005). Similarly, fishing communities and related industries are concentrated in coastal or low lying zones which are increasingly at risk from sea level rise, extreme weather events and wide range of human pressures (Nicholls et al., 2007a). While poverty in fishing communities or other forms of marginalization reduces their ability to adapt and respond to change, increasingly globalized fish markets are creating new vulnerabilities to market disruptions which may result from climate change.

Fisheries and fisher folk may have the impact in a wide range of ways due to climate change. The distribution or productivity of marine and fresh water fish stocks might be affected owing to the processes such as ocean acidification, habitat damage, changes in oceanography, disruption to precipitation and freshwater availability (Daw et al., 2009). Climate change, in particular, rising temperatures, can have both direct and indirect effects on global fish production. With increased global temperature, the spatial distribution of fish stocks might change due to the migration of fishes from one region to another in search of suitable conditions. Climate change will have major consequences for population dynamics of marine biota via changes in transport processes that influence dispersals and recruitment (Barange and Perry, 2009). These impacts will differ in magnitude and direction for populations within individual marine species whose geographical ranges span large gradients in latitude and temperature, as experimented by Mantzouni and Mackenzie (2010) in cod recruitment throughout the north Atlantic. The effects of increasing temperature on marine and freshwater ecosystems are already evident, with rapid pole ward shifts in distributions of fish and plankton in regions such as North East Atlantic, where temperature change has been rapid (Brander, 2007). Climate change has been implicated in mass mortalities of many aquatic species, including plants, fish, corals, and mammals (Harvell et al., 1999; Battin et al., 2007). Climate change will have impact on global biodiversity; alien species would expand into regions in which they previously could not survive and reproduce (Walther et al., 2009). Climate driven changes in species composition and abundance will alter species diversity and it is also likely to affect the ecosystems and the availability, accessibility, and quality of resources upon which human populations rely, both directly and indirectly through food web processes. Extreme weather events could result in escape of farmed stock and contribute to reduction in genetic diversity of wild stock affecting biodiversity. Climate variability and change is projected to have significant effects on the physical, chemical, and biological components of northern Canadian marine, terrestrial, and freshwater systems. According to a study conducted by Prowse et al. (2009), the northward migration of species and the disruption and competition from invading species are already occurring and will continue to affect marine, terrestrial, and freshwater communities. This will have implications for the protection and management of wildlife, fish, and fisheries resources; protected areas; and forests. Shifting environmental conditions will likely introduce new animal-transmitted diseases and redistribute some existing diseases, affecting key economic resources and some human populations. Stress on populations of iconic wildlife species, such as the polar bear, ringed seals, and whales, will continue as a result of changes in critical sea-ice habitat interactions. Where these stresses affect economically and culturally important species, they will have significant effects on people and regional economies. Further integrated, field-based monitoring and research programs, and the development of predictive models are required to allow for more detailed and comprehensive projections of change to be made, and to inform the development and implementation of appropriate adaptation, wildlife, and habitat conservation and protection strategies. Fisheries will also be exposed to a diverse range of direct and indirect climate impacts, including displacement and migration of human populations; impacts on coastal communities and infrastructure due to sea level rise; and changes in the frequency, distribution or intensity of tropical storms. Inland fisheries ecology is profoundly affected by changes in precipitation and run-off which may occur due to climate change. Lake fisheries in Southern Africa for example, will likely be heavily impacted by reduced lake levels and catches. The variety of different impact mechanisms, complex interactions between social, ecological and economic systems and the possibility of sudden and surprising changes make future effects of climate change on fisheries difficult to predict. In fact, understanding the ecological impacts of climate change is a crucial challenge of the twenty-first century. There is a clear lack of general rules regarding the impacts of global warming on biota. A study conducted by Daufresne et al. (2009) provided evidence that reduced body size is the third universal ecological response to global warming in aquatic systems besides the shift of species ranges toward higher altitudes and latitudes and the seasonal shifts in life cycle events. Apart from fisheries, global primary production (planktonic primary production) which is related to global fisheries catches at the scale of Large Marine Ecosystems appears to be declining, in some part due to climate variability and change, with consequences for the near future fisheries catches (Chassot et al., 2010). Other climatic change impacts on fisheries include surface winds, high $CO_2$ levels and variability in precipitations. While surface wind would alter both the delivery of nutrients in to the photic zone and strength and distribution of ocean currents, higher $CO_2$ levels can change the ocean acidity and variability in precipitation would affect sea levels. Global average sea level is rising at an average rate of 1.8 mm per year since 1961 and there is evidence of increased variability in sea level in recent decades. It is recently reported that ocean temperature and associated sea level increases between 1961 and 2003 were 50% larger than estimated in the 2007 IPCC Report. All coastal ecosystems are vulnerable to sea level rise and more direct anthropogenic impacts. Sea level rise may reduce intertidal habitat areas in ecologically important regions thus affecting fish and fisheries.

## Impact of climate change on the parasites and infectious diseases of aquatic animals

The potential trends of climate change on aquatic organisms and in turn in fisheries and aquaculture are less well documented and have primarily concentrated on coral bleaching and associated changes. An increase in the incidence of disease outbreaks in corals and marine mammals together with the incidence of new diseases has been reported. It was suggested that both the climate and human activities may have accelerated the global transport of species, bringing together of pathogens and previously unexposed populations (Harvell et al., 1999; De Silva and Sato, 2009). Climate changes could affect productivity of aquaculture systems and increase the vulnerability of cultured fish to diseases. All aquatic ecosystems, including freshwater lakes and rivers, coastal estuarine habitats and marine waters, are influenced by climate change (Parry et al., 2007; Scavia et al., 2002; Schindler, 2001).

**Table 1:** Impact of climate change on parasitic and other diseases of aquatic animals.

| Host | Disease /Parasite | Response to high temperature | Reference |
|---|---|---|---|
| Largemouth bass (*Micropterus salmoides*) | Red sore disease / bacterium *Aeromonas hydrophila* | Susceptibility to the disease increases | Esch and Hazen (1980) |
| Mosquitofish (*Gambusia affinis*) | Asian fish tapeworm (*Bothriocephalus acheilognathi*) | -do- | Granath and Esch (1983) |
| Trout (*Onchorhynchus spp.*) | Whirling disease / Myxozoan *Myxobolus cerebralis* | -do- | Hiner and Moffitt (2001) |
| Juvenile coho salmon (*O. kisutch*) | Blackspot disease/ trematode larvae (metacercariae) | Virulence is directly correlated with daily maximum temperature | Cairns et al., 2005 |
| A variety of reef fish | Ciguatera fish poisoning (CFP) caused by bioaccumulation of algal toxins | Increased incidence of CFP due to increased temperature | Tester et al., 2010 |
| Rainbow trout, *Oncorhynchus mykiss* | Infected with *Ichthyophonus* sp. | More rapid onset of disease, higher parasite load, more severe host tissue reaction and reduced mean-day-to-death at higher temperature | Kocan et al., 2009 |
| Freshwater bryozoans infected with myxozoan, *Tetracapsuloides bryosalmonae* | Spores released from sacs produced by the parasite during infection of freshwater bryozoans are infective to salmonid fish, causing the devastating Proliferative Kidney Disease (PKD) | Exacerbate PKD outbreaks and increase the geographic range of PKD as a result of the combined responses of T. bryosalmonae and its bryozoan hosts to higher temperatures. | Tops et al., 2009 |

Relatively small temperature changes alter fish metabolism and physiology, with consequences for growth, fecundity, feeding behavior, distribution, migration and abundance (Marcogliese, 2008). The general effects of increased temperature on parasites include, rapid growth and maturation, earlier onset of spring maturation, increased parasite mortality, increased number of generations per year, increased rates of parasitism and disease, earlier and prolonged transmission, the possibility of continuous, year-round transmission (Marcogliese, 2001).

Many diseases display greater virulence at higher temperatures that might be the result of reduced resistance of the host due to stress or increased expression of virulence factors/ increased transmission of the vectors. Some examples have been summarized in table 1.

As the emergence of disease is linked directly to changes in the ecology of hosts or pathogens, or both (Harvell et al., 1999), climate change will have a profound impact on the spread of parasites and disease in aquatic ecosystems (Harvell et al., 1999; Marcogliese, 2001; Harvell et al., 2002). Climate change will affect parasite species directly resulting from the extension of the geographical range of pathogens (Harvell et al., 2002). In addition, increased temperature may cause thermal stress in aquatic animals, leading to reduced growth, sub- optimal behaviors and reduced immunocompetence (Harvell et al., 1999; Harvell et al., 2002; Roessig et al., 2004) resulting in changes in the distribution and abundance of their hosts (Marcogliese, 2001). In the oceans, diseases are shown to increase in corals, sea urchins, molluscs, sea turtles and marine mammals, although not all can be linked unequivocally to climate alone (Lafferty et al., 2004). However, it was recently suggested that diseases may not increase with climate change, although distributions of parasites and pathogens will undoubtedly shift (Lafferty, 2009). Other factors may dominate over climate in controlling the distribution and abundance of pathogens, including: habitat alteration, invasive species, agricultural practices and human activities.

**Table 2:** General effects of increased temperature on parasite life cycles, their hosts and transmission processes (Marcogliese, 2008)

| Effects on parasites | Effects on hosts | Effects on transmission |
|---|---|---|
| Faster embryonic development and hatching | Altered feeding | Earlier reproduction in spring |
| Faster rates of development and maturation | Altered behavior | More generations per year |
| Decreased longevity of larvae and adults | Altered range | Prolonged transmission in the fall |
| Increased mortality of all stages | Altered ecology Reduced host resistance | Potential transmission year round |

Outbreaks of numerous water- borne diseases in both humans and aquatic organisms are linked to climatic events, although it is often difficult to disentangle climatic from other anthropogenic effects. In some cases, these outbreaks occur in foundation or keystone species, with consequences throughout whole ecosystems. There is much evidence to suggest that parasite and disease transmission, and possibly virulence, will increase with global warming. However, the effects of climate change will be superimposed on a multitude of other anthropogenic environmental changes. Climate change itself may exacerbate these anthropogenic effects. Moreover, parasitism and disease may act synergistically with these anthropogenic stressors to further increase the detrimental effects of global warming on animal and human populations, with debilitating social economic ramifications (Marcogliese, 2008). The repercussions of climate change are not limited solely to temperature effects on hosts and their parasites, but also have other possible effects such as: alteration in water levels and flow regimes, eutrophication, stratification, changes in acidification, reduced ice cover, changes in ocean currents, increased ultra-violet (UV) light penetration, run off, weather extremes (Cochrane et al., 2009).

## ANTICIPATED IMPACTS IN NEXT FEW DECADES

In addition to incremental changes of existing trends, complex social and ecological systems such as coastal zones and fisheries, may exhibit sudden qualitative shifts in behaviour when forcing variables past certain thresholds (Daw et al., 2009). For example, IPCC originally estimated that the Greenland ice sheet would take more than 1000 years to melt, but recent observations suggest that the process is already happening faster owing to mechanisms for ice collapse that were not incorporated into the projections (Lenton et al., 2008). The infamous collapse of the Northwest Atlantic northern cod fishery provides a non-climaterelated example where chronic over fishing led to a sudden, unexpected and irreversible loss in production from this fishery. Thus, existing observations of linear trends cannot be used to reliably predict impacts within the next 50 years (Daw et al., 2009). A study by Veron et al. (2009) also emphasizes impact of increasing atmospheric CO2 levels due to global warming on mass coral bleaching world-wide. According to this group, temperature-induced mass coral bleaching causing mortality on a wide geographic scale started when atmospheric $CO_2$ levels exceeded approximately

320 ppm. At today's level of approximately 387 ppm, allowing a lag-time of 10 years for sea temperatures to respond, most reefs world-wide are committed to an irreversible decline. Mass bleaching will in future become annual, departing from the 4 to 7 years return-time of El Niño events. Bleaching will be exacerbated by the effects of degraded water-quality and increased severe weather events. In addition, the progressive onset of ocean acidification will cause reduction of coral growth and retardation of the growth of high magnesium calcite-secreting coralline algae. If $CO_2$ levels are allowed to reach 450 ppm (due to occur by 2030-2040 at the current rates), reefs will be in rapid and terminal decline world-wide from multiple synergies arising from mass bleaching, ocean acidification, and other environmental impacts. Damage to shallow reef communities will become extensive with consequent reduction of biodiversity followed by extinctions. Reefs will cease to be large-scale nursery grounds for fish and will cease to have most of their current value to humanity. There will be knock-on effects to ecosystems associated with reefs, and to other pelagic and benthic ecosystems. This is likely to have been the path of great mass extinctions of the past, adding to the case that anthropogenic $CO_2$ emissions could trigger the Earth's sixth mass extinction (Veron et al., 2009).

## CLIMATE CHANGE IMPACTS ON INLAND FISHERIES - THE INDIAN SCENARIO

In recent years the climate is showing perceptible changes in the Indian subcontinent, where the average temperature is on the rise over the last few decades. In India, observed changes include an increase in air temperature, regional monsoon variation, frequent droughts and regional increase in severe storm incidences in coastal states and Himalayan glacier recession (Vass et al., 2009). In some states like West Bengal, the average minimum and maximum temperatures has increased in the range of 0.1 - 0.9 °C throughout the state. The average rainfall has decreased and monsoon is also delayed; consequently, the climate change impact is being felt on the temperature of the inland water bodies and on the breeding behavior of fishes. It is well known that temperature is an important factor which strongly influence the reproductive cycle in fishes. Temperature, along with rainfall and photoperiod, stimulate the endocrine glands of fishes which help in the maturation of the gonads.

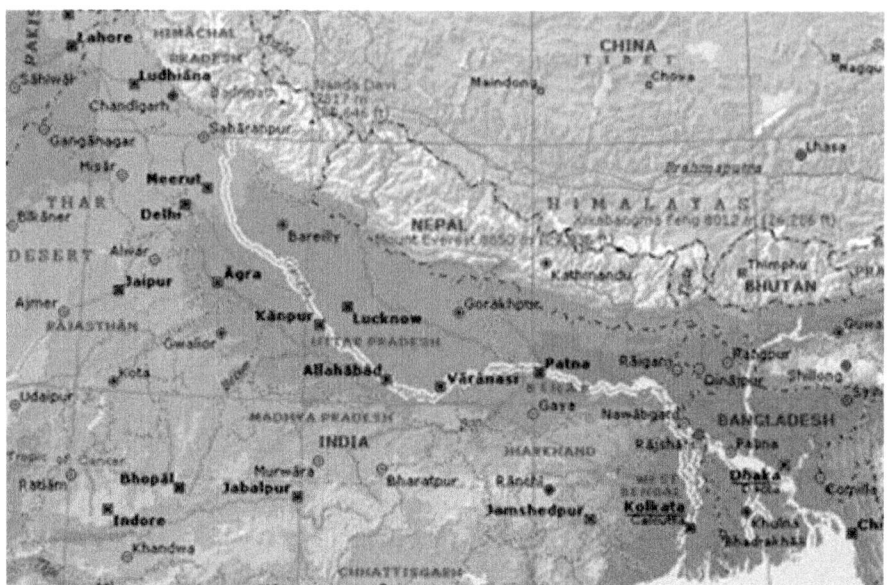

**Figure 1:** Course of the River Ganga showing different stretches (http://www.gits4u. com/ water/ganga1.gif)

In India, the inland aquaculture is centered on the Indian major carps, Catla catla, Labeo rohita and Cirrhinus mrigala and their spawning occurs during the monsoon (June-July) and extend till September. In recent years the phenomenon of IMC maturing and spawning as early as March is observed, making it possible to breed them twice a year. Thus, there is an extended breeding activity as compared to a couple of decades ago (Dey et al., 2007), which appears to be a positive impact of the climate change regime.

The mighty river Ganga forms the largest river system in India and not only millions of people depend on its water but it provides livelihood to a large group of fishermen also. The entire length of the river, with a span of 2,525 km from source to mouth is divided into three main stretches consisting of upper (Tehri to Kanauji), middle (Kanpur to Patna) and lower (Sultanpur to Katwa) (Figure 1). From analysis of 30 years' time series data on river Ganga and water bodies in the plains, Vass et al. (2009) reported an increase in annual mean minimum water temperature in the upper cold-water stretch of the river (Haridwar) by 1.5 °C (from 13 °C during 1970-86 to 14.5 °C during 1987-2003) and by 0.2- 1.6 °C in the aquaculture farms in the lower stretches in the Gangetic plains. This change in temperature clime has resulted in a perceptible biogeographically distribution of the Gangetic fish fauna. A number of fish species which were never reported in the upper stretch of the river and were predominantly available in the lower and middle stretches in the 1950s

(Menon, 1954) have now been recorded from the upper cold-water region. Among them, Mastocembelus armatus has been reported to be available at Tehri-Rishikesh and Glossogobius gurius is available in the Haridwar stretch (Sinha et al., 1998) and Xenentodon cancila has also been reported in the cold-water stretch (Vass et al., 2009). The predator-prey ratio in the middle stretch of the river has been reported to be declined from 1:4.2 to 1:1.4 in the last three decades. Fish production has been shown to have a distinct change in the last two decades where the contribution from IMCs has decreased from 41.4% to 8.3% and that from catfishes and miscellaneous species increased (Vass et al., 2009).

## ADAPTATION AND MITIGATION OPTIONS

Adaptation to climate change is defined in the climate change literature as an adjustment in ecological, social or economic systems, in response to observed or expected changes in climatic stimuli and their effects and impacts in order to alleviate adverse impacts of change, or take advantage of new opportunities. Adaptation is an active set of strategies and actions taken by peoples in response to, or in anticipation to the change in order to enhance or maintain their wellbeing. Hence adaptation is a continuous stream of activities, actions, decisions and attitudes that informs decisions about all aspects of life and that reflects existing social norms and processes (Daw et al., 2009). Many capture fisheries and their supporting ecosystems have been poorly managed, and the economic losses due to overfishing, pollution and habitat loss are estimated to exceed $50 billion per year (World Bank & FAO, 2008). The capacity to adapt to climate change is determined partly by material resources and also by networks, technologies and appropriate governance structures. Improved governance, innovative technologies and more responsible practices can generate increased and sustainable benefits from fisheries. There is a wide range of potential adaptation options for fisheries. To build resilience to the effects of climate change and derive sustainable benefits, fisheries and aquaculture managers need to adopt and adhere to best practices such as those described in the FAO 'Code of Conduct for Responsible Fisheries', reducing overfishing and rebuilding fish stocks. These practices need to be integrated more effectively with the management of river basins, watersheds and coastal zones. Fisheries and aquaculture need to be blended into National Climate Change Adaptation Strategies. In absence of careful planning, aquatic ecosystems, fisheries and aquaculture can potentially suffer as a result of adaptation measures applied by other sectors such as increased use of dams and hydro power in catchments with high rainfall, or the construction of artificial coastal defenses or marine wind farms (ftp://ftp.fao.org/FI/brochure/climate_

change/policy_brief.pdf). Mitigation solutions reducing the carbon footprint of Fisheries and Aquaculture will require innovative approaches. One example is the recent inclusion of Mangrove conservation as eligible for reducing emissions from deforestation and forest degradation in developing countries, which demonstrates the potential for catchment forest protection. Other approaches to explore include finding innovative but environmentally safe ways to sequester carbon in aquatic ecosystems, and developing low-carbon aquaculture production systems (ftp://ftp.fao.org/FI/brochure/climate_change/policy_brief.pdf). There is mounting interest in exploiting the importance of herbivorous fishes as a tool to help ecosystems recover from climate change impacts. Aquaculture of herbivorous species can provide nutritious food with a small carbon footprint. This approach might be particularly suitable for recovery of coral reefs, which are acutely threatened by climate change. Surveys of ten sites inside and outside a Bahamian marine reserve over a 2.5-year period demonstrated that increases in coral cover, including adjustments for the initial sizedistribution of corals, were significantly higher at reserve sites than those in non-reserve sites: macroalgal cover was significantly negatively correlated with the change in total coral cover over time. Reducing herbivore exploitation as part of an ecosystem-based management strategy for coral reefs appears to be justified (Mumby and Harborne, 2010). Furthermore, farming of shellfish, such as oysters and mussels, is not only good business, but also helps clean coastal water, while culturing aquatic plants help to remove waste from polluted water. In contrast to the potential declines in agricultural yields in many areas of the world, climate change opens new opportunities for aquaculture as increasing numbers of species are cultured (ftp://ftp.fao.org/FI/brochure/climate_change/policy_brief.pdf). Marine fish is one of the most important sources of animal protein for human use, especially in developing countries with coastlines. Marine fishery is also an important industry in many countries. The depletion of fishery resources is happening mainly due to anthropogenic factors such as overfishing, habitat destruction, pollution, invasive species introduction, and climate change. The most effective ways to reverse this downward trend and restore fishery resources are to promote fishery conservation, establish marineprotected areas, adopt ecosystem-based management, and implement a "precautionary principle." Additionally, enhancing public awareness of marine conservation, which includes eco-labeling, fishery ban or enclosure, slow fishing, and MPA (marine protected areas) enforcement is important and effective (Shao, 2009). The assessment report of the 4th International Panel on Climate Change confirms that global warming is strongly affecting biological systems and that 20-30% of species risk extinction from projected future increases in temperature. One of the widespread management strategies taken to conserve individual species and their constituent populations against climate-mediated declines has been the release of captive bred animals to wild in order to augment wild populations for many species. Using a regression model based on a 37-year study of wild and sea ranched Atlantic salmon (Salmo salar) spawning together in the wild, McGinnity et al. (2009) showed that the escape of captive bred animals into the wild can substantially depress recruitment and more specifically disrupt the capacity of natural populations to adapt to higher winter water temperatures associated with climate variability, thus increasing the risk of extinction for the studied population within 20 generations. According to them, positive outcomes to climate change are possible if captive bred animals are prevented from breeding in the wild. Rather than imposing an additional genetic load on wild populations by releasing maladapted captive bred animals, they propose that conservation efforts should focus on optimizing conditions for adaptation to occur by reducing exploitation and protecting critical habitats.

## MONITORING STRESS IN AQUATIC ANIMALS AND HSP70 AS A POSSIBLE MONITORING TOOL

Temperature above the normal optimum are sensed as heat stress by all organisms, Heat stress (HS) disturbs cellular homeostasis and can lead to severe retardation in growth and development and even death. Heat shock (stress) proteins (HSP) are a class of functionally related proteins whose expression is increased when cells are exposed to elevated temperatures or other stress. The dramatic up regulation of the HSPs is a key part of heat shock (stress) response (HSR). The accumulation of HSPs under the control of heat shock (stress) transcription factors (HSFs) play a central role in the heat stress response (HSR) and acquired thermo tolerance. HSPs are highly conserved and ubiquitous and occur in all organisms from bacteria to yeast to humans. Cells from virtually all organisms respond to different stress by rapidly synthesizing the HSPs and therefore, HSPs are widely used as biomarkers for stress response (Jolly and Marimoto, 2000). HSPs have multiple housekeeping functions, such as activation of specific regulatory proteins and folding and translocation of newly synthesized proteins. HSPs are usually produced in large amounts (induction) in response to distinct stressors such as ischemia, hypoxia, chemical/toxic insult, heavy metals, oxidative stress, inflammation and altered temperature or heat shock (Marimoto, 1998). Out of different HSPs, the HSP70 is unique in many ways; it acts as molecular chaperone in both unstressed and stressed cells. HSC70, the constitutive HSP70 is crucial for the chaperoning functions of unstressed cells, where as the inducible HSP70 is important for allowing cells to cope with acute stress, especially those affecting the protein machinery. HSP70 in marine mussels are widely used as a potential biomarker for stress response and aquatic environmental monitoring of the marine ecosystem (Li et al., 2000). The success of any organism depends not only on niche adaptation but also the ability to survive environmental perturbation from homeostasis, a situation generally described as stress (Clark et al., 2008a). Although species-specific mechanisms to combat stress have been described, the production of heat shock proteins (HSPs), such as HSP70, is universally described across all taxa. We have studied expression profile of the HSP70 proteins, in different tissues of the large riverine catfish Sperata seenghala (Mohanty et al., 2008), freshwater catfish Rita rita (Mohanty et al., 2010b), Indian catfish Clarias batrachus, Indian major carps Labeo rohita, Catla catla, Cirrhinus mrigala, exotic carp Cyprinus carpio var. communis and the murrel Channa striatus, the climbing perch Anabas testudineus (CIFRI, 2009; Mohanty et al., 2009). Out of these, the IMCs are the major aquaculture species and therefore are of much economic significance. Similarly, Anabas and Channa fetch good market value and their demand is increasing owing to their perceived therapeutic value (Mohanty et al., 2010a). The large riverine catfish S. seenghala comprises the major fisheries in majority of rivers and reservoirs and the freshwater catfish Rita rita has a good market demand and these two comprise a major share of the capture fisheries in India. Monoclonal anti-HSP70 antibody (H5147, Sigma), developed in mouse against purified bovine brain HSP70, in immunoblotting localizes both the constitutive (HSP73) and inducible (HSP72) forms of HSP70. The antibody recognizes brain HSP70 of bovine, human, rat, rabbit, chicken, and guinea pig. We observed immunoreactivity of this antibody with HSP70 proteins in different organs and tissues of a variety of fish species (Table 3). The strong immunoreactivity indicates that the HSP70 proteins of bovine and this riverine catfish Rita rita share strong homology although fish belong to a clade phylogenetically distant from the bovines. Persistent, high level of expression of HSP70 was observed in muscle tissues of Rita rita and for this reason, we have used and recommend use of white muscle tissue of Rita rita as a suitable positive control in analysis of HSP70 expression in tissues of other organisms (Mohanty et al., 2010b). Early studies on heat shock response in Antarctic marine ectoderms had led to the conclusion that both microorganisms and fish lack the classical heat shock response, i.e. there is no increase in HSP70 expression when warmed (Carratti et al., 1998; Hofmann et al., 2000). However, later it was reported that other Antarctic animals, show an inducible heat shock response, at a level probably set during their temperate evolutionary past (Clark et al., 2008 a, b); the bivalve (clam) Laternula elliptica and gastropod (limpet) Nacella concinna show an inducible heat shock response at 8 °C and 15 °C, respectively and these are temperatures in excess of that which is currently experienced by these animals, which can be attributed to the global warming (Waller et al., 2006). Permanent expression of the inducible HSP70 genes, species-specific high expression of HSC70 (N. concinna) and permanent expression of GRP78 (N concinna and L. elliptica) indicates that, as for fish, chaperone proteins form an essential part of the adaptation of the biochemical machinery of these animals to low but stable temperatures. High constitutive levels of HSP gene family member expression may be a compensatory mechanism for coping with elevated protein damage at low temperature analogous to the permanent expression of HSP70 in the Antarctic notothenoids (Clark et al., 2008 a). Such studies clearly indicate that both genetics and environment play important role in spatio-temporal gene expression.

**Table 3:** HSP70 expression profile in different tissues of some freshwater fishes, both aquacultured and wild stock.

| Fish species | Liver | Muscle | Kidney | Gill | Remarks |
|---|---|---|---|---|---|
| *Labeo rohita* | - | ++ | ++ | ++ | Mohanty et al. 2009 |
| *Cirrhinus mrigala* | ++ | - | - | ++ | CIFRI 2009; Mohanty et al. 2009 |
| *Cyprinous carpio var communis* | ++ | ++ | ++ | - | -do- |
| *Anabas testudineus* | ++ | - | - | ++ | -do- |
| *Channa punctatus* | - | - | ++ | | -do- |
| *Sperrata seenghala* | ++ | ++ | ++ | + | Mohanty et al. 2008 |
| *Rita rita* | ++ | ++ | ++ | + | Mohanty et al. 2010b |

There is need to standardize tools suitable for monitoring stress resulting from global warming and climate change impacts, in the aquatic animals from both aqua culture and capture fisheries systems. As HSP70 expression has been reported in many fish species (Table 3) it might serve as a suitable tool for monitoring impact of thermal stress/global warming; however, as HSP70 proteins are expressed under other conditions also, it is necessary to identify the heat shock (stress) transcription factors (HSFs) that can be specifically attributed to global warming (thermal stress) and climate change. It is also necessary to distinguish the constitutive and induced forms of the transcripts/ proteins by qPCR/proteomic analysis so that specific HSP70 forms suitable for monitoring performance of the farmed fishes can be monitored for better management of aquacultured animals. IPCC have predicted an average global warming between +2 and +6 °C, depending on the scenarios, within the next 90 years (IPCC 2007). The consequences of this increase in temperature are now well documented on both the abundance and geographic distribution of numerous taxa i.e. at population or community levels; in contrast, studies at the cellular level are still scarce. The study of the physiological or metabolic effects of such small increases in temperature is difficult because they are below the amplitude of the daily or seasonal thermal variations occurring in most environments. The underground water organisms are highly thermally buffered and thus are well suited for characterization of cellular responses of global warming. Colson-Proch et al. (2010) studied the genes encoding HSP70 family chaperones in amphipod crustaceans belonging to the ubiquitous subterranean genus Niphargus and HSP 70 sequence in 8 populations of 2 complexes of species of this genus (Niphargus rhenorhodanensis and Niphargus virei complexes). Expression profiles of HSP70 were determined for one of these populations by reverse transcription and quantitative polymerase chain reaction, confirming the inducible nature of this gene. An increase of 2 °C seem to be without any effect on N. rhenorhodanensis physiology whereas a heat shock of + 6 °C represented an important thermal stress for these individuals. Thus this study showed that although Niphargus individuals do not undergo any daily or seasonal thermal variations in underground water, they display an inducible HSP70 heat shock response (Colson-Proch et al., 2010).

## EPILOGUE

There are opposing viewpoints on the predicted impacts of 'global warming' also. Scientists warn against overselling climate change. Some experts feel that the data produced by models used to project weather changes, risk being over-interpreted by governments, organizations and individuals keen to make plans for a changing climate, with dangerous results. The point made is that the Global Climate Models (GCMs) help us understand pieces of the climate system, but that does not mean we can predict the details. Thus, indications of changes in the earth's future climate must be treated with the utmost seriousness and with the precautionary principle uppermost in our minds. Extensive climate change may alter and threaten the living conditions of much of mankind. They may induce large-scale migration and lead to greater competition for the earth's resources. Such changes will place particularly heavy burdens on the world's most vulnerable countries. There may be increased danger of violent conflicts and wars, within and between states. A wide array of adaptation options is available, but more extensive adaptation than is currently occurring is required to reduce vulnerability to climate change. Although the understanding of climate change has advanced significantly during the past few decades, many questions remain unanswered. The task of mitigating and adapting to the impacts of climate change will require worldwide collaborative input from a wide range of experts from various fields. The common man's contribution will play a major role in reducing the impacts of climate change and protecting the earth from climate change-related hazards. The impacts of climate change to freshwater aquaculture in tropical and subtropical region is difficult to predict as marine and freshwater populations are affected by synergistic effects of multiple climate and noncelibate stressors. If such noncelibate factors are identified and understood then it may be possible for local predictions of climate change impacts to be made with high confidence (De Silva and Soto, 2009). Coastal communities, fishers and fish farmers are profoundly affected by climate change. Climate change is modifying the distribution and productivity of marine and freshwater species and is already affecting biological processes and altering food webs, thus making the consequences for sustainability of aquatic ecosystems for fisheries and aquaculture, and for the people dependent on them, uncertain. Fisheries, aquaculture and fish habitats are at risk. Deltas and estuaries are in the fore front and thus, most vulnerable to climate change. Mitigation measures are urgently needed to neutralize and alleviate these growing threats, to adapt to their impacts and also to build our knowledge base on Complex Ocean and aquatic processes. The prime need is to reduce the global emissions of GHGs, which is the primary anthropogenic factor responsible for climate change (ProAct Network, 2008). Healthy aquatic ecosystems contribute greatly to food security and livelihoods. They are critical for production of wild fish and for some of the seed and much of the feed (trash fish) for aquaculture. Coastal ecosystems provide food, habitats and nursery grounds for fish. Estuaries, coral reefs, mangroves and sea grass beds are particularly important. Mangroves create barriers to destructive waves from storms and hold sediments in place with their extensive root systems thereby reducing coastal erosion. Healthy coral reefs, sea grass beds and wetlands provide similar benefits. Thus, these natural systems not only support fisheries, but help protect communities from the terrible impacts of natural hazards and disasters also (ProAct Network, 2008). In freshwater systems, ecosystem health and productivity is linked to water quality and flow and the health of wetlands. Ecosystem-based approaches to fisheries and coastal zone management are highly beneficial as such approaches recognize the need for people to use the ecosystem for their food security and livelihoods while enabling these valuable natural assets to adapt to the effects of climate change, and to reduce the threats from other environmental stresses (Hoegh-Guldberg et al., 2007). Fish and shellfish provide essential nutrition for 3 billion people and about 50% of animal protein and micronutrients to 400 million people in the poorest countries of the world. Fish is one of the cheapest sources of animal proteins and play important role in preventing protein-calorie malnutrition. The health benefits of eating fish are being increasingly understood by the consumers. Over 500 million people in the developing countries depend on fisheries and aquaculture for their livelihoods. Aquaculture is the world's fastest growing food production system, growing at 7% annually. Fish products are among the most widely traded foods internationally (ftp://ftp.fao.org/FI/brochure/climate_change/ policy_brief.pdf). Implementing adaptation and mitigation pathways for communities dependent on fisheries, aquaculture and aquatic ecosystems will need increased attention from policy-makers and planners. Sustainable and resilient aquatic ecosystems will benefit the fishers as well as the coastal communities and will provide good and services at national and global levels. Fisheries and aquaculture need specific adaptation and mitigation measures like: improving the management of fisheries and aquaculture as well as the integrity and resilience of aquatic ecosystems; responding to the opportunities for and threats to food and livelihood security due to climate change impacts; and helping the fisheries and aquaculture sector reduce GHG emissions. To conclude, the present generation is already facing the harmful effects of the climate change; however, the future generations will suffer most of the harmful effects of global climate change. So, the present generation need to decide, whether to aggressively reduce the chances of future harm at the cost of sacrificing some luxuries or to let our descendants largely fend for themselves (Broome, 2008). Thus, how we handle the issue of Climate Change is more of an ethical question and the global community must act sensibly and responsibly.

## REFERENCES

1. Barange, M., & Perry, R.I. (2009) Physical and ecological impacts of climate change relevant to marine and inland capture fisheries and aquaculture In: Climate change implications for fisheries and aquaculture overview of current scientific Knowledge, Cochrane, K., Young, C. De, Soto, D., & Bahri, T. (Eds). FAO Fisheries and Aquaculture Technical paper: No. 530, pp. 7-106, FAO, Rome.

2. Battin, J., Wiley, M. W., Ruckelshaus, M. H., Palmer, R. N,. Korb, E., Bartz, K. K., & Imaki, H. (2007) Projected impacts of climate change on salmon habitat restoration, Proc. Natl. Acad. Sci, USA, 104, 6720-6725.

3. Brander, K. M. (2007) Global fish production and climate change, Proc. Natl. Acad. Sci., USA, 104, 19709-19714.

4. Broome, J. (2008) The ethics of climate change, Sci. Am., 298, 96-100.

5. Cairns, M. A., Ebersole, J. L., Baker, J. P., Wigngton, P. J. Jr., Lavigne, H. R., & Davis, S. M. (2005) Influence of summer stream temperatures on black spot infestation of juvenile coho salmon in the Oregon Coast Range, Trans. Am. Fish. Soc., 134, 1471-1479.

6. Carrattù, L., Gracey, A. Y, B.uono, S., & Maresca, B. (1998) Do Antarctic fish respond to heat shock? In: Fishes of Antarctica. A Biological Overview. di Prisco, G., Pisano, E., Clarke, A. (Eds) Springer, Italy.

7.   Chassot, E., Bonhommeau, S., Dulvy NK, Mélin F, Watson R, Gascuel D, Le Pape O. (2010) Global marine primary production constrains fisheries catches. Ecol Lett., Feb 5. [Epub ahead of print]

8.   CIFRI (2009) Annual Report. Central Inland Fisheries Research Institute, Barrackpore, Kolkata, India. ISSN 0970 6267.

9.   Clark, M. S., Fraser, K. P. P., & Peck, L. S. (2008a) Antarctic marine molluscs do have an HSP70 heat shock response, Cell Stress Chaperon., 13, 39-49.

10.  Clark, M. S., Geissler, P., Waller, C., Fraser, K. P. P., Barnes, D. K. A., & Peck, L. S. (2008b) Low heat shock thresholds in wild Antarctic inter-tidal limpets (Nacella concinna). Cell Stress Chaperon., 13, 51-58

11.  Cochrane, K., Young, C. De, Soto, D., & Bahri, T. (2009) Climate change implications for fisheries and aquaculture: overview of current scientific knowledge. FAO Fisheries and Aquaculture Technical paper: No. 530,FAO, Rome.

12.  Colson-Proch, C., Morales, A., Hervant, F., Konecny, L., Moulin, C., & Douady, C. J. (2010) First cellular approach of the effects of global warming on groundwater organisms: a study of the HSP70 gene expression. Cell Stress Chaperon., 15, 3, 259-270.

13.  Daufresne, M., Lengfellner, K., & Sommer, U. (2009) Global warming benefits the small in aquatic ecosystems. Proc Natl Acad Sci USA., 106, 31, 12788-12793

14.  Daw, T., Adger, W. N., Brown, K., & Badjeck, M.-C. (2009) Climate change and capture fisheries: potential impacts, adaptation and mitigation. In: Climate change implications for fisheries and aquaculture overview of current scientific Knowledge, Cochrane, K., Young, C. De, Soto, D., & Bahri, T. (Eds). FAO Fisheries and Aquaculture Technical paper: No. 530, pp.107-150, FAO, Rome.

15.  De Silva, S. S. and Soto, D. 2009, Climate change and aquaculture: potential impacts, adaptation and mitigation In: Climate change implications for fisheries and aquaculture overview of current scientific Knowledge, Cochrane, K., Young, C. De, Soto, D., & Bahri, T. (Eds). FAO Fisheries and Aquaculture Technical paper: No. 530, pp. 151- 212, FAO, Rome.

16.  Dey, S., Srivastava, P. K., Maji, S., Das, M. K., Mukhopadhyay, M. K., & Saha, P. K. (2007) Impact of climate change on the breeding of Indian major carps in West Bengal. J. Inland Fish. Soc. India, 39, 1, 26-34

17.  Done, T., Whetton, P., Jones, R. et al. (2003) Global climate change and coral bleaching on the Great Barrier Reef. Final report to the State of

Queensland Greenhouse taskforce through the Department of Natural Resources and Mines, Queensland

18. Esch, G. W., & Hazen, T. C. (1980) Stress and body condition in a population of largemouth bass: implications for red-sore disease, Trans. Am. Fish. Soc., 109, 532-536.

19. FAO (2005) Increasing the contribution of small-scale fisheries to poverty alleviation and food security. FAO Technical Guidelines for Responsible Fisheries. No. 10, 79 p., FAO, Rome.

20. FAO (2007) The state of world fisheries and aquaculture – 2006, 162 p., FAO, Rome.

21. Granath, W. O. Jr., & Esch, G. W. (1983) Survivorship and parasite-induced host mortality among mosquio fish in a predator- free, North Carolina cooling reservoir, Am. Midland Naturalist, 110, 314-323

22. Harvell, C. D., Kim, K., Burkholder, J. M., Colwell, R. R., Epstein, P. R., Grimes, D. J., Hofmann, E. E., Lipp, E. K., & Osterhaus, A. D. Overstreet RM et al. (1999) Emerging marine diseases- climate links and anthropogenic factors, Science, 285, 1505-1510.

23. Harvell, C. D., Mitchell, C. E., Ward, J. R., Altizer, S., Dobson, A. P., Ostfeld, R. S & Samuel, M.D. (2002) Climate warming and disease risks for terrestrial and marine biota, Science, 296, 5576, 2158-2162.

24. Hiner, M., & Moffitt, C. M. (2001) Variation in infections of Myxobolus cerebralis in fieldexposed cutthroat and rainbow trout in Idaho, J. aquat. Anim. Hlth, 13, 124-132.

25. Hoegh-Goldberg, O. et al. (2007) Coral reefs under rapid climate change and ocean acidification. Science, 318, 1737-1742

26. Hofmann, G. E., Buckley, B. A., Airaksine, S., Keen, J. E., & Somero, G. N. (2000) Heat-shock protein expression is absent in the Antarctic fish Trematomus bernacchii family Nototheniidae. J Exp Biol., 203, 2331-2339

27. IPCC (2007) Fourth Assessment Report - Climate Change 2007:

28. Synthesis Report, 2007. IPCC (2001) Climate Change 2001. IPCC Third Assessment Report, 2001.

29. Jolly, C., & Marimoto, R. I. (2000) Role of the heat shock response and molecular chaperones in oncogenesis and cell death, J. Natl Cancer Inst. 92, 1564-1572.

30. Kocan, R., Hershberger, P., Sanders, G., & Winton, J. (2009) Effects of temperature on disease progression and swimming stamina in Ichthyophonus-infected rainbow trout, Oncorhynchus mykiss (Walbaum),

J Fish Dis., 32, 10, 835-43.

31. Lafferty, K. D. (2009) The ecology of climate change and infectious diseases. Ecology, 90, 888-900.

32. Lafferty, K. D., Porter, J. W & Ford, S. E. (2004) Are diseases increasing in the ocean? Ann. Rev. Ecol. Evol. Syst., 35, 31-54.

33. Lenton, T. M., Held, H., Kriegler, E., Hall, J. W., Lucht, W., Rahmstorf, S., & Schellnhuber, H. J. (2008) Tipping elements in the earth's climate system, Proc Natl Acad Sci, USA, 105, 6, 1786-1793.

34. Li, C. Y., Lee, J. S., Ko, Y. G., Kim, J. I. & Seo, J. S. (2000) Heat shock protein 70 inhibits apoptosis downstream of cytochrome c release and upstream of caspase-3 activation. J. Biol. Chem., 275, 25665-25671.

35. Ling, S. D., Johnson, C. R., Frusher, S. D., & Ridgway, K. R. (2009) Overfishing reduces resilience of kelp beds to climate-driven catastrophic phase shift, Proc Natl Acad Sci USA., 106, 52, 22341-22345.

36. Mantzouni, I., & Mackenzie, B. R. (2010) Productivity responses of a widespread marine piscivore, Gadus morhua, to oceanic thermal extremes and trends. Proc Biol Sci. Feb 10. [Epub ahead of print].

37. Marcogliese, D. J. (2008) The impact of climate change on the parasites and infectious diseases of aquatic animals, Rev. sci. tech. Off. int. Epiz., 27, 2, 467-484.

38. Marimoto, R.J. (1998) Regulation of the heat shock transcriptional response: cross talk between a family of heat shock factors, molecular chaperones and negative regulators. Genes Dev., 12, 3788-3796.

39. McGinnity, P., Jennings, E., DeEyto, E., Allott, N., Samuelsson, P., Rogan, G., Whelan, K., & Cross, T. (2009) Impact of naturally spawning captive-bred Atlantic salmon on wild populations: depressed recruitment and increased risk of climate-mediated extinction, Proc Biol Sci. 276, 1673, 3601-3610.

40. Menon, A. G. K. (1954) Fish geography of the Himalayas. Zoological Survey of India, Calcutta. 11, 4, 467-493.

41. Mohanty, B. P., Mondal, K., Bhattacharjee, S., & Vass, K. K. (2008) HSP 70 expression profile in tissues of the large riverine catfish Aorichthys seenghala (Sykes). P-GNB-58, p.153. 8th Indian Fisheries Forum 22-26 Nov 2008, Kolkata, India; jointly organized by CIFRI, Inland Fisheries Society of India and Indian Fisheries Forum. ISBN-81- 85482-14-4.

42. Mohanty, B. P., Bhattacharjee, S, Mondal, K., & Das, M. K. (2009) HSP 70 expression in different tissues of some important tropical freshwater fishes. 96th Indian Science Congress, 3-7 January 2009, organized by

NEHU, Shillong, India.

43. Mohanty, S., & Mohanty, B. P. (2009) Global climate change: a cause of concern, Natl Acad Sci Lett, 32, 5 & 6, 149-156.

44. Mohanty, B. P., Behera, B. K., & Sharma, A. P. (2010a) Nutritional significance of small indigenous fishes in human health. Bulletin No. 162, Central Inland Fisheries Research Institute, Barrackpore, Kolkata, India. ISSN 0970-616X.

45. Mohanty, B. P., Bhattacharjee, S., Mondal, K., & Das, M. K. (2010b) HSP70 expression profiles in white muscles of riverine catfish Rita rita show promise as biomarker for pollution monitoring in tropical rivers. Natl Acad Sci Lett., 33, 5 & 6, 177-182.

46. Mumby, P. J., & Harborne, A. R. (2010) Marine reserves enhance the recovery of corals on Caribbean reefs, PLoS One 5, 1, e8657.

47. National Climate Data Centre, National Oceanic and Atmospheric Administration. Global warming: frequently asked questions. Available at: www.ncdc.noaa.gov/oa/ climate/globalwarming.html. Accessed December 9, 2008.

48. Nicholls, R. J., Wong, P. P., Burkett, V. R., Codignotto, J. O., Hay, J. E., McLean, R. F., Ragoonaden, S., & Woodroffe, C. D. (2007) Coastal systems and low-lying areas. In: Climate Change 2007: impacts, adaptation and vulnerability, Parry, M. L., Canziani, O. F., Palutikof, J. P., Linden, V. D. & Hanson, C. E., (Eds.), pp. 315-356. Contribution of working group II to the Fourth Assessment Report of the Intergovernmental Panel on Climate Change, Cambridge University Press, Cambridge, UK.

49. Orr, J. C., Fabry, V. J., Aumont, O., Bopp, L., Doney, S. C., Feely, R. A., Gnanadesikan, A., Gruber, N., Ishida, A., Joos, F., Key, R. M., Lindsay, K., Maier-Reimer, E., Matear, R., Monfray, P., Mouchet, A., Najjar, R. G., Plattner, G-K, Rodgers, K. B., Sabine, C. L., Sarmiento, J. L., Schlitzer, R., slater, R. D., Totterdell, I. J., Weirig, M-F., Yamanaka, Y., & Yool, A. (2005) Anthropogenic ocean acidification over the twenty-first century and its impact on calcifying organisms. Nature, 437, 681-686.

50. ProAct Network (2008) The role of environmental management and eco-engineering in disaster risk reduction and climate change adaptation.

51. Prowse, T. D., Furgal, C., Wrona, F. J., & Reist, J. D. (2009) Implications of climate change for northern Canada: freshwater, marine, and terrestrial ecosystems, Ambio, 38, 5, 282-289.

52. Regional Framework for action to protect human health from effects of climate change in the South East Asia and Pacific Region. 2007. Available at http://www.searo.who.int/ en/Section260/Section2468_14335.htm.

Accessed December 9, 2008.

53. Scott, M. A., Locke, M., & Buck, L. T (2003) Tissue- specific expression of inducible and constitutive Hsp70 isoforms in the western painted turtle, J Exptl. Biol., 206, 303-311.

54. Shao, K. T. (2009) Marine biodiversity and fishery sustainability. Asia Pac J Clin Nutr., 18, 4, 527-531.

55. Shea, K. M., & the Committee on Environmental Health. (2007) Global Climate Change and children's health. Pediatrics, 120, e1359-e1367.

56. Sinha, M., De, D. K., & Jha, B. C. (1998) The Ganga- Environment and Fishery. Central Inland Fisheries Research Institute, Barrackpore, Kolkata, India.

57. Tester, P. A., Feldman, R. L., Nau, A. W., Kibler, S. R., & Wayne Litaker, R. (2010) Ciguatera fish poisoning and sea surface temperatures in the Caribbean Sea and the West Indies. Toxicon. Mar 3. [Epub ahead of print]

58. Thorpe, A., Reid, C., Anrooy, R. V., Brugere, C., & Becker, D. (2006) Poverty reduction strategy papers and the fisheries sector: an opportunity forgone?, J Intl. Dev., 18, 4, 487-517.

59. Tops, S., Hartikainen, H. L., & Okamura, B. (2009) The effects of infection by Tetracapsuloides bryosalmonae (Myxozoa) and temperature on Fredericella sultana (Bryozoa). Int J Parasitol., 39, 9, 1003-1010.

60. Understanding and responding to Climate Change. 2008 Edn. pp. 1-24. The National Academies, USA (http://www.national-academies.org)

61. Vass, K. K., Das, M. K., Srivastava, P. K. & Dey, S. (2009) Assessing the impact of climate change on inland fisheries in River Ganga and its plains in India. Aqu Ecosys Health & Management., 12, 2, 138-151.

62. Veron, J. E., Hoegh-Guldberg, O., Lenton, T. M., Lough, J. M., Obura, D. O., Pearce-Kelly, P., Sheppard, C. R., Spalding, M., Stafford-Smith, M. G., & Rogers, A. D. (2009) The coral reef crisis: the critical importance of<350ppm CO2. Mar Pollut Bull., 58, 10, 1428-1436.

63. Waller, C., Barnes, D. K. A., & Convey, P. (2006) Ecological contrasts across an Atlantic landsea interface, Austral Ecol, 31, 656-666.

64. Walther, G. R., Roques, A., Hulme, P. E., Sykes, M. T., Pysek, P., Kühn, I., Zobel, M., Bacher, S., Botta-Dukát, Z., Bugmann, H., Czúcz, B., Dauber, J., Hickler, T., Jarosík, V., Kenis, M., Klotz, S., Minchin, D., Moora, M., Nentwig, W., Ott, J., Panov, V. E., Reineking, B., Robinet, C., Semenchenko, V., Solarz, W., Thuiller, W., Vilà, M., Vohland, K., & Settele, J. (2009) Alien species in a warmer world: risks and opportunities. Trends Ecol Evol., 24, 12, 686-693

65.   WMO World Data Centre for Greenhouse Gases. Greenhouse gas bulletin: the state of greenhouse gases in the atmosphere using global observations up to December 2004. Vol.1, March 14, 2006.

66.   World Bank & FAO (2008) The sunken billions: the economic justification for fisheries reform. Agriculture and Rural Development Dept. The World Bank: Washington DC. www.worldbank.org.sunkenbillions

# Chapter 3

## ANTIBACTERIAL DRUGS IN FISH FARMS: APPLICATION AND ITS EFFECTS

Selim Sekkin and Cavit Kum

Adnan Menderes University Turkey

## INTRODUCTION

Antibacterial chemotherapy has been applied in aquaculture for over 60 years. The discovery of antibacterials changed the treatment of infectious diseases, leading to a dramatic reduction in morbidity and mortality, and contributing to significant advances in the health of the general population. Antibacterials are used both prophylactically, at times of heightened risk of disease and therapeutically, when an outbreak of disease occurs in the system. The removal of antibacterials from fish medicine would cause great welfare problems. There are many antibacterial drugs for animal health. However, pharmacological research on aquaculture drugs has focused mainly on a few antibacterials widely used in aquaculture. It is well recognised that the issues relating to antibacterial use in animal food are of global concern. Currently, there is a general perception that veterinary medicines, and in particular antibacterials, have not always been used in a responsible manner. In some cases, rather than providing a solution, chemotherapy may complicate health management by triggering toxicity, resistance, residues and occasionally public health and environmental consequences. As a result, authorities have introduced national regulations on the use of antibacterials.

## ANTIBACTERIAL USE IN AQUACULTURE

Aquaculture continues to be the fastest growing animal food producing sector, and aquaculture accounted for 46% of total food fish supply (FAO, 2011). The intensification of aquaculture has led to the promotion of conditions that favour the development of a number of diseases and problems related to biofouling. It is worth remembering the age-old adage that "prevention is better than cure?" and certainly it is possible to devote more attention to preventing the occurrence of disease in fish. Fish may be reared under ideal conditions, in which case the stock are inevitably in excellent condition and without signs of disease (Austin & Austin, 2007). However, disease is a component of the overall welfare of fish (Bergh, 2007). Consequently, a wide range of chemicals are used in aquaculture, including antibacterials, pesticides, hormones, anaesthetics, various pigments, minerals and vitamins, although not all of them are antibacterial agents. As is the case in terrestrial animal production, antibacterials are also used in aquaculture in attempts to control bacterial disease (Burka et al., 1997; Horsberg, 1994; Defoirdt et al., 2011). Usage patterns also vary between countries and between individual aquaculture operations within the same country.

Antibacterials are among the most-used drugs in veterinary medicine (Sanders, 2005). The principal reasons behind the control of infectious diseases in hatcheries are to prevent losses in production; to prevent the introduction of pathogens to new facilities when eggs, fry, or broodstock are moved; to prevent the spread of disease to wild fish via the hatchery effluent or when hatchery fish are released or stocked out; and to prevent the amplification of pathogens already endemic in a watershed (Phillips et al., 2004; Winton, 2001; Lupin, 2009). Antibacterial usage requires veterinary prescription in aquaculture as with usage in terrestrial animals (Sanders, 2005; Prescott, 2008; Rodgers & Furones, 2009). We have limited data about antibacterial use in world aquaculture. For most of the species farmed, we also lack adequate knowledge of the pharmacokinetics (PK) and pharmacodynamics (PD) of administration (Smith et al., 2008). Along with widespread use comes a growing concern about irresponsible use, such as the covert use of banned products, misuse because of incorrect diagnose and abuse owing to a lack of professional advice. That said, there are still not enough approved products for a range of species and diseases in aquaculture (FAO, 2011). Antibacterials are drugs of natural or synthetic origin that have the capacity to kill or to inhibit the growth of micro-organisms. Antibacterials that are sufficiently non-toxic to the host are used as chemotherapeutic agents in the treatment of infectious diseases amongst humans, animals and plants (Table 1). Drug choices for the treatment of common infectious diseases are becoming increasingly limited and expensive and, in some cases, unavailable due to the emergence of drug resistance in bacteria that is threatening to reverse much of the medical progress of the past 60 years (FAO, 2005). In aquaculture, antibacterials have been used mainly for therapeutic purposes and as prophylactic agents (Shao, 2001; Sapkota et al., 2008; FAO, 2005). The voluntary use of antibacterials as growth promoters in any aspect of aquaculture is generally rare. Prophylactic treatments, when they are employed, are mostly confined to the hatchery, the juvenile or larval stages of aquatic animal production. Prophylactic treatments are also thought to be more common in small-scale production units that cannot afford, or cannot gain access to, the advice of health care professionals. There are no antibacterial agents that have been specifically developed for aquacultural use and simple economic considerations suggest that this will always be the case (Smith et al., 2008; Rodgers & Furones, 2009). Despite the widespread use of antibacterials in aquaculture facilities, limited data is available on the specific types and amounts of antibacterials used (Sapkota et al., 2008; Heuer et al., 2009). General considerations in the selection and use of antibacterial drugs are given by Figure 1. (Walker & Giguére, 2008). Treatment options will be different for animals that are held in net pens at sea as opposed to those held in an indoor facility or an aquarium. A treatment must also be feasible: an appropriate treatment route for aquarium fish or selected broodstock individuals may be cost- or labour-prohibitive in commercial aquaculture ventures. The stress associated with treatments must be balanced with the need for and the expected benefits of treatment (Smith et al., 2008). Also, before making a decision to treat a group of fish, the following questions should be asked (Winton, 2001):

1.  Does the loss-rate, severity or nature of the disease warrant treatment?

2.  Is the disease treatable, and what is the prognosis for successful treatment?

3.  Is it feasible to treat the fish where they are, given the cost, handling, and prognosis?

4.  Is it worthwhile to treat the fish or will the cost of treatment exceed their value?

5.  Are the fish in a good enough condition to withstand the treatment?

6.  Will the treated fish be released or moved soon, and is adequate withdrawal or recovery time available?

a   Resistance to antibacterials belonging to different classes in at least one of the isolates. bRepresents a generalisation only, the actual relationship can be variable when an individual drug is involved.

**Table 1:** Properties of the major classes of antibacterial agents (Modified from (Yan & Gilbert, 2004) and (Defoirdt et al., 2011)).

| Antibacterial class | Mode / mechanism of action | Mechanisms of resistance | Multiple resistance[a] | PK-PD relationship[b] |
|---|---|---|---|---|
| ß-Lactams (penicillins, cephalosporins, and carbapenems) | Bactericidal. Inhibition of the penicillin binding proteins (PBPs) located on the cytoplasmic membrane | ß-lactamase production, PBPs modifications, reduced permeability, and efflux | Yes | Time-dependent |
| Aminoglycosides (streptomycin and neomycin) | Bactericidal. Protein synthesis inhibition through binding to the 30s subunit of the ribosome | Decreased permeability, efflux, modification of enzymes, and target (ribosome) modification | Yes | Concentration-dependent |
| Macrolides (erythromycin, tylosin and spiramycin) | Bacteriostatic. Protein synthesis inhibition through binding to the 50s subunit of the ribosome | Target (ribosome) modification of enzymes, decreased permeability and efflux | Yes | Time-dependent |
| Fluoroquinolones (enrofloxacin and ciprofloxacin) | Bactericidal. Inhibition of DNA gyrase and topoisomerase | Target point mutations decreased permeability, efflux and plasmid mediated mechanism | Yes | Concentration-/time-dependent |
| Tetracyclines (oxytetracycline and chlortetracycline) | Bacteriostatic. Protein synthesis inhibition at the ribosomal level (interference with peptide elongation) | Efflux, drug detoxification, and ribosome modification | Yes | Time-dependent |
| Folate synthesis inhibitors (sulphonamides and ormetoprim) | Single bacteriostatic, combination bactericidal. Inhibition of dihydro-pteroate synthase and dihydrofolate reductase | Decreased permeability, formation of enzymes with reduced sensitivity to the drugs | Yes | Concentration-dependent |
| Phenicols (florfenicol and chloramphenicol) | Bacteriostatic. Inhibit the peptidyltransferase reaction at the 50s subunit of the ribosome | Decreased target binding, reduced permeability, efflux and modifying enzymes | Yes | Time-dependent |

## The pathogen and the host

Organisms responsible for disease in aquatic species include fungi, bacteria, nematodes, cestodes and trematodes as well as parasitic protozoans, copepods and isopods. Some can cause death, while others may stress the affected animal to the point that it becomes more susceptible to additional diseases (Stickney, 2005). Disease forms a part of the lives of wild fish and farmed fish. Often, it is not cultured fish that are most susceptible, due to efficient prophylactic strategies and good culture practices. Unprotected wild fish, as exemplified in the case of salmon lice, will be more susceptible to infections and mortality (Bergh, 2007). The difference between health and disease typically depends on the balance between the pathogen and the host, and that balance is greatly influenced by environmental factors, such as temperature and water chemistry (Winton, 2001). The diagnostic techniques for pathogens that are used range from gross observation to highly technical bimolecular-based tools. Pathogen screening is another health management technique, which focuses on the detection of pathogens in subclinical or apparently healthy hosts (Subasinghe, 2009).

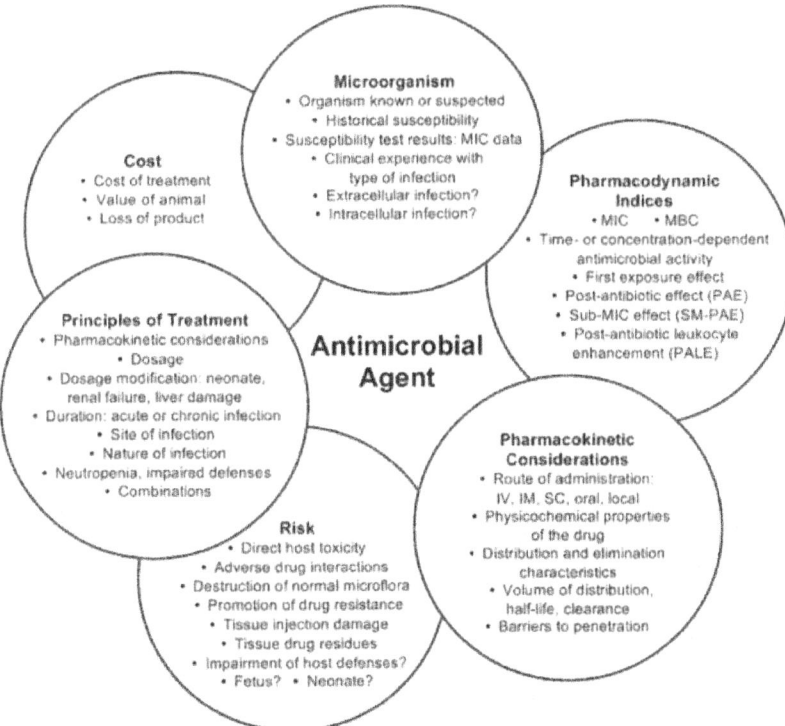

**Figure 1:** Some considerations in selecting and using antibacterial drugs (Walker & Giguére, 2008).

The primary pathogens in aquaculture are bacteria and viruses (Shao, 2001). More than 100 bacterial pathogens of fish and shellfish have been reported (Alderman & Hastings, 1998; Winton, 2001). The artificially high host-densities associated with aquaculture are evolutionarily beneficial for pathogens (Bergh, 2007). Bacterial pathogens probably cause more disease problems overall than all other causes combined. In virtually every type of aquaculture, bacterial diseases rank number one amongst aetiological agents. In each type of culture, and for virtually every species, specific bacterial pathogens are responsible for serious disease problems. Gram-negative bacilli are the most frequent cause of bacterial diseases in finfish. Although only a few Gram-positive forms affect finfish, such bacteria cause serious diseases among crustaceans (Meyer, 1991; Rodger, 2010; Roberts, 2004). Whereas similar types of pathogens affect freshwater and marine fish, relatively few pathogens are transmissible from freshwater fish to marine fish, and vice versa (i.e. most pathogens affect either marine or freshwater fish, but not both). This is the rationale for why many freshwater pathogens can be treated with salt, and many marine pathogens can be treated with freshwater (Noga, 2010). Choosing the right drug depends in part on such factors as age, size and the housing of the animal. Common bacterial fish diseases, their definition, aetiology and treatment, as well as control issues, are resumed in Table 2.

**Table 2:** Common bacterial fish diseases, their aetiology and treatment – control issues (Modified from (Rodger, 2010)).

| Disease / Aetiology | Treatment and control |
|---|---|
| **Mycobacteriosis.** Mycobacteriosis in fish is a chronic progressive disease caused by certain bacterial species within the genus Mycobacterium. *Mycobacterium marinum, M. fortuitum, M. salmoniphilum and M. chelonae are all considered* pathogenic for fish. All are aerobic, acid-fast, Gram-positive and non-spore forming. | There is no fully effective treatment. Therefore, the best course is to cull and disinfect the premises. Rifampicin in combination with tetracycline (Boos et al., 1995) and clarithromycin may reduce infection (Collina et al., 2002). |
| **Coldwater Diseases & Rainbow Trout Fry Syndrome (RTFS).** Bacterial coldwater disease is a serious septicaemic infection of hatchery-reared salmonids, especially young coho salmon, which has also been referred to as peduncle disease. *Flexibacter psychrophilus, Cytophaga psychrophila* and *Flavobacterium psychrophilum* are all terms that have been used for the causal agents of these diseases. Most of the recent classificatory work indicates that *Flavobacterium psychrophilum* is the correct name for these bacteria. These bacteria are Gram-negative. | Broad-spectrum antibacterials have been partially ineffective in controlling an outbreak, but the improvement of the environment and using 3-4 times the recommended doses of antibacterials have shown benefits (Bebak et al., 2007). Florfenicol also appears to be effective for recommended dose regimes. |
| **Bacterial Kidney Disease (BKD).** A serious disease of freshwater and seawater fish, farmed and wild salmonids, that results in an acute to chronic systemic granulomatous disease. *Renibacterium salmoninarum* is a Gram-positive diplococcus that grows best at 15-18°C, is causative agent of BKD and a significant threat to the healthy and sustainable production of salmonid fish worldwide (Wiens et al., 2008). | Chemotherapy (erythromycin) provides limited and only temporary relief. The bacteria can survive and multiply within phagocytic cells. |

| | |
|---|---|
| **Enteric Redmouth Disease (ERM).** A bacterial septicaemic condition of farmed salmonids, and in particular rainbow trout. There are recent reports amongst channel catfish. *Yersinia ruckeri* is the causal agent and the Gram-negative, motile rod-shaped bacterium is catalase positive and oxidase negative. Several serotypes have been identified. | Broad-spectrum antibacterials are effective in controlling an outbreak, although increasing antibacterial resistance is observed. The bacteria can survive and multiply within phagocytic cells (Tobback et al., 2009; Rykaert et al., 2010). |
| **Furunculosis.** Furunculosis is a fatal epizootic disease, primarily of salmonids. It also causes clinical diseases in other fish species, where it is named ulcer disease or carp erythrodermatitis. *Aeromonas salmonicida* is a Gram-negative bacteria. Atypical furunculosis is caused by a slower growing non-pigmenting isolate, *A. salmonicida achromogenes.* | Broad-spectrum antibacterials are effective in controlling an outbreak, but increasing antibacterial resistance is observed. |
| **Piscirickettsiosis.** A disease of salmonids caused by *Piscirickettsia salmonis* and a significant disease problem in farmed marine salmonids. *Piscirickettsia salmonis* is a Gram-negative, acid-fast, non-motile, spherical to coccoid, non-capsulated (although often pleomorphic) organism. | Broad-spectrum antibacterial therapy is used, although some resistance is developing. Outbreaks are usually associated with stressful events, such as algal blooms, sudden changes in the environment or grading. |
| **Bacterial Gill Disease (BGD).** BGD is an important disease in farmed freshwater salmonids. The bacterium *Flavobacterium branchiophila* causes a chronic, proliferative response in gill tissue. *Flavobacterium branchiophila* is a Gram-negative, long, thin, filamentous rod. | BGD usually responds well to antiseptic and surfactant baths, such as chloramine T and benzalkonium chloride. Providing adequate oxygen is a useful supportive therapy. |
| **Vibriosis.** Vibriosis is the term most commonly used to describe infections associated with *Vibrio spp*. In recent years, vibriosis has become one of the most important bacterial diseases in marine-cultured organisms (Stabili et al., 2010). *Vibrio anguillarum, V. ordalii* and other *Vibrio sp.* may cause similar clinical signs in wild and farmed fish. It is Gram-negative, and has straight or slightly curved rods which are motile. | Broad-spectrum antibacterials are effective in controlling an outbreak, but increasing antibacterial resistance is observed. Vaccines are widely used. Caprylic acid may be helpful as an alternative or as an adjunct to antibacterial treatment (Immanuel et al., 2011). |
| **Epitheliocystis.** Epitheliocystis is a chronic and unique infection caused by the *Chlamydia spp.* organism and which results in hypertrophied epithelial cells - typically of the gills but sometimes also of the skin - or certain freshwater and marine fishes. | Broad-spectrum antibacterials have been used with some degree of success, though the avoidance of infected fish should be adhered to at all costs. |

| | |
|---|---|
| **Tenacibaculosis.** An infection of marine fish by *Tenacibaculum maritimum* is common in farmed fish and many species. The bacteria appear to be opportunistic, commonly infecting fish after minor epidermal or epithelial trauma or irritation, and they can rapidly colonise such tissue. It is Gram negative, with slender bacilli which multiply in mats on the damaged tissue. | Oral treatment with broad-spectrum antibacterials is generally successful if the fish are maintained in a low-stress environment. |
| **Francisellosis.** Francisellosis is the term used to describe infection associated with *Francisella philomiragia subspecies noatunensis*, which has emerged as a major pathogen of farmed cod. *Francisella spp.* is also a major pathogen of farmed tilapia. It is Gram-negative, with intracellular coccobacilli. | There is no effective treatment due to the intracellular nature of the infection. Removal of the affected fish and the disinfection of the premises and equipment, and fallowing. |
| **Rainbow Trout Gastro-Enteritis (RTGE).** RTGE is an enteric syndrome of freshwater farmed rainbow trout, reported in several European countries, and which results in significant economic loss and daily mortalities. It is not fully established and the role of the segmented filamentous bacteria remains unclear, as they are also found in apparently healthy fish. However, *Candidatus arthromitis* may have some role to play in the disease. This bacterium have not yet been cultured in vitro. | Changing the diet-type or the addition of salt to the diet as well as broad-spectrum antibacterials all appear to be effective, once the disease is present. However, none appear to be preventative. Biosecurity is important in preventing the disease from entering a farm. |
| **Red Mark Syndrome (RMS) or Cold Water Strawberry Disease.** RMS is an infectious dermatitis of rainbow trout which does not cause mortality but which presents as dramatic haemorrhagic marks on the skin. It is not fully established, although *Flavobacterium psychrophilum* and rickettsia-like organisms have been associated with it. | The lesions will resolve eventually without treatment; however, broad-spectrum antibacterials do induce the rapid healing of the condition. The avoidance of any livestock from infected farms reduces the chance of the introduction of RMS onto a site. |

## Antibacterial susceptibility testing of aquatic bacteria

Resistance is a description of the relative insusceptibility of a microorganism to a particular treatment under a particular set of conditions. Therefore, it should be noted that resistance or at least the resistance level depends strongly upon the test type and test conditions, as well as the type of compound and its mode of action (Kümmerer, 2008). The empirical use of antibacterials should be avoided. The use of antibacterials should always be based upon an examination of the clinical case, the diagnosis of a bacterial infection and the selection of a clinically efficacious antibacterial agent. However, in certain situations (such as when the animal is seriously ill or where there is an outbreak with a high mortality or a rapid spread) therapy may be initiated on the basis of clinical signs (Guardabassi & Kruse, 2008; Smith et al., 2008). The target organisms must be known or shown to be susceptible, and adequate concentrations must be shown to reach the target (Phillips et al., 2004). A definitive diagnosis requires the isolation and identification of the causative organism, preferably from three to five infected fish (Smith et al., 2008). Samples for a bacteriologic culture should be collected from the actual site of infection before administering an antibacterial drug (Walker & Giguére, 2008). Currently, a wide range of standardised methods are available (Smith et al., 2008; Guardabassi & Kruse, 2008; CLSI, 2006a, 2006b). It should be expected that there will be differences between the bacteria isolated and their antibacterial sensitivities, between freshwater and saltwater fish, between different taxa of fish, and possibly even between different species of fish (Mulcahy, 2011). Furthermore, due to the varying activity spectrum of the different compounds in some tests, microbial population dynamics may overrule their effects in some populations. They may thereby mask effects (Kümmerer, 2008). The discrepancies between testing methods may also require further studies (Kum et al., 2008). Differences in the measurement of zone sizes by individual scientists also represent a possible source of inter-laboratory variation (Nic Gabhainn et al., 2004).

## The treatment route

In intensive fish farming, the antibacterials used to treat bacterial infections are administrated generally by either water-borne or oral means, or else through injection (Shao, 2001; Treves-Brown, 2001; Zounkova et al., 2011). Agents that are intended to treat diseases must reach therapeutic levels in target tissues. It is always advisable to perform a bioassay of a small number of individuals before treating any fish species without a known history of response to the treatment. A bioassay can be performed by placing five or six fish in an aquarium that has the treated pond water. The fish should be observed for 1-2 days before treatment so as to be sure that none have died from the stress of collection. Fish should never be left unattended during treatment and, if an adverse response occurs, the drug should be immediately removed by transferring the fish to clean water or diluting the treatment water. It is necessary to take the presence of these additives into account when calculating the active drug quantity required for any treatment (Noga, 2010; Rodgers & Furones, 2009). Adequate plans for detoxification and the removal and disposal of used drugs must be in place before treatment is begun (e.g., ammonia and nitrite levels must be monitored closely during therapy). When hospitalisation is completed, the aquarium, the filter, and all other materials in contact with the hospitalisation aquarium should be disinfected before re-use. Used drugs must be disposed of responsibly. Deteriorated or otherwise uncontrolled water quality poses particular challenges to farmed fish and their surroundings. Outputs from these systems can further harm local wildlife and the ecosystem (Cottee, 2009). There is no one specific drug application method which is better than other; rather, the method of treatment should be based on the specific situation encountered. Here, experience is exceptionally valuable. A fish health professional or other knowledgeable source should be consulted if one is unfamiliar with the disease or treatment proposed (Winton, 2001). Methods for the application of antibacterials to fish are resumed in Table 3.

**Table 3:** Methods for application of antibacterials to fish (Haya et al., 2005).

| Method of application | Comments |
|---|---|
| Oral route (on food) | Needs palatable components; minimal risk of environmental pollution |
| Bioencapsulation | Needs palatable compounds; minimal risk of environmental pollution |
| Bath | Need for a fairly lengthy exposure to the compound, which must be soluble or capable of being adequately dispersed; problem of the disposal of spent drug |
| Dip | Brief immersion in a compound, which must be soluble or capable of being adequately dispersed; problem of disposal of the dilute compound |
| Flush | Compound added to a fish holding facility for brief exposure to fish; must be soluble or capable of being adequately dispersed; poses a problem of environmental pollution |
| Injection | Feasible for only large and/or valuable fish; usually requires prior anaesthesia; slow; negligible risk of environmental pollution |
| Topical application | Feasible for the treatment of ulcers on valuable/pet fish |

## *Water medication*

The water-borne route is the most common method for administering treatments to fish and it has distinct advantages, such as being relatively non-stressful and easy to administer. Drugs are added to water for two distinct purposes. The first and most obvious one is so that the drug will be absorbed by, and so medicate, the fish; the second is to kill the freeliving and, hence, transmissible stages of parasites (Treves-Brown, 2001). Seawater fish drink significant amounts of water and may absorb large amounts of a drug via the gastrointestinal tract (Noga, 2010). Application by the water-borne route becomes necessary if the fish refuse to eat, and, therefore, would be unlikely to consume any medicated food. With these methods, the fish are exposed to solutions/suspensions of the drug for a predetermined period. This may be only briefly, i.e., a few seconds duration (a "dip") or for many minutes to several hours (a "bath") (Haya et al., 2005). Waterborne antibacterial treatments will vary depending upon the animal and its holding conditions. Treating fish by applying the drug to the water avoids stressing the fish by handling (Reimschuessel & Miller, 2006). However, there are disadvantages. Relative to other treatment routes, dosing is less precise (too little or too much). Baths and dips are not as effective as some of the other treatment methods – particularly for systemic infections – because of generally poor internal absorption of the antibacterial being used. Water-borne treatments are mainly used for surface-dwelling (skin and gill) pathogens, including parasites, bacteria and water moulds. Certain species, such as scaleless fish are often especially sensitive to water-borne treatments (Rodgers & Furones, 2009). Antibacterials which are absorbed from the water include chloramines, dihydrostreptomycin, enrofloxacin, erythromycin, flumequine, furpyrinol, kanamycin, oxolinic acid, oxytetracycline, nifurpirinol, sulphadimethoxine, sulphadimidine, sulpha-monomethoxine, sulphanilamide, sulphapyridine, sulphisomidine and trimethoprim. Antibacterials that are absorbed poorly or not at all include chloramphenicol and gentamicin (Reimschuessel & Miller, 2006). With bath-type treatments, more antibacterials are required when compared with oral (feed) treatments or injections. Bath treatments are also not recommended for recirculation systems or aquarium systems using biological filters. The accurate calculation of the volume of water in the tank, pond or cage is also required (Rodgers & Furones, 2009). If both short- and long-term exposures are probably equally feasible and effective, it is preferable to use a short-duration drug exposure. The advantages of this type of treatment lie in reduced waste (and thus reduced expense) and less environmental contamination (Reimschuessel & Miller, 2006). Even where absorption is known to occur, the technique does have some important disadvantages. In particular, in most cases less than 5% of the administered dose will be absorbed by the fish. In this case, the technique is wasteful, expensive (at least twenty times the dose required by the fish must be provided) and environmentally undesirable (Treves-Brown, 2001).

## *Oral medications*

In food fish or ornamental aquaculture, many of the bacterial diseases of fish can be successfully treated with medicated feeds, and it is usually the preferred method of treatment. However, care must be taken because some of the causes of disease – such as stress – can lead to treatment failures or the recrudescence of disease after the completion of treatment (Rodgers & Furones, 2009). Fish in ponds are best treated using oral medications. However, sick fish may not eat, and withholding food for 12-24 hours may increase the acceptance of a medicated feed (Reimschuessel & Miller, 2006). The incorporation of an antibacterial in the feed is usually via a powdered premix in conjunction with a binder, such as gelatine (up to 5%), fish or vegetable oil (Shao, 2001). The dosage required for treatment with a medicated feed depends upon the original level of active ingredient/kg fish body weight. The dosage rates used in medicated feeds will vary according to the specific antibacterial used, but usually the rate is based on a number of grams per 100 kg of fish per day. The exact dosage will also require the number and average weight of the fish to be treated, as well as a daily feeding rate and consideration of whether the fish are marine or fresh water species. It is also important that treated fish must not be harvested for food use until a specified withdrawal period has elapsed (Rodgers & Furones, 2009). One problem for the treatment of marine species is that antibacterials have been shown to be less effective in seawater, which is related to their reduced bioavailability, e.g., tetracycline has a low bioavailability in fish ($< 10\%$) due to binding with sea-water-borne divalent cations such as $Mg^{2+}$ and $Ca^{2+}$. It is noteworthy that non-bioavailable tetracyclines contaminate the environment (Toutain et al., 2010). The bioavailability of some aquaculture drugs in salmon held in seawater is shown in Table 4. (Rodgers & Furones, 2009).

**Table 4:** Examples of reduced bioavailability for some aquaculture drugs in seawater (Rodgers & Furones, 2009).

| Antibacterial | Bioavailability (%) |
|---|---|
| Oxytetracycline | 1 |
| Amoxicillin | 2 |
| Sarafloxacin | 2 |
| Oxolinic acid | 30 |
| Flumequine | 45 |
| Sulfadiazine | 50 |
| Trimethoprim | 96 |
| Florfenicol | 97 |

The dosage can vary within certain limits and depending upon the feeding rate. It is usually best to use a feed that has enough medication so that feeding at a rate of 1% of body weight per day will give the needed dosage (Noga, 2010). Absorption from the intestinal tract may vary between species. Saltwater fish will drink and, therefore, drugs may bind cations in the water in their intestinal tracts, affecting bioavailability (Reimschuessel & Miller, 2006). For particular applications, like the treatment of young larvae and fry, some success has been obtained with the bio-encapsulation of drugs in live feeds, especially with artemias. Other innovative methods of oral delivery – like microspheres or coated beads – offer the possibility of protecting fragile molecules from deterioration in the gastric juices, carrying them up to their target sites in the intestine. Though still quite recent, large developments are expected from these innovative technologies in the near future (Daniel, 2009).

## Injection

The injection of antibacterials can be a more effective treatment for bacterial infections than the use of a medicated feed, particularly for advanced infections and as the best way of being sure of the given dose (Douet et al., 2009). However, it is usually only practical for valuable individual fish, such as brood stock or ornamental fish, rather than for fish in large-scale production facilities. Injection quickly leads to high blood and tissue levels of antibacterials (Yan & Gilbert, 2004; Haya et al., 2005). Normally, an individual fish will also need to be anaesthetised before treatment. Typical injection sites include the intraperitoneal (IP) cavity and the intramuscular route (IM) (Rodgers & Furones, 2009; Treves-Brown, 2001). Disadvantages include the stress imposed by capturing the fish and, for aquarium fish, the need to bring the fish to the clinic for every injection, since the owner is usually unable to perform the treatment (Noga, 2010). The IP route is the widely used route for injection. Fish should be fasted for 24 hours prior to injection. The landmarks for an IP injection are the pelvic fins and the anus. All fish should be at least 35g. Improper injection can lead to peritoneal adhesion, ovulation problems, mortality from injection, reduced efficacy, side effects (local reactions), reduced carcass quality and therapy failure (Treves-Brown, 2001; Noga, 2010). IM injection is best used only on fish more than 13 cm long or else more than 15g. The best site is the dorsal musculature just lateral to the dorsal fin. Only relatively small amounts can be injected (0.05 ml/50 grams of fish). Injections should be done slowly. The IM route has the disadvantage of causing damage to carcass quality and has the potential of forming sterile abscesses (Noga, 2010). The volume required for the injection of antibacterials is based on the weight of fish to be treated, the recommended dosage for the antibacterial being used and its supplied concentration (Rodgers & Furones, 2009). This is usually expressed as:

Volume of antibacterial required = recommended dosage (mg/kg) x weight of fish (kg) /supplied solution concentration (mg/ml)

## Topical application

The topical application of drugs to fish is rare. Anaesthesia is an essential preliminary procedure. Topical treatments are usually only necessary for more valuable individual fish, such as ornamental varieties or brood stock. Ointments containing antibacterials have sometimes been used in fish surgeries, applied to the sutures and incision site. Commercial antibacterial ointments are most commonly used (Mulcahy, 2011). Open sores or ulcers that are secondarily infected by bacteria or water moulds can be treated. A cotton swab should be dipped in a drug solution and then used to gently touch the lesion, allowing the solution to soak the lesion via capillary action. Nevertheless, it is possible that ulcers may heal themselves with improved water quality and the elimination of parasites (Treves-Brown, 2001; Haya et al., 2005; Noga, 2010).

## Water treatment

Disinfection can reduce the risk of disease transmission within aquaculture facilities, and from facilities to the environment, by deactivating or destroying pathogens with disinfecting agents. Disinfection can be done routinely, but also in response to the outbreak of specific diseases (Winton, 2001). In this procedure the drug is applied to all the water in the aquarium. It is, therefore, not applicable to antibacterial drugs, as these would inactivate the filter (Treves-Brown, 2001).

## Dosage

PK and PD data has allowed the design of therapeutic regimens, with the PK/PD variable providing the most appropriate surrogate for drug effectiveness being dependent upon several factors (Rigos & Troisi, 2005; Martinez & Silley, 2010; Toutain et al., 2010). Within some species there may be considerable differences both within and between breeds in PK and PD profiles; veterinary pharmacogenetics aims to identify genetic variations (polymorphisms) as the origin of differences in the drug response of individuals within a given species. These between- and within-species differences in drug response are largely explained by variations in drug PK and PD, the magnitude of which varies from drug to drug (Toutain et al., 2010). We will not be able to apply the full power of the PK/PD approach to either the design of treatment regimens that minimise the development of resistance or the setting of clinical breakpoints that provide an empirical definition of resistance (Smith et al., 2008). There is a considerable amount of information available on the PK of various antibacterials delivered by different routes to different species of fish, but little information about the plasma levels of antibacterials that are required to be of benefit for the implanted fish or the calculation of the dosage of antibacterial required to obtain a positive benefit (Mulcahy, 2011). It is important to remember that it is the host immune system that is ultimately responsible for success in combating bacterial disease (Martinez & Silley, 2010). If one is unsure about the dose to use, it is usually best to start with the lower recommended dose. If the disease does not respond adequately, repeat the treatment with a higher dose. For oral medications, dosage varies with feed intake. Fish that are eating less need a higher percentage of the drug in their diet, but there are limits on the legally allowable amount as well as practical considerations, since some drugs are unpalatable at high doses (e.g., many antibacterials) (Noga, 2010; Winton, 2001). Drug dosage regimens also are host-dependent. Fish species reared in warm water may absorb, metabolise and excrete drugs at a different rate (often faster) than those in cold water.

**Table 5:** Half-lives and dosages of antibacterials in fish (Reimschuessel et al., 2005).

| Drug | Species | t[1/2] (hr) | Dosage | Route[b] | °C |
|---|---|---|---|---|---|
| Amoxicillin | Atlantic salmon | 120 | 12.5 mg/kg sd[a] | IM | 13 |
| | Atlantic salmon, sea bream | 14-72 | 40-80 mg/kg sd | IV/PO | 16-22 |
| Chloramphenicol | Carp | 48-72 | 40 mg/kg sd | IP | 9 |
| Ciprofloxacin | Carp, rainbow trout, African catfish | 11-15 | 15 mg/kg sd | IM/IV | 12-25 |
| Difloxacin | Atlantic salmon | 16 | 10 mg/kg sd | PO | 11 |
| Enrofloxacin | Atlantic salmon, red pacu, rainbow trout, sea bass, sea bream | 24-105 | 5-10 mg/kg sd | IM/IV/PO | 10-26 |
| Erythromycin | Chinook salmon | 120 | 0.1 g/kg 21 d | PO | 10 |
| Florfenicol | Atlantic salmon | 12-30 | 10 mg/kg sd | IV/PO | 10-11 |
| | Cod | 39-43 | 10 mg/kg sd | IV/PO | 8 |
| Flumequine | Eel | 255 | 9 mg/kg sd | IM | 23 |
| | Atlantic halibut, brown trout, corkwing wrasse, Atlantic halibut, Atlantic salmon, cod, goldsinny wrasse, sea bass, sea bream, turbot | 21-96 | 5-25 mg/kg sd | IP/IV/PO | 5-25 |
| | Eel | 208-314 | 10 mg/kg sd | IV/PO | 23 |
| | Rainbow trout | 285-736 | 5 mg/kg sd | IV/PO | 13 vs 3 |
| Furazolidone | Channel catfish | 1-24 | 1 mg/kg sd | IV/PO | 24 |
| Gentamicin | Channel catfish, brown shark, goldfish | 12-54 | 1-3.5 mg/kg sd | IC/IM | 20-25 |
| | Toadfish | 602 | 3.5 mg/kg sd | IM | 19 |
| Miloxacin | Eel | 35 | 30-60 mg/kg sd | IV/PO | 27 |
| Nalidixic acid | Rainbow trout, amago salmon | 21-46 | 5-40 mg/kg sd | IV/PO | 14-15 |
| Nifurstyrenate | Yellowtail | 2 | 100 mg/kg sd | PO | 23 |
| Ormetoprim | Atlantic salmon, channel catfish, rainbow trout, hybrid striped bass | 4-25 | 4-50 mg/kg sd | IV/PO | 10-28 |
| Oxolinic acid | Atlantic salmon, corkwing wrasse, channel catfish, cod, rainbow trout, red sea bream, sea bass | 15-87 | 4-20 mg/kg sd | IP/IV | 8-24 |
| | Atlantic salmon, cod, rainbow trout | 82-146 | 25-75 mg/kg sd | PO | 5-8 |

| | | | | | |
|---|---|---|---|---|---|
| | Atlantic salmon, gilthead sea bream, rainbow trout, sharpsnout, sea bream, turbot | 13-48 | 10-40 mg/kg up to 10d | PO | 9-19 |
| Oxytetracycline | African catfish, carp, rainbow trout, red pacu, sockeye salmon | 63-95 | 5-60 mg/kg sd | IM | 12-25 |
| | African catfish, Atlantic salmon, ayu, carp, Chinook salmon, eel, rainbow trout, red pacu, sea bass, sea bream, sharpsnout, sea bream | 6-167 | 5-60 mg/kg sd | IV | 8-25 |
| | Arctic charr | 266-327 | 10-20 mg/kg sd | IV | 6 |
| | Atlantic salmon, ayu, black sea bream, carp, channel catfish, eel, perch, rainbow trout, sea bass, sea bream, hybrid striped bass, summer, flounder, walleye | 43-268 | 10-100 mg/kg up to 10d | PO | 7-27 |
| | Arctic charr, sockeye salmon, Chinook salmon | 428-578 | 10-100 mg/kg sd | PO | 6-11 |
| Piromidic acid | Eel, goldfish | 24 | 5 mg/kg sd | PO | 26 |
| Sarafloxacin | Atlantic salmon, cod | 12-45 | 10-15 mg/kg sd | IV/PO | 8-24 |
| Streptozotocin | Toadfish | 24 | 50 uCi | IV | |
| Sulphachlor-pyridazine | Channel catfish | 4-5 | 60 mg/kg sd | IC/PO | 22 |
| Sulfadiazine | Atlantic salmon, carp, rainbow trout | 26-96 | 25-200 mg/kg sd | IV/PO | 8-24 |
| Sulfadimethoxine | Atlantic salmon, channel catfish, rainbow trout, hybrid striped bass | 1-48 | 25-200 mg/kg sd | IV/PO | 10-20 |
| Sulphadimidine | Carp, rainbow trout | 18-57 | 100-200 mg/kg sd | IV/PO | 10-20 |
| Sulphametho-xypyridazine | Rainbow trout | 72 | 200 mg/kg sd | PO | 13 |
| Sulphamono-methoxine | Rainbow trout, yellowtail | 5-33 | 100-400 mg/kg sd | IV/PO | 15-22 |
| Sulphanilamide | Rainbow trout | 36 | 200 mg/kg sd | PO | 13 |
| Sulfathiazole | Rainbow trout | 60 | 200 mg/kg sd | PO | 13 |
| Thiamphenicol | Sea bass | 21 | 30 mg/kg 5d | PO | 19 |
| Tobramycin | Brown shark | 48 | 1-2.5 sd | IM | 25 |
| Trimethoprim | Atlantic salmon, carp, rainbow trout | 21-50 | 1-100 mg/kg sd | IV/PO | 8-24 |
| Vetoquinol | Cod | 79 | 25 mg/kg sd | PO | 8 |
| | Atlantic salmon | 16 | 40 mg/kg sd | PO | 10 |

asd: single dose. bAbbreviations, IM: intra muscular, IV: intra venous, PO: per os (oral), IP: intra peritoneal, IC: intra coelom, uCi: a unit of radioactivity (Curie= 3,7x1010 disintegration per second).

The salinity of the holding water also affects drug kinetics. Fish kept in saltwater drink the water while freshwater fish do not. Thus, antibacterials in the gastrointestinal tract of fish species held in saltwater may bind cations, which can reduce their uptake (Smith et al., 2008; Toutain et al., 2010). This is especially true for antibacterials – such as the tetracyclines – that have low bioavailability even in freshwater. The half-lives of drugs in fish are highly dependent upon the dosage regimen, the route and the temperature. Therefore, these parameters are included in the Phish-Pharm Database and should be considered when administering antibacterials to fish. Table 5 shows some drug dosages that have been reported for fish (Reimschuessel et al., 2005). It is important to realise that the dosages listed in Table 5 may not have been shown to be safe or effective in all fish species. No generalisations are possible. Successful therapy often depends on maintaining adequate blood levels over a course of seven to ten days. Temperature is a very important factor in deciding on the dose and treatment intervals (Toutain et al., 2010).

## Drug metabolism in fish

Liver is the primary organ for the detoxification of drugs in fish. Similarities exist in the metabolism of drugs by fish and mammals. The metabolism of aquaculture antibacterials by the cytochrome P450 system could affect their activation, tissue distribution and elimination rates, and determine the persistence of residues as well as the length of the withdrawal period before the fish can be used for human consumption (Moutou, 1998). The elimination rate of antibacterials from fish tissues varies greatly with the temperature. The temperature dependency of drug PK is an important consideration for drug residues. The elimination half-life of antibacterial drugs increases significantly as the temperature decreases. Ideally, the drug dose should be adjusted according to the water temperature, but in clinical practise the dose is normally fixed (Toutain et al., 2010). However, unmetabolised oxytetracycline can be passed unabsorbed through the body of treated sparids and then excreted via the faeces into the local marine environment (Rigos & Troisi, 2005; Rigos et al., 2004).

## Duration of antibacterial treatment

It is universally recognised that a drug must be present in a sufficient concentration for an adequate length of time at the site of the infection, although the variables affecting the length of treatment have not yet been fully defined (Walker & Giguére, 2008). The responses of different types of infections to antibacterial drugs vary, and clinical experience with many infections is important in assessing the response to the treatment. For serious acute infections, treatment should last at least 7 to 10 days. If no response is seen by that time, both the diagnosis and treatment should be reconsidered (Walker & Giguére, 2008). It is important to remember that it is the host immune system that is ultimately responsible for any success in combating bacterial diseases (Martinez & Silley, 2010).

## Failure of antibacterial therapy

Treatment failure has many causes. The selected antibacterial may be inappropriate because of misdiagnosis, poor drug diffusion at the site of the infection, inactivity of a given drug at the site of infection, failure to identify the aetiological agent including inaccurate results of laboratory tests, resistance of pathogens, intra-cellular location of bacteria, metabolic state of the pathogen, or errors in sampling. Other factors that may contribute are inadequate dosage or the use of drugs with low bioavailability. When failure occurs, diagnose must be reassessed and proper samples collected for laboratory analysis. Patient factors such as the persistence of foreign bodies, neoplasia, and impairment of host defences are important to consider. It is important also to ensure that persons medicating their own animals comply with dosing instructions (Walker & Giguére, 2008; Winton, 2001; Treves - Brown, 2001; Noga, 2010).

# TREATMENT OPTIONS IN VARIOUS AQUACULTURE SYSTEMS

Another important factor influencing treatment is the type of culture system. The four major types of culture system are aquaria, ponds, cages and flow-through systems (Noga, 2010). The main factors that may influence a treatment's success are given in Table 6.

**Table 6:** Major types of culture systems influencing the diseases' treatment (Table established from (Noga, 2010)).

| Aquaria | Ponds | Cages | Raceways |
|---|---|---|---|
| The most highly controllable culture systems for maintaining temperature, biological filtration and oxygen. Amenable to various waterborne treatments. Ease of manipulability. | Influenced by natural factors such as light, temperature and rainfall. Natural biological cycles are less controllable. Interventional strategies are more limited compared with aquaria. | Susceptible to the vagaries of natural environmental changes. Water-borne treatments are possible in such systems, but are much more difficult. The fish that need to be treated in such systems must have their cage enclosed. Alternatively, the fish must be treated in a closed system (e.g., a bath treatment) or the medications must be delivered orally. | Raceways and other flow-through systems are the least manipulable systems by virtue of the constant and rapid water turnover. Similar adverse environmental consequences can follow such treatments. Flow-through systems are even more limited than cages in the ability to use water-borne treatments. |

# LEGAL USE OF ANTIBACTERIALS

A number of international and regional codes of practice, agreements and technical guidelines exist for aquatic animals (Subasinghe, 2009). The drugs available for use and their treatment protocols are tightly regulated. The consumers of fish – and particularly in the world's richer economies – are increasingly demanding that retailers guarantee that the fish which they offer are not only of a high quality and safe to eat, but also that they derives from fisheries that are sustainable (FAO, 2011). As health threats have appeared, management practices have evolved and fish husbandry has greatly improved over the past 20 years, resulting in a reduction in the use of some chemicals, and particularly the use of antibacterials in most jurisdictions (Burridge et al., 2010). The banning of any antibacterial usage in animals based upon the "precautionary principle" in the absence of a full quantitative risk assessment is likely to be wasted at best, and even harmful at worst, both to animal and human health (Phillips et al., 2004). The antibacterials used in veterinary medicine are only prescribed by veterinarians in the European Union (EU). The prescription scheme could be discussed and improved, and non-approved and even banned antibacterials are purchased "over-the-counter" (without the need for a prescription) or their use is undeclared in fish feed formulations. The use of specifically banned antibacterials in aquaculture is a violation of regulations (Lupin, 2009). The user safety data included on labelling and packaging inserts should provide sufficient information for such occupational safety assessments to be made (Alderman & Hastings, 2009). Before approval, drugs are assessed for the definition of their maximum residue limits (MRLs) (Table 7), and their environmental impact and efficacy (Sanders, 2005). MRLs are generated by a number of bodies, such as the EU, and more globally within the framework of the FAO/WHO Codex Alimentarius Commission, which is advised scientifically by the JECFA (Joint FAO/WHO Expert Committee on Food Additives). The use of antibacterial agents in food animal species, including fish, is controlled by regulations, particularly in Europe and the USA.

**Table 7:** Main antibacterial compounds having fixed MRLs (Modified from (Daniel, 2009)).

| Antibacterial | Species* | Tissue** | MRL | Comments |
|---|---|---|---|---|
| Amoxicillin | All FPS | Muscle | 50 pg/kg | |
| Ampicillin | All FPS | Muscle | 50 pg/kg | |
| Benzylpenicillin | All FPS | Muscle | 50 pg/kg | |
| Chlortetracycline | All FPS | Muscle | 100 pg/kg | |
| Cloxacillin | All FPS | Muscle | 300 pg/kg | |
| Colistine | All FPS | Muscle | 150 pg/kg | |
| Danofloxacin | All FPS | Muscle | 100 pg/kg | |
| Dicloxacillin | All FPS | Muscle | 300 pg/kg | |
| Difloxacin | All FPS | Muscle | 300 pg/kg | |
| Enrofloxacin | All FPS | Muscle | 100 pg/kg | Enro.+ciprofloxacin |
| Erythromycin | All FPS | Muscle | 200 pg/kg | Erythromycin A |
| Florfenicol (Fish) | Fish | Muscle+skin | 1000 pg/kg | |
| Flumequine | Fish | Muscle+skin | 600 pg/kg | |
| Lincomycin | All FPS | Muscle | 100 pg/kg | |
| Neomycin (Incl. Framycetin) | All FPS | Muscle | 500 pg/kg | Neomycin B |
| Oxacillin | All FPS | Muscle | 300 pg/kg | |
| Oxolinic Acid | Fish | Muscle+skin | 100 pg/kg | |
| Oxytetracycline | All FPS | Muscle | 100 pg/kg | |
| Paromomycin | All FPS | Muscle | 500 pg/kg | |
| Sarafloxacin (Fish & Poultry) | Salmonids | Muscle+skin | 30 pg/kg | |
| Spectinomycin | All FPS | Muscle | 300 pg/kg | |
| Sulphonamides (All) | All FPS | Muscle | 100 pg/kg | |
| Tetracycline | All FPS | Muscle | 100 pg/kg | |
| Thiamphenicol | All FPS | Muscle | 50 pg/kg | |
| Tilmicosine | All FPS | Muscle | 50 pg/kg | |
| Trimethoprim | All FPS | Muscle | 50 pg/kg | |
| Tylosin | All FPS | Muscle | 100 pg/kg | |

*All FPS: all food producing species (with some exclusions and depending on each compound). **For all fish MRLs, the target tissues "muscle" or "muscle and skin" shall be understood as "muscle and skin in natural proportions."

The approval process is very costly and time consuming, and the sales potential for the aquaculture market in global terms is limited, which in some cases has meant a certain lack of interest on the part of pharmaceutical companies for developing new antibacterials and registering them (Alderman & Hastings, 2009; Rodgers & Furones, 2009). In the USA, the regulatory authority for the approval of Veterinary Medicinal Products (VMPs) is the FDA (FDA, 2009, 2011). The body that is responsible for the authorization procedure in the EU is the European Medicines Agency (EMEA) and the European Commission, or else the national competent authorities in the

EU Member States (depending on the procedure chosen for the marketing authorisation application). The EMEA's Committee for Medicinal Products for Veterinary Use (CVMP) carries out the scientific evaluation (Sanders, 2005; Prescott, 2008; Valois et al., 2008; Alderman, 2009). Compared to agricultural use and medicinal use, the market for aquaculture antibacterials is fairly small and the approval process can be expensive. The availability of antibacterial agents for aquacultural use is affected by the setting of MRLs. However, these withdrawal times are based on studies that are mainly performed on fish held in temperate freshwater. The excretion of a drug by a fish can vary greatly with its environmental conditions, and especially the temperature (Daniel, 2009; Noga, 2010). Because of the variability of drug excretion, especially with temperature, a rule of thumb called "degree days" has been advocated for estimating the required withdrawal time. If the data does not indicate a temperature effect on depletion, then a day-based withdrawal can be accepted (Alderman & Hastings, 2009).

## PROBLEMS ASSOCIATED WITH ANTIBACTERIAL USE IN AQUACULTURE

Consumers demand guarantees that their food has been produced, handled and sold in a way that is not dangerous to their health, and which respects the environment and addresses various other ethical and social concerns (FAO, 2011). Even if the occurrence, effects and fate of antibacterials have been considered from the perspective of scientific interest, little is still known about the actual risk to both humans and the environment (Kemper, 2008). However, medicines legislation requires that user-safety be assessed in the safety package and that the product label must include advice and warnings to the user, giving guidance for safe use. Any hazards associated with feed medication – whether in feed mills or on farms – must be considered, as must any hazards to the final user (the fish farm staff) (Alderman & Hastings, 2009). The majority of fishers and aquaculturists are in developing countries, and mainly in Asia which has experienced the largest increases over recent decades, reflecting the rapid expansion of aquacultural activities (Sapkota et al., 2008; FAO, 2011; Smith et al., 2008). Fish diseases are generally coupled with cultured fish and viewed as a result of aquaculture (Bergh, 2007). As a consequence, it is probable that the majority of antibacterial use in world aquaculture is not associated with any classification of the target bacterium or of its susceptibility to the range of available antibacterials (Smith et al., 2008). There is also a need for assurance that the usage will not harm animals or humans (Phillips et al., 2004). With an increase in consumers' recognition of the health benefits associated with seafood consumption, the volume of fisheries and aquaculture products consumed is expected to rise (Storey, 2005). There is little doubt that aquaculture production will continue to grow (Asche et al., 2008). The world food supply will probably have to double in quantity and increase in quality over the next 30–50 years as populations and incomes rise. The demand for fish as food will probably double and could even more than double (Pullin et al., 2007). Consequently, an increase in the number of problems associated with aquaculture production may be expected.

## Toxicity to the host

Antibacterials that are sufficiently non-toxic to the host are used as chemotherapeutic agents in the treatment of the infectious diseases of humans, animals and plants. Direct host toxicity is the most important factor limiting drug dosage. Tolerance studies must be carried out to determine the safety of the product to the target fish species (Alderman & Hastings, 2009). Also, it is important for the clinician to report adverse drug events to legal authorities. Antibacterial agents can have a wide variety of damaging effects on the host, including: (1) direct host toxicity; (2) adverse interactions with other drugs; (3) interference with the protective effect of normal host microflora or the disturbance of the metabolic function of microbial flora in the digestive tract of herbivores; (4) the selection or promotion of antibacterial resistance; (5) tissue necrosis at injection sites; (6) drug residues in animal products that are intended for human consumption; (7) impairment of the host's immune or defence mechanisms; and (8) damage to foetal or neonatal tissues (Guardabassi & Kruse, 2008; Mulcahy, 2011). Nonetheless, the most used aquaculture antibacterial agent oxytetracycline may have genotoxic and ecotoxic effects in aquatic ecosystems (Zounkova et al., 2011). The selective toxicity of antibacterials is variable. Some agents, such as betalactams, are generally considered to be safe, whereas others, such as the aminoglycosides, are potentially toxic (Guardabassi & Kruse, 2008).

## Resistance of aquatic bacteria

The capacity of bacteria to adapt to changes in their environment and thus survive is called resistance. Drug choices for the treatment of common infectious diseases are becoming increasingly limited and expensive and, in some cases, unavailable due to the emergence of drug resistance in bacteria (FAO, 2005). In general, aquatic bacteria are not different from other bacteria in their responses to exposure to antibacterial agents, and they are capable of transferring antibacterial resistance genes to other bacteria (Heuer et al., 2009). The WHO has long recognised that antibacterial use in food animals – which seems to outweigh antibacterial use for human therapy in many countries –

contributes importantly to the public health problem of antibacterial resistance (WHO, 2011). The resistance of pathogenic bacteria to antibacterials is a growing problem in human and veterinary medicine, and antibacterial use in fish – especially in aquaculture – is an area of increasing concern over health risks (Kemper, 2008; Mulcahy, 2011). The fact that some of the bacteria that cause infections in fish belong to the same genera as the bacteria causing infections in humans is likely to increase the probability of the spread of antibacterial resistance from aquaculture to humans (Heuer et al., 2009). The continued use of subtherapeutic levels of antibacterials to prevent disease increases the likelihood of establishing populations of multiply resistant strains of pathogenic bacteria. These may ultimately result in outbreaks of disease which cannot be controlled by antibacterial therapy (Mulcahy, 2011; Roberts, 2004). Also, the selection and use of inappropriate antibacterials, and the use of insufficient dosages, incorrect routes of application, incorrect dosing frequencies and administering antibacterials for an insufficient time period, are ways to select for antibacterial-resistant bacteria (Mulcahy, 2011). The excessive use of antibacterials in fish aquaculture is increasing the resistance in bacteria that can infect both humans and animals (Burridge et al., 2010; Kümmerer, 2010; Defoirdt et al., 2011). It is not only the direct therapeutic use of antibacterials, but also their indirect contact with them which might enhance the resistance of bacteria: not taking into account the bacteria's origins, resistance genes have been isolated from human pathogens, bacteria of animal origin and even environmental bacteria (Kemper, 2008; Martinez & Silley, 2010; Martinez, 2009). The consequences of increasing resistance in bacteria and the diminishing impact of therapeutic drugs reach far beyond the geographic origins of antibacterial compounds and are, therefore, of global concern (Kemper, 2008). Antibacterials exhibit different activity spectra and mechanisms of action. It has been recognised for some time that susceptibility to antibacterials varies markedly both between different groups of organisms and within these groups (Kümmerer, 2008). A large variety of antibacterial resistance mechanisms have been identified in bacteria and several different mechanisms may be responsible for resistance to a single antibacterial agent in a given bacterial species. Antibacterial resistance mechanisms can be classified into four major categories (shown in Table 3.1 by the asterisk): (1) the antibacterial agent can be prevented from reaching its target by reducing its penetration into the bacterial cell; (2) general or specific efflux pumps may expel antibacterial agents from the cell; (3) the antibacterial agent can be deactivated by modification or degradation, either before or after penetrating the cell; or (4) the antibacterial target may be modified so that the antibacterial cannot act on it anymore, or else the microorganism's acquisition or activation of an alternate pathway may render the target dispensable (see Table 8) (Boerlin & White, 2008; Nikaido, 2009). Drug resistance may be natural or acquired (Roberts, 2004; Douet et al., 2009). Some organisms have always been resistant to a particular agent by the nature of their physiology or biochemistry (i.e., inherent or intrinsic resistance); others have acquired resistance as a result of the application of antibacterials by humans (i.e., acquired resistance) (Kümmerer, 2008). Resistance to antibacterials may be acquired by the mutation of a chromosomal gene which modifies the structure of the ribosomal target or by the infection of the cell with a resistant R-factor plasmid. Plasmids are extrachromosomal circular DNA molecules capable of autonomous replication (Alderman & Hastings, 1998; Boerlin & White, 2008; Defoirdt et al., 2011). Once they are integrated in successful gene-transmission elements, antibacterial resistance genes can persist and spread even in the absence of antibacterials (Martinez, 2009). Multidrug resistance in bacteria occurs with the accumulation – on resistant R plasmids or transposons – of genes, with each coding for resistance to a specific agent, and/or by the action of multidrug efflux pumps, each of which can pump out more than one drug-type (Nikaido, 2009). The demonstration of R-factor transfer to fish pathogens was first shown with certain strains of Aeromonas salmonidae.

**Table 8:** Examples of resistance mechanisms (Boerlin & White, 2008).

| Antibacterial agent | Resistance mechanism | * | Examples of genetic determinant |
|---|---|---|---|
| Tetracycline | Inducible efflux of tetracycline in *E. coli* and other *Enterobacteriaceae* | 2 | tetA, tetB, tetC |
| | Ribosomal protection in Gram-positive bacteria | 4 | tetO, tetM |
| Chloramphenicol | Efflux in *Enterobacteriaceae* | 2 | cmlA, floR |
| | Acetylation in *Enterobacteriaceae* | 3 | catA |
| ß-lactams | ß-lactamases in *Enterobacteriaceae*, *Staphylococcus aureus* | 3 | bla$_{TEM}$, bla$_{SHV}$, bla$_{CMY-2}$, bla$_Z$ |
| Oxacillin, methicillin | Alternate penicillin-binding proteins in *Staphylococcus aureus* | 4 | mecA |
| Imipenem | Decreased porin formation in *Enterobacter aerogenes* and *Klebsiella spp.* | 1 | Mutations |
| Aminoglycosides | Phosphorylation, adenylation, and acetylation of aminoglycosides in Gram-negative and -positive bacteria | 3 | Numerous genes with a broad variety of specificities |
| Streptomycin | Modification of ribosomal proteins or of 16s rRNA in *Mycobacterium spp.* | 4 | Mutations |
| Macrolides, lincosamides, streptogramins | Methylation of ribosomal RNA in Gram-positive organisms | 4 | ermA, ermB, ermC |
| Macrolides, streptogramins | *Staphylococcus spp.* | 2 | vga(A), msr(A) |
| Fluoroquinolones | DNA topoisomerases with low affinity to quinolones | 4 | Mutations in gyrA, gyrB, parC, parE |
| Sulphonamides | Bypass of blocked pathways through additional resistant dihydropteroate synthase in Gram-negative bacteria | 4 | Sul1, sul2, sul3 |
| Trimethoprim | Bypass of blocked pathways through additional resistant dihydrofolate reductase | 4 | Diverse dfr genes |

Note: This is by no means a comprehensive list of all the resistance mechanisms for each category of antibacterials listed. *: Numbers 1, 2, 3 and 4 refer to mechanisms listed in the text.

Also, transferable R-factor plasmids in drug-resistant strains were shown with Aeromonas hydrophila, Vibrio anguillarum, marine Vibrio sp., Edwardsiella tarda and Patteurella piscicida (Alderman & Hastings, 1998). Tetracycline-resistance genes are found even in small farms which rarely use antibacterials. The copy numbers of tetA, tetC, tetH, and tetM genes (tetR reported by Seyfried et al., (2010)) remain elevated at farms over the surveillance period of four years in the absence of any selection pressure from tetracycline or even other antibacterials (Schmitt & Römbke, 2008; Tamminen et al., 2010). The continued introduction of tetracycline-resistant organisms from the hatchery to the stream, even after a significant time period had elapsed since the use of antibacterials, indicates the presence of reservoirs of organisms or unknown sources of resistance (Stachowiak et al., 2010) as well as other aquatic bacteria, and also illustrates that these bacteria can act as reservoirs of resistance genes that can be further disseminated. Ultimately, resistance genes in the aquatic environment may reach human pathogens and thereby add to the burden of antibacterial resistance in human medicine (Heuer et al., 2009).

Aquaculture is thought to stimulate the spread and stability of antibacterial resistance in the environment (Sapkota et al., 2008). Commercial fish production facilities could be a source of antibacterial-resistant microorganisms to receiving waters at times when there is no active use of antibacterials as a result of cross-resistance induced by biocides (Stachowiak et al., 2010). It has been shown that antibacterial-resistant bacteria are more likely to occur in the water and sediment associated with aquaculture. Already, in several areas of the world, this is beginning to take place. The comparison of predicted antibacterial concentrations to published minimum inhibitory concentrations suggests that antibacterials in wastewater – but probably not antifungals – may select for low-level antibacterial resistance (Kostich & Lazorchak, 2008). Also, the presence of R-factor-infected populations of bacteria in aquaculture systems may lead to the transfer of antibacterial resistance to other micro organisms, including potential human pathogens (Roberts, 2004; Cabello, 2006). Both the percentage and level of bacterial resistance to drugs was higher when drugs were administered as medicated feed. In addition, the duration of the resistance was longer when medicated feed was the mode of administration. The presence of feed residue in the aquatic system would have an important effect for the generation and maintenance of the drug resistance of bacteria in sediment (Yu et al., 2009). These results call for the development of better management strategies for fish farming so as to prevent the emergence of resistant gene pools in the sediments of aquaculture facilities, and to promote the disappearance of established resistant gene pools (Tamminen et al., 2010). Principles for the prudent use of antibacterials should be developed and awareness of the problem of antibacterial resistance should be raised by informing the public (FAO, 2005). The most effective and direct approach is thought to be the reasonable use of antibacterials in health protection and agriculture production (Zhang et al., 2009). Without a doubt, a promising approach for proper risk-assessment and management would be the reduction of the emission of antibacterials into the environment, whether of human or veterinary medical origin. In either case, it may not be appropriate to assume that terminating the use of antibacterials will lead to a rapid decrease in resistant organisms (Stachowiak et al., 2010). The appropriate use of antibacterials in livestock production will preserve the long-term efficacy of existing antibacterials, support animal health and welfare, and limit the risk factors of transferring antibacterial resistance to animals and humans (Kemper, 2008). Whatever is done, the competent surveillance of disease and antibacterial resistance, as well as the repeated refinement of risk analyses, are a necessity if we are to concentrate our efforts to limit the effects of antibacterial resistance on what is shown to work in practice (Phillips et al., 2004; Sanders, 2005). In general, the emergence of resistance to antibacterials is a highly complex process, which is not yet fully understood with respect to the significance of the interaction of bacterial populations and antibacterials, even in a medicinal environment (Kümmerer, 2010). In the EU, the EMA works for the development of a harmonised approach to the surveillance of antibacterial usage in animals and the collection of data from EU Member States (WHO, 2011). Also, research projects should be encouraged which aim at the better understanding of the mechanisms of the emergence and spread of resistance within a species, and from animal to human and the environment (FAO, 2005).

## Aquatic food residues

The case of the residues of antibacterial substances in fish and fish products represents, in practice, a complex problem for society and regulators, and particularly in developing countries where regulations and the possibilities for enforcing them are scarce (Cabello, 2006; Lupin, 2009). In addition to selecting for antibacterial resistance, the heavy prophylactic and therapeutic use of antibacterials in aquaculture environments can lead to elevated antibacterial residues in ponds, marine sediments, aquaculture products, wild fish and other natural aquatic environments that are impacted by aquaculture facilities (Sapkota et al., 2008). Also, the use of large amounts of antibacterials that have to be mixed with fish food creates problems for industrial health and increases the opportunities for the presence of residual antibacterials in fish meat and fish products (Cabello, 2006). Withdrawal times are recommended, and in many countries they are legally enforced for some drugs, and especially antibacterials. The Food Animal Drug Avoidance Databank (FARAD) assists veterinarians in estimating residue-depletion times for antibacterial agents that are administered at doses in excess of label recommendations (Walker & Giguére, 2008). A good rule of thumb for the withdrawal time is 500 degree days. Thus, if the mean daily water temperature after treatment is 10°C, the withdrawal period should be at least 50 days (10 x 50 = 500), while at 25° C, the withdrawal period would be 20 days (Noga, 2010). Obviously, this can only be a rough estimate of the elimination-rate because temperatures fluctuate diurnally and from day-to-day and other factors besides temperature affect eliminationrates. Note also that 500 degree days might not be sufficient in some cases (Treves - Brown, 2001). Therefore, the accurate and sensitive determination of antibacterial residues is now a necessity. In order to protect human health, the EU and other regulatory authorities worldwide have established MRLs for antibacterial residues in animal products entering the human food chain (Cañada-Cañada & Pena, 2009; Lupin, 2009). Research projects should be promoted on pharmacology and the PK of antibacterials in aquatic species in order to provide a more exact approach to establishing MRLs' values (Table 7) (FAO, 2005).

## Environmental Impact of Antibacterial Use in Aquaculture

Aquaculture is so integrally linked to the surrounding environment that if sustainable practices are not employed, the degradation of the surrounding environment will ultimately lead to the degradation of the industry itself (Bergh, 2007). The wellbeing of the environment – in cases of disease and treatment – is related to two aspects of biota conservation; the transmission of microbial pathogens to wild populations and the pollution from chemotherapeutics (Grigorakis, 2010). The extensive use of veterinary pharmaceuticals is supposed to represent a daunting public health risk, resulting not only in the emergence and spread of resistant bacteria, but also in other human, animal and environmental impairments (Kemper, 2008). The input of resistant bacteria into the environment from different sources seems to be the most important basis of resistance in the environment. The possible impact of resistant bacteria on the environment is not yet known and the health risks of active pharmaceutical ingredients remain poorly understood (Kümmerer, 2010). The physicochemical fate and environmental concentrations of antibacterials in soil has been the subject of a number of recent studies. During recent years, significant attention has been paid to the occurrence of drugs in the environment. Several classes of antibacterials have been detected in field soils, and their sorption behaviour and degradation have been studied to a large extent (Schmitt & Römbke, 2008; Zounkova et al., 2011). In general, farmed fish is as safe and nutritious as wild-caught species, but there are public health hazards associated with ignorance, abuse and the neglect of aquaculture technology. Numerous small fish ponds increase the shoreline of ponds, causing higher densities of mosquito larvae and cercaria, which can increase the incidence and prevalence of lymphatic filariasis and schistosomiasis (Lessenger, 2006). Fish production can generate considerable amounts of dissolved effluents, which potentially affect water quality in the vicinity of the farms and, due to rapid dilution, also at larger scales (km-scale) (Costanzo et al., 2005; Holmer et al., 2008). High antibacterial load in sediments and in concentrations potent enough to inhibit the growth of bacteria have been reported for aquaculture (Kümmerer, 2008).

Tetracycline has a low bioavailability in fish ($< 10\%$), due to binding with sea-water-borne divalent cations such as $Mg_{+2}$ and $Ca_{+2}$. It is noteworthy that non-bioavailable tetracyclines contaminate the environment (Rigos & Troisi, 2005). However, it has been shown that residues of oxytetracycline in marine sediments were very stable over a period of months (Toutain et al., 2010). Often, the existing data used to assess the environmental effects of antibacterials is not adequate for the establishment of how long bacteria maintain antibacterial resistance in the absence of continued selective pressure for that resistance (Kümmerer, 2008). Also, in order to minimise the possible risks of antibacterials in dust, the use of antibacterials in livestock farming should be strictly reduced to therapeutic use (Hamscher & Hartung, 2008). In one study (Wei et al., 2011) conducted in China (Jiangsu Province) – the biggest aquaculture producer (FAO, 2011) – contamination with antibacterials indicated that ten veterinary antibacterials around farms were found in animal wastewaters, eight antibacterials were detected in pond waters, and animal farmeffluents and river water samples were contaminated by nine antibacterials. The most frequently detected antibacterials were sulphamethazine (75%), oxytetracycline (64%), tetracycline (60%), sulfadiazine (55%) and sulphamethoxazole (51%). This research has demonstrated that animal wastewater is a major source of pollution of veterinary antibacterials. By applying the animal wastewater to agricultural soils, the antibacterials might contaminate the soils and surrounding water systems, thus posing a serious threat to humans and wildlife (Figure 2) (Boxall et al., 2004; Wei et al., 2011). Antibacterials may be detected in effluent entering receiving waters and be detectable 500m from the source (Costanzo et al., 2005). There is very little information about the chronic toxicity or the bioaccumulation potential of pharmaceuticals in biota and food chains (Christen et al., 2010). Not much is known about the occurrence, fate and activity of metabolites (Kümmerer, 2010). Another study showed that more than 30 antibacterial substances have been found in sewage influent and effluent samples, in surface waters and even in ground and drinking water (Kemper, 2008). At the same time, with antibacterials, disinfectants, and heavy metals being released into water, they might exert selective activities as well as ecological damage in water communities, resulting in antibacterial resistance (Baquero et al., 2008). For example, the exposure of eels to pollution during their development is inducing changes on the biomarkers involved in physiological functions that are determinants for the survival and performance of the eels, namely biotransformation enzymes and antioxidative stress defences, and these alterations may have negative effects on sexual development. In addition, the mechanisms used to face chemical stress need energy which is probably allocated from other functions, such as tissue repair, growth and weight increase, and which are determinants for a successful migration into the reproduction area (Gravato et al., 2008).

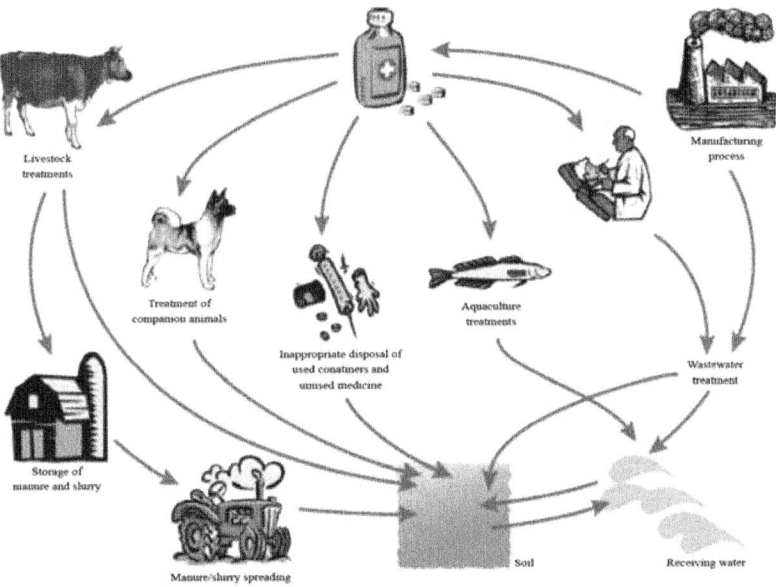

**Figure 2:** Routes of pharmaceuticals entering the environment (Boxall et al., 2004).

Ideally, aquaculture operations would be planned with background knowledge of the ecosystems in which the facilities will operate as well as knowledge of the potential environmental, social, and economic effects (both positive and negative) that could be incurred, and the cost:benefit ratio associated with operating, given knowledge of that background (Bergh, 2007). Environmental observations and models can then be combined with effective aquaculture husbandry practices so as to manage environmental risks from all sources (Hargrave et al., 2005). Hopefully, research on aquaculture-environment interactions has progressed remarkably during recent years, particularly in the framework of EU-funded projects, which have provided useful information for the understanding of various ecosystem processes affected by the presence and operation of fish farms (Holmer et al., 2008).

# ANTIBACTERIAL USAGE SUGGESTIONS IN AQUACULTURE

When it is apparent that a treatment is necessary, the following check-lists may be useful (Winton, 2001):

Before treating:

1. Accurately determine the water-volume, flow-rate, and temperature.
2. Accurately determine the number and total weight of fish in the rearing unit.
3. Confirm the identity, expiration date, and active ingredient concentration of the regulated product to be applied.
4. Double-check treatment calculations. Beware of confusion from mixing metric and standard units.
5. Have aeration devices ready for use if needed.
6. If treated water is to be discharged, make sure all appropriate permits are in place and regulatory authorities have been notified.
7. If possible, conduct a bioassay on a small group of fish before treating the entire population in the rearing unit.

When treating:

1. Dilute the regulated product with rearing water before applying it (or follow product directions).
2. Ensure that the regulated product is well-mixed and evenly applied in the rearing units.
3. Observe the fish closely and frequently during treatment for signs of distress.
4. Monitor the temperature and dissolved-oxygen levels in the rearing unit during treatment.
5. Except for oral treatments, discontinue feeding during treatment. Fish are unlikely to feed during treatment, and uneaten feed will foul the system and may reduce the efficacy of some treatments.
6. Discontinue treatment and restore normal culture conditions if the fish

become distressed.

After treating:

1.    Observe the fish frequently for at least 24 hours following the treatment.

2.    Do not stress the treated fish for at least 48 hours.

3.    Recheck the fish to determine the efficacy of the treatment

Judicious antibacterial use principles for veterinarians are discussed and concluded in Table 9 (FDA, 2009).

**Table 9:** Judicious antibacterial use principles for veterinarians (resumed from (FDA, 2009)).

| The food fish veterinarian should: |
| --- |
| 1.   Accept responsibility for helping clients design management, immunization, production unit and nutritional programmes that will reduce the incidence of disease and the need for antibacterial treatment. |
| 2.   Use antibacterial drugs only within the confines of a valid veterinarian-client-patient relationship, including both the dispensing and issuing of prescriptions and veterinary feed directives. Extra-label usage should be consistent with regulatory agency laws, regulations and policies. |
| 3.   Properly select and use antibacterial drugs. Veterinarians should participate in continuing education programmes that include therapeutics and the emergence and/or development of antibacterial resistance. |
| 4.   Have strong clinical evidence of the identity of the disease's aetiology, based upon history, clinical signs, necropsy, laboratory data, and/or past experience before recommending an antibacterial drug treatment. |
| 5.   Treat food fish with antibacterial drugs according to the product label recommendations (including indication, dosage, duration, fish species and environmental conditions). |
| 6.   Choose an antibacterial drug and treatment regimen based on the available laboratory and label (including package insert) information, additional data in the literature, and consideration of the pharmacokinetics, spectrum of activity and pharmacodynamics of the drug. |
| 7.   Use antibacterial drugs with a specific clinical outcome(s) in mind, including a specific target for population morbidity and/or mortality-rate reduction. |
| 8.   Specific outcome criteria will prevent an unnecessarily long therapy and indicate when the current therapy is no longer effective. |

9. Determine the production population pathogen susceptibility at the first indication of increasing morbidity or mortality, and monitor the therapeutic response so as to detect changes in microbial susceptibility and in order to evaluate antibacterial selections.
10. Routine necropsy examination of fish populations should be periodically performed, including antibacterial susceptibility testing and update historical information for developing treatment and control protocols.
11. Use products that have the narrowest spectrum of activity and known effectiveness *in vivo* against the pathogen causing the disease problem.
12. Choose antibacterial drugs of lesser importance in human medicine, if these receive future food fish use approval, and do not choose an antibacterial for which the emergence of resistance is expected in an advanced stage.
13. Use, whenever possible, an antibacterial drug labelled to treat the condition diagnosed.
14. Do not use combination antibacterial drug therapy unless there is information to show that this decreases or suppresses the target organism resistance development.
15. Do not compound antibacterial drug formulations.
16. Do not use antibacterial drugs to treat cases with a poor chance of recovery.
17. Do not use antibacterial drugs prophylactically.
18. Ensure proper on-farm drug use and protect antibacterial drug integrity through proper handling, storage and observation of the expiration date.
19. Prescribe, dispense or write a Veterinary Feed Directive for drug quantities appropriate to the production-unit size and expected need using the approved formulation.
20. Work with producers and/or facility fish health management personnel so as to ensure that farm personnel receive adequate training on the use of antibacterial drugs, including indications, diagnoses, dosages, withdrawal times, the route of administration, storage, handling and accurate record-keeping.
21. Work closely with all other fish health experts involved in fish population health management at the fish production facility.

## CONCLUSION

The presence of disease in farmed fish populations has severe welfare implications for the affected fish, and poses a threat to the welfare of unaffected fish. Large quantities of antibacterials are used in aquaculture in some countries, often without professional consultation or supervision. Consequently, many problems are associated with the use of antibacterials in aquaculture. More research is needed in order to determine the consequences of the application of large quantities of antibacterials. Considering the rapid growth and importance of the aquaculture industry in many regions of the world and the widespread, intensive, and often unregulated use of antibacterial agents for animal production, additional efforts are required to prevent the development and spread of antibacterial resistance in aquaculture. Also, safer, more effective medicines are necessary, along with improvements in husbandry and management which will reduce the need for those medicaments. However, without the use of veterinary medicines, aquaculture food production would be impaired. Furthermore, fish farmers and their veterinary surgeons must confirm that fish are kept in the best state of health and welfare. Governments, farmers and veterinary surgeons all have a shared responsibility to ensure that medicines are used judiciously

## REFERENCES

1.  Alderman, D. J. (2009). Control of the Use of Veterinary Drugs and Vaccines in Aquaculture in the European Union. In The Use of Veterinary Drugs and Vaccines in Mediterranean Aquaculture, edited by Basurco Rogers C.J., B. Zaragoza: CIHEAM - IAMZ.

2.  Alderman, D. J. & Hastings, T. S. (1998). Antibiotic Use in Aquaculture: Development of Antibiotic Resistance - Potential for Consumer Health Risks. International Journal of Food Science and Technology no. 33:139 -155.

3.  Asche, F., Roll, K. H. & Tveterås, S. (2008). Future Trends in Aquaculture: Productivity Growth and Increased Production. In Aquaculture in the Ecosystem, edited by Marianne Holmer, Kenny Black, Carlos M. Duarte, Nuria Marbà and Ioannis Karakassis, 271-292. Springer Netherlands.

4.  Austin, B. & Austin, D. A. (2007). Control. In Bacterial Fish Pathogens, 337-404. Springer Netherlands.

5.  Baquero, F., Martínez, J. L. & Cantón, R. (2008). Antibiotics and antibiotic resistance in water environments. Curr Opin Biotechnol no. 19 (3):260-265.

6.  Bebak, J. A., Welch, T. J., Starliper, C. E. Baya, A. M. & Garner, M.M. (2007). Improved husbandry to control an outbreak of rainbow trout fry syndrome caused by infection with Flavobacterium psychrophilum. J Am Vet Med Assoc no. 231 (1):114-6.

7.  Bergh, O. (2007). The dual myths of the healthy wild fish and the unhealthy farmed fish. Dis Aquat Organ no. 75 (2):159-64.

8.  Boerlin, P. & White, D. G. (2008). Antimicrobial Resistance and Its Epidemiology. In Guide to Antimicrobial "Use": in Animals, edited by Lars Bogø Jensen Luca Guardabassi, Hilde Kruse, 27 - 43. Oxford, UK: Blackwell Publishing Ltd.

9.  Boos, S., Schmidt, H., Ritter, G. & Manz, D. (1995). Effectiveness of oral rifampicin against mycobacteriosis in tropical fish. Berl Munch Tierarztl Wochenschr no. 108 (7):253-5.

10. Boxall, A. B., Fogg, L. A., Blackwell, P. A., Kay, P., Pemberton, E. J.

& Croxford, A. (2004). Veterinary medicines in the environment. Rev Environ Contam Toxicol no. 180:1- 91.

11. Burka, J. F., Hammell, K. L., Horsberg, T. E., Johnson, G. R., Rainnie, D. J. & Speare, D. J. (1997). Drugs in salmonid aquaculture – A review. Journal of Veterinary Pharmacology and Therapeutics no. 20 (5):333-349.

12. Burridge, L., Weis, J. S., Cabello, F., Pizarro, J. & Bostick, K. (2010). Chemical use in salmon aquaculture: A review of current practices and possible environmental effects. Aquaculture no. 306 (1-4):7-23.

13. Cabello, F. C. (2006). Heavy use of prophylactic antibiotics in aquaculture: a growing problem for human and animal health and for the environment. Environmental Microbiology no. 8 (7):1137-1144.

14. Cañada-Cañada, F., Muñoz de la Peña, A. & Espinosa-Mansilla, A. (2009). Analysis of antibiotics in fish samples. Analytical and Bioanalytical Chemistry no. 395 (4):987- 1008.

15. Christen, V., Hickmann, S., Rechenberg, B. & Fent, K. (2010). Highly active human pharmaceuticals in aquatic systems: A concept for their identification based on their mode of action. Aquat Toxicol no. 96 (3):167-81.

16. (CLSI), Clinical and Laboratory Standards Institute. (2006a). Methods for Antimicrobial Disk Susceptibility Testing of Bacteria Isolated From Aquatic Animals. In Approved Guideline. Wayne, Pennsylvania, USA.

17. (CLSI). 2006b. Methods for Broth Dilution Susceptibility Testing of Bacteria Isolated From Aquatic Animals. In Approved Guideline. Wayne, Pennsylvania, USA.

18. Collina, G., Morandi, L., Lanzoni, A. & Reggiani, M. (2002). Atypical cutaneous mycobacteriosis diagnosed by polymerase chain reaction. Br J Dermatol no. 147 (4):781-4.

19. Costanzo, S. D., Murby, J. & Bates, J. (2005). Ecosystem response to antibiotics entering the aquatic environment. Mar Pollut Bull no. 51 (1-4):218-23.

20. Cottee, S. & Petersan, P. (2009). Animal Welfare and Organic Aquaculture in Open Systems. Journal of Agricultural and Environmental Ethics no. 22 (5):437-461.

21. Daniel, P. (2009). Drugs and Chemicals in Aquafeeds: The problems and solutions. In The Use of Veterinary Drugs and Vaccines in Mediterranean Aquaculture, edited by Basurco Rogers C.J., B. Zaragoza: CIHEAM - IAMZ.

22.  Defoirdt, T., Sorgeloos, P. & Bossier, P. (2011). Alternatives to antibiotics for the control of bacterial disease in aquaculture. Current Opinion in Microbiology no. 14 (3):251- 258.

23.  Douet, D.G., Le Bris, H. & Giraud, E. (2009). Environmental Aspects of Drug and Chemical Use in Aquaculture: An Overview. In The Use of Veterinary Drugs and Vaccines in Mediterranean Aquaculture, edited by Basurco Rogers C.J., B. Zaragoza: CIHEAM -IAMZ.

24.  FAO. (2005). Responsible Use of Antibiotics in Aquaculture. edited by FAO Fisheries Technical Paper. No.469. Rome, Italy.

25.  FAO. (2011). The State of World Fisheries and Aquaculture 2010. edited by FAO Fisheries and Aquaculture Department. Rome, Italy.

26.  FDA. Judicious Use of Antimicrobials for Aquatic Veterinarians. U.S. Food and Drug Administration 2009 [cited 11.07.2011]. Available from http:// www.fda.gov/AnimalVeterinary/SafetyHealth/ AntimicrobialResistance/ JudiciousUseofAntimicrobials/ucm095473.htm

27.  FDA. (2011). Aquaculture Drugs. In Fish and Fishery Products Hazards and Controls Guidance, 183 -208. Rockville, USA: Center for Food Safety and Applied Nutrition.

28.  Gravato, C, Melissa, F., Anabela, A., Joana, S. & Lúcia, G. (2008). Biomonitoring Studies Performed with European Eel Populations from the Estuaries of Minho, Lima and Douro Rivers (NW Portugal). In Advanced Environmental Monitoring, edited by Young J. Kim and Ulrich Platt, 390-401. Springer Netherlands.

29.  Grigorakis, K. (2010). Ethical Issues in Aquaculture Production. Journal of Agricultural and Environmental Ethics no. 23 (4):345-370.

30.  Guardabassi, L. & Kruse, H. (2008). Principles of Prudent and Rational Use of Antimicrobials in Animals. In Guide to Antimicrobial Use in Animals, edited by Lars Bogø Jensen Luca Guardabassi, Hilde Kruse, 1-12. Oxford, UK: Blackwell Publishing Ltd.

31.  Hamscher, G. & Hartung, J. (2008). Veterinary Antibiotics in Dust: Sources, Environmental Concentrations, and Possible Health Hazards. In Pharmaceuticals in the Environment, edited by Klaus Kümmerer, 95-102. Springer Berlin / Heidelberg.

32.  Hargrave, B., William, S. & Paul, K. (2005). Assessing and Managing Environmental Risks Associated with Marine Finfish Aquaculture. In Environmental Effects of Marine Finfish Aquaculture, edited by Barry Hargrave, 433-461. Springer Berlin / Heidelberg.

33.  Haya, K., Burridge, L., Davies, I. & Ervik, A. (2005). A Review and

Assessment of Environmental Risk of Chemicals Used for the Treatment of Sea Lice Infestations of Cultured Salmon. In Environmental Effects of Marine Finfish Aquaculture, edited by Barry Hargrave, 305-340. Springer Berlin / Heidelberg.

34.  Heuer, O. E., Kruse, H., Grave, K., Collignon, P., Karunasagar, I. & Angulo, F. J. (2009). Human Health Consequences of Use of Antimicrobial Agents in Aquaculture. Clinical Infectious Diseases no. 49 (8):1248-1253.

35.  Holmer, M., Hansen, P.K., Karakassis, I., Borg, J.A. & Schembri, P.J. (2008). Monitoring of Environmental Impacts of Marine Aquaculture. In Aquaculture in the Ecosystem, edited by Marianne Holmer, Kenny Black, Carlos M. Duarte, Nuria Marbà and Ioannis Karakassis, 47-85. Springer Netherlands.

36.  Horsberg, T. E. (1994). Experimental methods for pharmacokinetic studies in salmonids. Annual Review of Fish Diseases no. 4:345-358.

37.  Immanuel, G., Sivagnanavelmurugan, M. & Palavesam, A. (2011). Antibacterial Effect of Medium-chain fatty Acid: Caprylic Acid on Gnotobiotic Artemia franciscana nauplii Against Shrimp Pathogens Vibrio harveyi and V. parahaemolyticus. Aquaculture International no. 19 (1):91-101.

38.  Kemper, N. (2008). Veterinary antibiotics in the aquatic and terrestrial environment. Ecological Indicators no. 8 (1):1-13.

39.  Kostich, M. S. & Lazorchak, J.M. (2008). Risks to aquatic organisms posed by human pharmaceutical use. Science of The Total Environment no. 389 (2-3):329-339.

40.  Kum, C., Kirkan, S., Sekkin, S., Akar, F. & Boyacioglu, M. (2008). Comparison of in vitro antimicrobial susceptibility in Flavobacterium psychrophilum isolated from rainbow trout fry. J Aquat Anim Health no. 20 (4):245-51.

41.  Kümmerer, K. (2008). Effects of Antibiotics and Virustatics in the Environment. In Pharmaceuticals in the Environment, edited by Klaus Kümmerer, 223-244. Springer Berlin /Heidelberg.

42.  Kümmerer, K. (2010). Pharmaceuticals in the Environment. Annual Review of Environment and Resources no. 35 (1):57-75.

43.  Lessenger, J. E. (2006). Diseases from Animals, Poultry, and Fish. In Agricultural Medicine, edited by James E. Lessenger, 367-382. Springer New York.

44.  Lupin, H. M. (2009). Human Health Aspects of Drug and Chemical

Use in Aquaculture. In The Use of Veterinary Drugs and Vaccines in Mediterranean Aquaculture, edited by Basurco Rogers C.J., B. Zaragoza: CIHEAM - IAMZ.

45. Martinez, J. L. (2009). Environmental pollution by antibiotics and by antibiotic resistance determinants. Environmental Pollution no. 157 (11):2893-2902.

46. Martinez, M. & Silley, P. (2010). Antimicrobial Drug Resistance. Handb Exp Pharmacol (199):231-268.

47. Meyer, F. P. (1991). Aquaculture disease and health management. J Anim Sci no. 69 (10):4201-8.

48. Moutou, K. A., Burke, M. D. & Houlihan, D. F. (1998). Hepatic P450 monooxygenase response in rainbow trout (Oncorhynchus mykiss [Walbaum]) administered aquaculture antibiotics. Fish Physiol Biochem no. 18 (1):97-106.

49. Mulcahy, D. (2011). Antibiotic use during the intracoelomic implantation of electronic tags into fish. Reviews in Fish Biology and Fisheries no. 21 (1):83-96.

50. Nic Gabhainn, S., Bergh, O., Dixon, B., Donachie, L., Carson, J., Coyne, R., Curtin, J., Dalsgaard, I., Manfrin, A., Maxwell, G. & Smith, P. (2004). The precision and robustness of published protocols for disc diffusion assays of antimicrobial agent susceptibility: an inter-laboratory study. Aquaculture no. 240 (1-4):1-18.

51. Nikaido, H. (2009). Multidrug Resistance in Bacteria. Annual Review of Biochemistry no. 78 (1):119-146.

52. Noga, E. J. (2010). Fish Disease Diagnose and Treatment. Second ed. Iowa, USA: Wiley - Blackwell.

53. Phillips, I., Casewell, M., Cox, T., De Groot, B., Friis, C., Jones, R., Nightingale, C., Preston,

54. R. & Waddell, J. (2004). Does the use of antibiotics in food animals pose a risk to human health? A critical review of published data. J Antimicrob Chemother no. 53 (1):28-52.

55. Prescott, J.F. (2008). Antimicrobial use in food and companion animals. Animal Health Research Reviews no. 9 (Special Issue 02):127-133.

56. Pullin, R., Froese, R. & Pauly, D. 2007. Indicators for the Sustainability of Aquaculture. In Ecological and Genetic Implications of Aquaculture Activities, edited by Theresa M. Bert, 53-72. Springer Netherlands.

57. Reimschuessel, R. & Miller, R. (2006). Antimicrobial Drug Use in Aquaculture. In Antimicrobial Therapy in Veterinary Medicine, edited

by J.F. Prescott, Baggot, J.D., Walker, R.D., Dowling, P.M., 593-606. Iowa, USA: Blackwell Publishing Professional.

58. Reimschuessel, R.L., Stewart, E., Squibb, K., Hirokawa, T., Brady, D., Brooks, B., Shaikh & Hodsdon, C. (2005). Fish drug analysis--Phish-Pharm: a searchable database of pharmacokinetics data in fish. AAPS J no. 7 (2):E288-327.

59. Rigos, G., Nengas, I., Alexis, M. & Athanassopoulou, F. (2004). Bioavailability of oxytetracycline in sea bass, Dicentrarchus labrax (L.). J Fish Dis no. 27 (2):119-22.

60. Rigos, G. & Troisi, G. (2005). Antibacterial Agents in Mediterranean Finfish Farming: A Synopsis of Drug Pharmacokinetics in Important Euryhaline Fish Species and Possible Environmental Implications. Reviews in Fish Biology and Fisheries no. 15 (1):53-73.

61. Roberts, R.J. (2004). The Bacteriology of Teleosts. In Fish Pathology, edited by R.J. Roberts, 297 - 331. Philadelphia, USA: W.B. Saunders.

62. Rodger, H. D. (2010). Fish Disease Manual. Marine Institute and the Marine Research SubProgramme of the National Development Plan. Original edition, PBA/AF/08/003.

63. Rodgers, C.J. & Furones, M.D. (2009). Antimicrobial Agents in Aquaculture: Practice, needs and issues. In The Use of Veterinary Drugs and Vaccines in Mediterranean Aquaculture, edited by Basurco Rogers C.J., B. Zaragoza: CIHEAM - IAMZ.

64. Ryckaert, J., Bossier, P., D'Herde, K., Diez-Fraile, A., Sorgeloos, P., Haesebrouck, F. & Pasmans, F. (2010). "Persistence of Yersinia ruckeri in trout macrophages." Fish Shellfish Immunol no. 29 (4):648-55.

65. Sanders, P. (2005). Antibiotic Use in Animals—Policies and Control Measures Around Europe. In Antibiotic Policies, edited by Ian Gould and Jos Meer, 649-672. Springer US.

66. Sapkota, A., Sapkota, A. R., Kucharski, M., Burke, J., McKenzie, S., Walker, P. & Lawrence, R. (2008). Aquaculture practices and potential human health risks: current knowledge and future priorities. Environ Int no. 34 (8):1215-26.

67. Schmitt, H. & Römbke, J. (2008). The Ecotoxicological Effects of Pharmaceuticals (Antibiotics and Antiparasiticides) in the Terrestrial Environment – a Review. In Pharmaceuticals in the Environment, edited by Klaus Kümmerer, 285-303. Springer Berlin /Heidelberg.

68. Seyfried, E., Newton, R., Rubert, K., Pedersen, J. & McMahon, K. (2010). Occurrence of Tetracycline Resistance Genes in Aquaculture

Facilities with Varying Use of Oxytetracycline. Microbial Ecology no. 59 (4):799-807.

69.  Shao, Z. J. (2001). Aquaculture pharmaceuticals and biologicals: current perspectives and future possibilities. Adv Drug Deliv Rev no. 50 (3):229-43.

70.  Smith, P.R., Le Breton, A., Horsberg, T.E. & Corsin, F. (2008). Guidelines for Antimicrobial in Aquaculture. In Guide to Antimicrobial Use in Animals, edited by Lars Bogø Jensen Luca Guardabassi, Hilde Kruse, 207- 218. Oxford, UK: Blackwell Publishing Ltd.

71.  Stabili, L., Gravili, C., Boero, F., Tredici, S. & Alifano, P. (2010). Susceptibility to Antibiotics of Vibrio sp. AO1 Growing in Pure Culture or in Association with its Hydroid Host Aglaophenia octodonta (Cnidaria, Hydrozoa) Microbial Ecology no. 59 (3):555-562.

72.  Stachowiak, M., Clark, S., Templin, R. & Baker, K. (2010). Tetracycline-Resistant Escherichia coli in a Small Stream Receiving Fish Hatchery Effluent Water, Air, & Soil Pollution no. 211 (1):251-259.

73.  Stickney, R.R. (2005). Aquaculture. In Encyclopedia of Coastal Science, edited by Maurice L. Schwartz, 33-38. Springer Netherlands.

74.  Storey, S. (2005). Challenges with the development and approval of pharmaceuticals for fish. AAPS J no. 7 (2):E335-E343.

75.  Subasinghe, R. (2009). Disease Control in Aquaculture and the Responsible use of Veterinary Drugs and Vaccines: The Issues, Prospects and Challenges. In The Use of Veterinary Drugs and Vaccines in Mediterranean Aquaculture, edited by Basurco Rogers C.J., B. Zaragoza: CIHEAM - IAMZ.

76.  Tamminen, M., Karkman, A., Lõhmus, A., Muziasari, W. I., Takasu, H., Wada, S., Suzuki, S. & Virta, M. (2010). Tetracycline Resistance Genes Persist at Aquaculture Farms in the Absence of Selection Pressure. Environmental Science & Technology no. 45 (2):386- 391.

77.  Tobback, E., Decostere, A., Hermans, K., Ryckaert,J., Duchateau, L., Haesebrouck, F. & Chiers, K. (2009). Route of entry and tissue distribution of Yersinia ruckeri in experimentally infected rainbow trout Oncorhynchus mykiss. Dis Aquat Organ no. 84 (3):219-28.

78.  Toutain, P. L., Ferran, A. & Bousquet-Melou, A. (2010). Species differences in pharmacokinetics and pharmacodynamics. Handb Exp Pharmacol (199):19-48.

79.  Treves - Brown, K.M. (2001). Applied Fish Pharmacology. Edited by G. Poxton Michael. 3 vols. Vol. 3, Aquaculture. The Netherlands: Kluwer

Academic Publishers.

80.  Valois, A.A., Endoh, Y.S., Grein, K. & Tollefson, L. (2008). Geographical Differences in Market Availability, Regulation and Use of Veterinary Antimicrobial Products. In Guide to Antimicrobial Use in Animals, edited by Lars Bogø Jensen Luca Guardabassi, Hilde Kruse, 59 - 76. Oxford, UK: Blackwell Publishing Ltd.

81.  Walker, R.D. & Giguére, S. (2008). Principles of Antimicrobial Drug Selection and Use. In Antimicrobial Therapy in Veterinary Medicine, edited by J.F. Prescott, Baggot, J.D.,

82.  Walker, R.D., Dowling, P.M., 107-117. Iowa, USA: Blackwell Publishing Professional.

83.  Wei, R., Ge, F., Huang, S., Chen, M. & Wang, R. (2011). Occurrence of veterinary antibiotics in animal wastewater and surface water around farms in Jiangsu Province, China. Chemosphere no. 82 (10):1408-1414.

84.  WHO. (2011). Tackling Antibiotic Resistance From a Food Safety Perspective in Europe. Edited by World Health Organization. Copenhagen, Denmark.

85.  Wiens, G. D., Rockey, D. D., Wu, Z., Chang,J., Levy,R., Crane,S., Chen,D.S., Capri,G.R., Burnett,J.R., Sudheesh, P.S., Schipma,M.J., Burd,H., Bhattacharyya,A., Rhodes, L.D., Kaul,R. & Strom,M.S. (2008). Genome sequence of the fish pathogen Renibacterium salmoninarum suggests reductive evolution away from an environmental Arthrobacter ancestor. J Bacteriol no. 190 (21):6970-82.

86.  Winton, J. R. (2001). Fish Health Management. In Fish Hatchery Management, edited by G. Wedemeyer, 559 - 640. Bethesda, USA: American Fisheries Society.

87.  Yan, S. S. & Gilbert, J.M. (2004). Antimicrobial drug delivery in food animals and microbial food safety concerns: an overview of in vitro and in vivo factors potentially affecting the animal gut microflora. Adv Drug Deliv Rev no. 56 (10):1497-521.

88.  Yu, D., Yi,X., Ma,Y., Yin,B., Zhuo,H., Li, J. & Huang, Y. (2009). Effects of administration mode of antibiotics on antibiotic resistance of Enterococcus faecalis in aquatic ecosystems. Chemosphere no. 76 (7):915-920.

89.  Zhang, X. X., Zhang, T. & Fang, H. (2009). Antibiotic resistance genes in water environment. Applied Microbiology and Biotechnology no. 82 (3):397-414.

90. Zounkova, R., Klimesova, Z., Nepejchalova, L.,Hilscherova, K. & Blaha, L. (2011). Complex evaluation of ecotoxicity and genotoxicity of antimicrobials oxytetracycline and flumequine used in aquaculture. Environ Toxicol Chem no. 30 (5):1184-9.

# Chapter 4

## EJACULATE ALLOCATION AND SPERM COMPETITION IN ALTERNATIVE REPRODUCTIVE TACTICS OF SALMON AND TROUT: IMPLICATIONS FOR AQUACULTURE

Tomislav Vladić

Department of Zoology, Stockholm University, Stockholm, Sweden

## INTRODUCTION

Aquacultural production has increased globally and today most of salmon consumed in Sweden originates from hatcheries. It is predicted that aquaculture will produce more food for human consumption than capture fisheries (Anon 2009). Freshwater aquaculture contributes to 48 percent by value and mariculture contributes 36 percent by value globally. Norway and Chile are the leading nations producing farmed salmonids, accounting for 33 and 31 percents of aquacultural production (Anon 2009). A common way of salmonid propagation in hatcheries involves mixing of milt from several adult males for fertilization eggs of single or several females. Such procedure invariably involves sperm competition between milt from several males to fertilize eggs of individual females. Atlantic salmon (Salmo salar) and brown trout (Salmo trutta) are cold water stenotherm fish species with pronounced population differences in time of gonad maturation and life history patterns. Ecological conditions and genetic liability fuel seasonal variation in the reproductive cycle dynamics. This cycle is characterized by a feed-back mechanism: gonadal development in smolt (sub-adults migrating from the nascent river to the sea) determines time of sea/lake period before fish return to freshwater spawning grounds. Northern, cold-climate populations may mature as precocially mature parr in the second year of life (Dalley et al 1983; Myers 1984), whereas southern salmonid populations were found to mature precocially already in the first year of life (Bagliniere & Maisse 1999). Sperm competition is the competition between sperm from several males for fertilization of the female's eggs during a single reproductive cycle (Parker 1970). Since males are capable of repeated matings at a much higher rate than females, they have evolved adaptations to prevent competitor males from ejaculating with the same female, securing thereby paternity. Adaptations to competition for securing reproductive success have created alternative reproductive strategies, which are genetically based life history allocation and behavioural rules affecting the manner an individual spreads its reproduction over the lifetime (Brommer 2000). Atlantic salmon and brown trout exhibit alternative male maturation phenotypes (tactics): anadromous males and preociously mature parr males, which commonly engage in sperm competition at spawning grounds. Alternative reproductive tactics are phenotypes that are an expression of the life history strategy, selected to maximize individual reproductive success, even if this involves reduced survival. In many fish species, alternative reproductive tactics are characterized by a conspicuous difference in age and size at maturity and differ in relative investment to gonad and somatic tissue (Taborsky 1994). Difference in age and size at sexual maturity has created alternative spawning behavioural tactics that are tools for securing reproductive success, such as "guarder" behaviour by dominant males and "sneaker" behaviour by subordinates. This review looks at energy allocation strategies of the alternative mating phenotypes of Atlantic salmon and brown trout and places this selective pressure in the context of interbreeding between escaped farmed fish and their wild conspecifics. Its objective is to review the proximate mechanisms of sperm competition and its evolutionary implications in the two sympatrically occurring salmonid species exhibiting alternative male maturation tactics and connect these to increased aqua-cultural production today. Possible effects of these procedures on genetic population structure as a consequence of escaped farmed fish from aquacultural production facilities will be considered in this chapter.

## SPERM CELL

In the salmonid spermatozoon, the following parts are morphologically distinct:

1.  Sperm plasma membrane which is the mediator of the signals for sperm motility,

2.  Sperm head with a nucleus containing the haploid paternal genetic material,

3.  Sperm mid-piece with a circular mitochondrion, where glycogene, phospholipids and phosphocreatine are the substrates for ATP production that provides energy for sperm motility,

4.  sperm flagellum with a central bundle of microtubules with a 9+2 organisational pattern, the axoneme, which is a locomotory component of the sperm cell.

## Sperm Membrane

Sperm plasma membrane is the semipermeable barrier that defines sperm body, about 10 nm thick (Baccetti 1985). Water surrounding the cell membrane tends to enter the cell, which would eventually cause it to burst. Freshwater fishes have evolved mechanisms to keep water outside the cytoplasm, in order to maintain cellular stability. Sperm plasma membrane functions as the main receptor of the environmental signals for motility, such as the hypotonicity in freshwater after ejaculation, which initiates sperm motility of freshwater teleosts (Morisawa and Suzuki 1980). High potassium concentrations in the seminal plasma of salmonid fishes are responsible for the inhibition of sperm motility (Stoss 1983). In contrast, changes in the external divalent cation concentrations and in osmolality initiate sperm motility concomitantly (Morisawa and Suzuki 1980). After ejaculation, potassium leaves sperm cell through ion channels hyperpolarizing thereby the cell membrane. This membrane hyperpolarization event is the trigger for the initiation of sperm motility (Morisawa 1994). Simultaneous increase in intracellular calcium levels causes activation of the enzyme adenylyl cyclase, which catalyzes the synthesis of cyclic AMP (cAMP) from ATP (Morisawa and Okuno 1982). cAMP is an intracellular signal, which activates the enzyme protein kinase, the enzyme activating tyrosine kinase with the function to phosphorylate a 15K protein (Morisawa and Hayashi 1985). Change in intracellular pH is not a primary factor for regulation of sperm motility in freshwater fishes (Morisawa et al 1999). Signal transmission traversing the sperm plasma membrane results in the cascade of events with an ultimate function to maintain the communication between a mature sperm cell and its environment.

## Sperm head

Ellipsoid trout sperm head is 2, 5 μm long and 1,5-2 μm in diameter, containing the cell nucleus (Billard 1983). Teleost fish spermatozoa have no acrosome at the anterior of the sperm head, a structure containing the enzymes that hydrolyse the egg envelopes, which is coupled with the presence of an orifice, the micropyle, on the teleostean egg (Ginsburg 1972). Sperm nucleus is transcriptionally inactive. The nucleoplasm of fish spermatozoa consists of nucleoprotamines, which after fertilization, when the hereditary material is activated, are substituted for histones (Figure 1). Sperm movement ensues as the result of the viscous interactions of sperm flagella with the surrounding medium (Taylor 1951). Gray and Hancock (1955) calculated that the viscous drag of the sperm head was small relative to the viscous drag of the flagellum itself. Thus, sperm head has only a negligible effect on the sperm cell locomotion (Gray and Hancock 1955; Humphries at al. 2008).

## Mid-piece

The salmon sperm middle piece is 0.30- 0.95 µm long (Vladić et al 2002). Sperm movement commences from the base of the flagellum, and is performed by sliding movements between flagellar proteins. A single ring-shaped mitochondrion surrounds the midpiece in salmonid fishes (Jamieson 1991; Figure 1). The mitochondrial function is to synthesize ATP by the process of oxidative phosphorylation from endogenous phospholipids, glycolipids and glycogene (Stoss 1983). ATP produced in the mitochondrial oxidative phosphorylation prior to ejaculation is the main energetic source for sperm motility. It is hydrolysed by the molecular motor, dynein ATPases, in the course of motlilty. An ATP molecule contains two phosphoanhydride bonds, which liberate free energy when hydrolysed to ADP or AMP. This ATP/ADP cycle is the fundamental mode of energy conversion in living systems. Atkinson (1968) proposed that the energy charge: regulates the energy metabolism in all living systems. In salmon, sperm energy charge increased with sperm tail length (Vladić et al 2002), which agrees with the finding of greater ADP concentrations in shorter sperm cells (Vladić 2001).

$$EC = \frac{2(ATP) + (ADP)}{2[(ATP) + (ADP) + (AMP)]}$$

$$(1)$$

In the process of energy transduction, the free energy of respiration is the driving force for ATP production by the $F_1 F_0$ ATP synthase (Harold 1986; Kinosita et al 2000). It was suggested that trout mitochondria have a low oxidative phosphorylation capacity, as ATP stores are quickly depleted in the course of sperm motility due to hydrolysis of ATP by dynein ATPase (Christen et al. 1987). Thus, a rate of mitochondrial respiration in fish spermatozoa is insufficient to maintain endogenous ATP reserves for prolonged motility (Cosson at al. 1999). In the Atlantic salmon, length of the sperm mid-piece was positively associated with the sperm ATP concentrations confirming thereby mitochondrial origin of ATP (Vladić et al 2002).

## Sperm Flagellum

The salmon sperm flagellum is 35- 45 µm long (Vladić et al 2002). Sperm flagella contain a uniform microtubular structure of the axoneme, comprised of nine peripheral and two central microtubules, surrounded by the plasma membrane (Afzelius 1959). The two central tubules are comprised of single microtubules, while the nine peripheral tubules are comprised of a complete A-tubule and an incomplete B-tubule. The molecular motor enzyme complex, adenosine triphosphatase (ATPase) activity involves a "dynein", which is an ATPase protein that drives the sliding of the outer doublet microtubules in sperm flagella (Harrison & King 2000). Some dynein molecules are assembled together with other proteins into macromolecular complexes called dynein arms. The peripheral microtubules have two rows of dynein arms along the length of the principle part of the flagellum. Guanosine nucleotides instead of adenosine nucleotides as in actin were detected in an additional isolated microtubule protein, which was named "tubulin" (Mohri and Ogawa 1975). This dynein-tubulin system is the molecular motor, which drives the flagellar movement of spermatozoa (Woolley 2000). Dynein ATPase proteins are bound to the A-tubule. The dynein arms of one peripheral doublet will walk on its neighbour to produce force for the flagellar movement, under hydrolysis of ATP. Thus, the beating of sperm flagella is the result of an active sliding between adjacent doublets of the axoneme powered by the ATP-driven mechanochemical cycle (Omoto 1991; Harrison & King 2000). A thin filament devoid of dynein arms is present at the terminal end of the sperm flagellum (Vladić et al 2002). This cell structure is an universal feature of animal spermatozoa (Retzius 1904, Franzén 1956). Length of this flagellar portion in salmon spermatozoa was 2-4 µm and contains only the two central microtubular pair (Vladić et al 2002). Since this region of the sperm flagellum experiences an increased viscous drag relative to the rest of the sperm flagellum an improvement in sperm propulsive effectiveness as an adaptation to viscous resistances has been proposed (Omoto and Brokaw 1982). Importantly, sperm fertility was positively related to the size of the thin sperm tail end piece fillament, suggesting adaptation to the viscosity of the aquatic medium (Vladić et al 2002; Figure 2).

# SPERM COMPETITION IN SALMONIDS

Success in sperm competition depends on behaviour of the competing males and of the contested female, as well as on the frequencies with which different behavioural tactics are played in the population: therefore sperm competition models are analysed in the framework of the evolutionary game theory (Maynard Smith 1982). These games are designed to find an "unbeatable" or evolutionary stable strategy (ESS), strategy which, after adopted by all individuals in the population, cannot be invaded by a mutant playing an alternative strategy. The game-theoretic models are phenotypic and do not include genetics of the participants (Maynard Smith 1982)

The assumptions of the sperm competition games are (sensu Parker 1998):

1. Males in competition have specified information about their physiological state, and a tactical decision is made about the behaviour that yields the highest pay-off for that individual. Indeed, phenotypic plasticity may be viewed as an individual response to its physiological state (McNamara and Houston 1996);

2. There is a range of possible ejaculation decisions that are dependent on his state, which includes resource-holding potential, fighting ability, age and size of the individual;

3. Several male ejaculates must compete for fertilization of a single egg clutch; piece of a salmon. Sperm nucleus contains densely packed chromatin, which is transcriptionally inactive. Below, a transverse section of the middle piece. Note the cell membrane around the sperm head and a single circular mitochondrion in the middle piece.

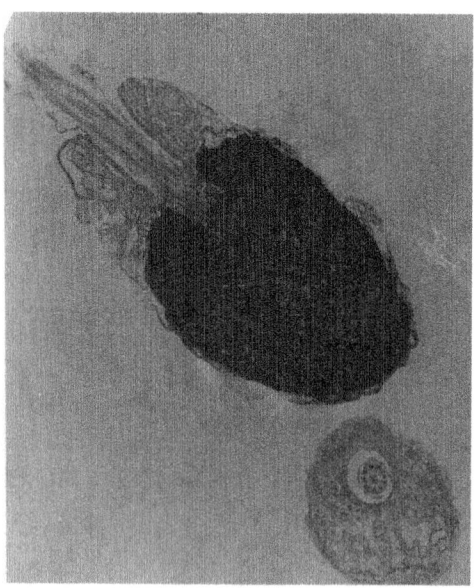

**Figure 1:** Electron micrograph showing a longitudinal section through sperm head and middle

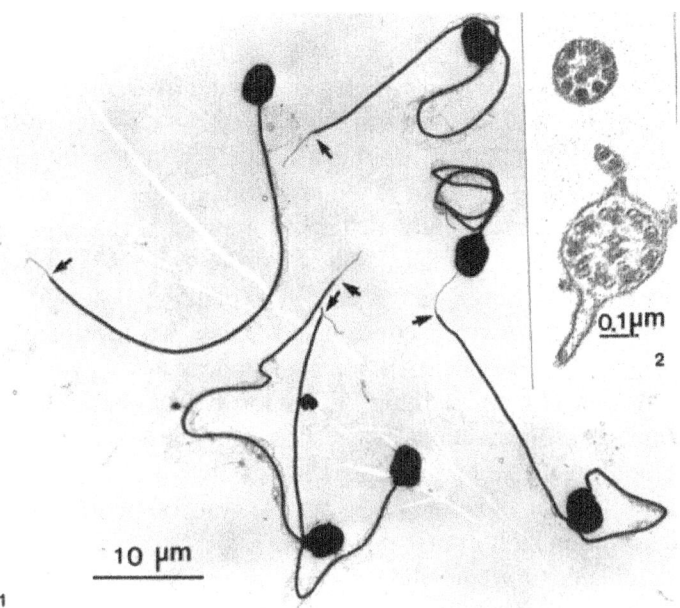

**Figure 2:** Whole mount of spermatozoa from a salmon parr. X2200. At the end of flagella, tail tip, containing only two centrally positioned, inner microtubules are indicated (arrows). 2) Sections through sperm tails from an anadromous salmon. The transected main piece (bottom) has nine microtubular doublets carrying inner and outer dynein arms and the two inner microtubules; the cell membrane forms two side fins. The end piece near the upper side fin contains only the two inner microtubules. The upper sperm transect contains 8 + 2 singlets; it is probably close to the transition region. X80 000. (from Vladić, Afzelius & Bronnikov, Biology of Reproduction, 66: 98-105).

4.  Sperm may be used either randomly for fertilization depending on their densities in ejaculates (i.e. fair raffle), or spermatozoa originating from competing ejaculates may differ in some features determining mating order or ejaculate quality, leading to different propensities for a success (i.e. loaded raffle). In the fair raffle, all sperm compete on equal terms, that is competing ejaculates are physiologically and energetically equal. Thus, only an increased number of spermatozoa relative to competitor's spermatozoa in the competition will yield a greater chance for success in the fair raffle situation. When the physiological quality of spermatozoa differs between competing males, it is expected that sperm quality will determine the outcome of competition. In the loaded raffle, a compensatory mechanism is predicted by which males, the sperm of which are devalued, compensate for this disadvantage by expending a greater proportion of reproductive effort on sperm (Parker 1990a,b).

5.    In externally fertilizing fish, a female does not exert a direct preference for the competing ejaculates.

If males have an imperfect information about the role in sperm competition (eg. mating hierarchy), sperm expenditure should increase proportionally with the risk of sperm competition (Parker 1990a). The mating role (dominant or sub-ordinate) doesn´t have to be randomly assigned, like in the mating systems in which male characters correlated with fitness, like body size, determine dominance and access to females. Males in disfavoured role are here expected to compensate for mating disadvantage by expending a greater proportion of reproductive effort on sperm production than males in dominant role. Dominant males are likely to suffer an informational handicap making them uncertain of engagement in sperm competition. This informational handicap of dominant males favours sneaking tactic, which attains significantly greater paternity than dominant adult tactic when the probability of cuckolding (mating out of pair bond) is low, but not when the intensity of sperm competition is high (Parker 1990b). The reason for the latter expectation is that an increase in the number of players (ejaculates) in competition results in reduced payoffs when the number of players is greater than 2 (Parker 1998). Therefore, disfavoured males ought to be selected to expand more energy in sperm production and/or quality than dominant males (Parker 1998). In all salmonid species, there are at least two distinct life histories in males (Jones, 1959; Fleming 1996): one, with dominant anadromous males with variable degrees of fighting ability that have developed linear dominance hierarchy, and second, with small precociously mature males-parr that do not migrate to sea to acquire food for prolonged growth, but stay in the stream of their hatching or "grilse", who return to the spawning ground after single season in the sea. Because of the smaller amounts of food in the river, and different ecological conditions of the freshwater habitat (reviewed by Gibson, 1993) these males are miniature in size relative to dominant males (Gage et al 1995) (Figure 3). In salmonids, genes are propagated into future generations by means of alternative life history strategies. Thus, younger, precociously mature males are using "sneaking" tactic in the vicinity of the spawning salmon pair - after it have swam unnoticed into the spawning territory where dominant male have already courted a female, parr is trying to "sneak" into the sperm cloud of dominant male ejaculates, when it will also ejaculate spermatozoa in what is to be a trial to fertilize females eggs in the red (Figures 4 and 5). This strategy can be very costly for pre-cocious parr if discovered by dominant male, because dominant will not hesitate to retaliate. Such retaliatory behaviour can sometimes incure injuries, or even death of precocious salmon parr (Hutchings & Myers, 1987). Reproductive strategies can be defined as genetically based behavioural programmes influencing individual allocation decisions to reproductive effort between alternative tactics within a sex (Gross

1996). Phenotypes that became as coevolved responses of life history traits to ecological problems are alternative life history tactics (Stearns 1976). The alternative strategy involves a genetically based life history program, which has evolved under environmental, usually frequency-dependent selective pressure (Gross 1996). When average individuals have reduced fitness compared to individuals on extremes of phenotypic distribution, disruptive selection might select for extreme individuals (Rueffler et al 2006). Variability in response to environmental pressure among individuals within salmonid populations is shaped by differences in survival and reproductive success. Atlantic salmon (Salmo salar) is an anadromous species, which spawn in freshwater, a feature characteristic for all salmonids. During its life history, two ecological environments are inhabited, a freshwater environment, in which salmon hatch and spawn, and a marine environment, where fast growth is achieved. The seaward migration is preceded by one to eight years. Before returning to the river of hatching for spawning the anadromous males and females stay between one and five years in the sea. Immature fish in the river (parr), become smolt in the spring of the second, third or fourth year. The process of smoltification involves various morphological, behavioural and physiological changes, as adaptations to marine environment. Salmon that return to the spawning river after only a single year are called 'grilse' (Mills 1971). Salmon do not feed during spawning migration, when males develop conspicuous lower hook and red bodily coloration, viz. secondary sexual characters (Figure 3). In northern latitudes, spawning may last from October to February, depending on duration of returning time to the nascent river. Males contribute no nest guarding or offspring tending to the female, but only their spermatozoa. Although after the spawning most of the males (called 'kelts') die due to high energetic costs that are paid in terms of intra-sexual fights for female acquisition and metabolic demands of sexual maturation (Jonsson et al 1991), relatively small proportion of females return to spawn in the following season (Mills 1971). Sea trout (Salmo trutta) males pursue shorter migratory routes and have greater iteroparity (i.e. mating in several consecutive seasons) than Atlantic salmon males, which undertake long migrations to the feeding grounds far off the coast and have an increased mortality rate after single spawning season (i.e. semelparity) (Belding 1934, Jones 1959, Mills 1989). Alternative male sexual maturation strategies are apparent both in salmon and brown trout, with adult, anadromous males which shed sperm simultaneously with precociously mature parr, a situation which results in sperm competition (Fig. 3). Number of males that may compete over fertilization of a single female eggs can vary between one and ten males (Hutchings 1986, Petersson and Järvi 1997). Gonad maturation was proposed to be determined by a genetical threshold (Thorpe 1986), which in concert with environmental control of maturation (shortening photoperiods; Lundqvist 1980; higher-than-average temperatures; Adams and Thorpe 1989; and food in excess enabling good growth, Alm 1959) determines male maturation pattern. Males that have a good growth rate tend to mature precociously (Alm 1959); they are maturing as precocious parr (Jones 1959, Mills 1989). A genetic component in male reproductive strategies was found in the Atlantic salmon (Glebe and Saunders 1986; Garant et al 2003). Individual precociously mature Atlantic salmon parr males have very variable fertilization success and may fertilize up to 65% of female eggs in the redd (Hutchings and Myers 1988, Thomaz et al 1997; GarciaVazquez et al 2002).

Precocious parr invest relatively more into gonadal tissue for their body mass than anadromous adults which invest more into secondary sexual traits which are frequently used in aggressive intrasexual interactions (Vladić & Järvi 2001). Also, higher metabolic demands of the parr reproductive strategy may be the reason that the relative heart weight in precocious parr is greater than this in immature fish (Armstrong and West 1994). This life history strategy is also associated with the impaired sea-water adaptability and reduced smoltification (Myers 1984, Lundqvist et al 1989). Male success is more dependent on social environment than is female success, which is dependent on the allometric relation between body size and gonad mass; therefore no single optimal life history is expected (Thorpe et al. 1998). Several decisions about the number of winters in the sea before returning to spawning may be exhibited within a single cohort. This results in the variable proportions of precociously mature parr, "grilse" and anadromous males that have spent a varying number of years in sea (Thorpe et al. 1998). In reproductive biology, male quality equals individual reproductive success. Sperm movement in externally fertlizing fish is dependent on the cellular energy, produced in the mitochondria located in the sperm mid-piece. The synthesis of ATP is coupled to respiratory electron transport requiring the expression of mitochondrial genes. As the sperm cytoskeletal microtubular assembly, the axoneme, is extending throughout the sperm flagellum, sperm size is mainly related to sperm tail length (reviews in Gibbons 1981, Witman 1990). Reduction in sperm size with the increase in time between ejaculation and fertilization of the egg was predicted when the sperm tail size is positively correlated with sperm velocity at the expense of sperm longevity (Parker 1993). This prediction is applicable typically to internally fertilizing species. The logic is that it is difficult to adjust sperm size before given mating, since sperm have matured in the reproductive tract before the information about the role in competition could influence male ejaculation tactic. In externally fertilizing species, like salmon, it was found that sperm size decreases with sperm competition across fish species (Stockley et al 1997). In the Atlantic salmon, positive associations between different sperm length parameters and sperm energy charge, ATP concentrations and fertilization ability were found (Vladić et al 2002). In addition, salmonid sperm show adaptation to natural spawning temperatures (ie 3-4 °C), whereas trout eggs exhibit higher thermotolerance than salmon eggs, possibly reflecting the southern origin of trout (Vladić & Järvi 1997). Sperm density was higher in both brown trout and salmon precocious parr, whereas salmon sperm are containing greater ATP concentrations than trout sperm (Vladić 2001). These features may be connected to the greater semelparity of salmon as compared to trout (Vladić 2000). Jonsson and Jonsson (2005) discuss greater energy allocation in reproduction of the precocially mature parr of Atlantic salmon than this in precocially mature trout parr in connection to conspicuous body size difference between the species and relatively longer migration distances to the feeding areas at sea in the Atlantic salmon as compared to the brown trout. Female eggs were found to be fertile after 512 s in water, significantly longer than sperm were mobile, i.e. 100-300 s at 2-4 C° (Vladić & Järvi 1997). Salmon parr have greater sperm vigour (percentage of motile cells in ejaculates) (Vladić & Järvi 2001) and trade-off between sperm velocity and longevity after one-third of time since sperm activation (Vladić 2001), the result in agreement with the result published by Levitan (2000) (see Rosengrave et al 2009 for the discussion of effect of ovarian fluid on sperm behaviour). Therefore, studies on sperm traits should not imply contention of individual male quality unless these traits are tested in fertilization experiments. Besides sperm ATP content, sperm velocity was found to be the most important determinant of success in sperm competition (Gage et al 2004; Burness et al 2004; Yeates et al 2007). In addition, salmon parr were found to produce more ATP per sperm cell and are beter in fertilizing eggs both in the non-competitive situation (Vladić & Järvi 2001) and in sperm competition (Vladić et al 2010) confirming thereby the loaded model of sperm competition (Parker 1998). Sperm density in the competition is high; therefore all eggs in externally fertilizing fish are expected to be fertilized instantaneously. At ESS, there will be a natural level of egg loss due to sperm death rate (Ball and Parker 1996). This "adaptive infertility" is opposed by an increase in sperm competition intensity; it benefits females to tolerate group spawning promoting thereby conflict between males and thereby sperm competition (Ball and Parker 1997). In addition, trade-off between offspring quality and quantity might be expected. This emanates from the fact that in structured populations, natural selection does not maximise short-term individual reproductive success (quality) but rather longterm value associated with genotype distribution in the population (McNamara et al 2011).

## HUMAN IMPACT ON SALMONID EJACULATE ALLOCA-
## TION AND HERITABILITY OF PHENOTYPIC PLASTICITY

Human impact on wild habitats has proven devastating in many instances, as a consequence of the extensive hydroelectric power plants dam construction in most Swedish rivers. Occurrence of interspecific hybrids between the salmon and trout, which are called "laxing" in Sweden, was attributed to the shrinkage of the spawning area and destruction of natural spawning habitats caused by new hydroelectric plants (Jansson and Öst 1997). Such inter-specific hybrids are sterile. In addition, wild-farmed Atlantic salmon hybrids were found to have lowered fitness in comparison to wild fish, cautioning that frequent escapes from hatchery facilities may potentially reduce fitness for wild populations (Fleming et al 2000; McGinnity et al 2003; Araki et al 2007). Therefore, hatchery rearing for compensatory purposes has created new demographic pressures on endangered wild populations of these fish. Some traits are artificially selected for in hatchery environment, like genes for pathogen resistance (reviewed by Fjaelstad et al 1993), for instance genes of the Major Histocomapatibility Complex (MHC) that confer resistence to disease in vertebrates (Reusch et al 2001). However, considering the fact that local populations contain long-term adapted genomes to local environments, gene flow between farmed and wild fishes is likely to erode local adaptation and possibly lead to their extinction (McGinnity et al 2003). Hereditary basis for age at maturity in salmonids cautions that the compensatory breeding programs could have a significant demographic influence on wild Atlantic salmon populations (see Garant et al 2003), since the proportion of early maturing males may be substantial in hatcheries from where the supplementary fish are recruited, especially under favourable conditions of culture (i.e. higher-than-average temperature, food in excess) (reviewed by Jonsson & Jonsson 2006). Therefore a knowledge about ejaculate quality from alternative male morphotypes commonly engaging in sperm competition at spawning grounds has comprehensive areas of application, from those related to basic evolutionary questions to application of the findings in species management context.

**Figure 3:** Differences in size between andromous and preciously mature parr of sea trout (above) and salmon from river Dalälven (below), showing the asymmetry in roles that these male morphotypes experience during sperm competition. Below, at the top, a female (81 cm long, 5.2 kg total weight), beneath a male (92 cm long, 6.7 kg total weight) with a large kype at the lower jaw, elongated nose and big adipose fin. The three smaller salmon are mature parr. Note differences in body morphology between parr males and anadromous adults, and differences in the expression of secondary sexual traits between the anadromous male, the female and precocious parr. Photo by Erik Petersson and Anna Löf. (Lower photo reprinted by permission from Vladić 2001, with permission).

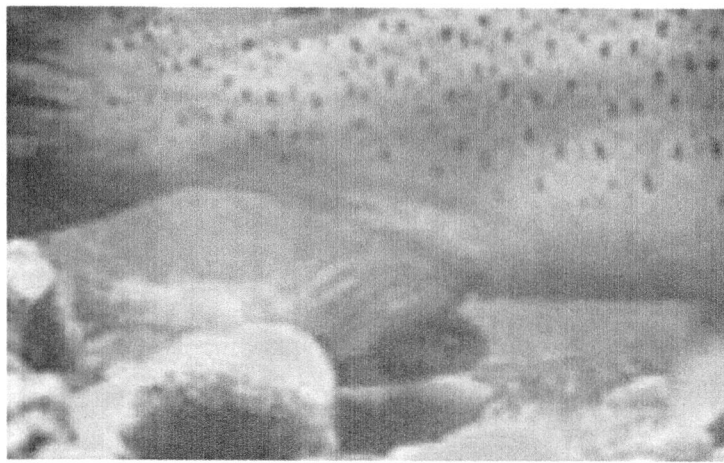

**Figure 4:** Sea trout precocious parr assumes an advantageous"sneaking" position beneath the female genital vent. Photo: Tomislav Vladić.

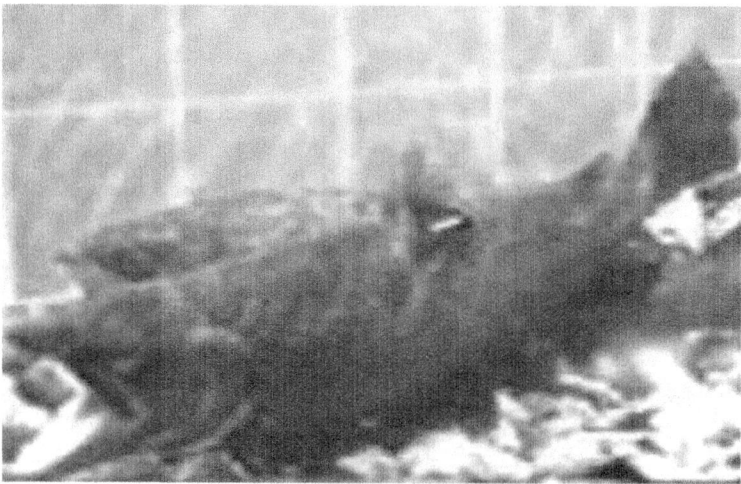

**Figure 5:** Orgasm and ejaculation of an anadromous trout male. The female that expulsed eggs is behind the ejaculating male. Note the precocious parr in the sperm cloud beneath the anadromous male; it released sperm in sperm competition. Photo: Tomislav Vladić.

Since behaviourally sub-ordinate males can acquire fitness benefits by exploiting sexual investment of behaviourally dominant males, male-biased operational sex ratio selects for alternative reproductive tactics, possibly through frequency-dependent selection. Conditional variation (variation depending on environmental influence) may produce different phenotypes from single genotype (Gross 1996). Reaction norms are genotype responses describing the manner individuals respond to environmental change within a population (Woltereck 1909). In salmonid fishes, growth rates influence choice of male mating tactic; males in good condition tend to mature early whereas males in poor condition tend to postpone reproduction in favour of prolonged body growth (Hutchings & Myers 1994; Thorpe et al 1998). This choice is a "threshold trait", where a liability toward maturation-age decision depends on the phenotype's position relative to some physiological (or environmental, see below) threshold value. It is evident that human impact exercised through high food ratios supplemented to hatchery fish aimed at compensatory releases changes male maturation pattern in the population, since conditional variation produces different phenotypes from the same genotype over environmental gradient (i.e. phenotypic plasticity). Therefore, decrease in egg size is expected when the food supplementation is in excess in hatcheries (Heath et al 2003), as offspring from larger eggs could have an advantage early in life (Einum & Fleming 1999). Interestingly, brown trout sea-running males were found to preferentially fertilize eggs of intermediate sizes, on the contrary to precocially maturing parr, which fertilized all eggs sizes indiscriminately; this mechanism was proposed as an expression of cryptic male choice (Vladić 2006). Recently, phenotypic plasticity in sperm production as a response to sperm competition risk was emphasized (Rudolfsen et al 2006; Cornwallis & Birkhead 2007; Pizzari et al 2007; Ota et al 2010). Importantly for conditional strategies to evolve, environmental cue affecting gonad maturation must be reliable, whereas finesses of the alternative maturation life histories are not necessarily equal (Tomkins & Hazel 2008). Each reaction norm can be understood as different conditional strategy. A cue property of the genetic response in reaction norms as a result of selection, G, is that the selection differential before and after the episode of selection measured by mean mortality in different environments, Sk, and heritability of the plastic maturation trait, $h_2$, will vary in the function of environmental cue distribution, e, (e.g. feeding, temperature, photoperiod) within a given generation k:

$$\Delta G_k = \frac{S_k \, h^2}{e}$$

(2)

Evolutionary changes arise through change in frequency distribution of the environmental cue, each side of this distribution favoring different maturation conditional tactic (eg. early versus late maturing phenotypes) (Tomkins & Hazel 2008). Above equation implies that stronger additive genetic variation for the maturation trait, the stronger response to selection should be over an environmental gradient. Heritability of phenotypic plasticity depends on the extent of genotype x environment interaction and strength of sexual selection. A genotype x environment reaction exists in the population if slopes of alternative fitness functions over environmental gradient cross. If slopes of fitness functions do not cross, male tactic may depend on individual male condition, which is commonly influenced by developmental constraints (Tomkins & Hazel 2008). These conditional reproductive tactics are central in sexual selection. Thus, genetic variance in phenotypic plasticity in the population, $\delta2_{PL}$, can be defined by partitioning total variance,

$$\delta^2_{PL} = \delta^2_E + \delta^2_{GxE} + \delta^2_S$$

(3)

where $\delta^2_E$ is environmental variance, $\delta2_{GxE}$ is genotype x environment variance and $\delta^2_S$ is conditional variance due to sexual selection. Reduced phenotypic plasticity due to interbreeding between escaped farmed and wild fish may reduce capacity of population to cope with environmental change and thereby disrupt population dynamics. Genetic diversity is crucial in small, isolated wild populations with increased levels of genetic drift, whereby chance events may cause random fixation of deleterious or invasive alleles. Maintaining genetic diversity and minimizing potential bottlenecks due to genetic drift requires minimum relatedness among individuals (Ohta 1982). In the generation k, genetic diversity can be defined as

$$GDk = 1 - \frac{1}{2Nef}$$

(4)

where $2N_{ef}$ is the effective allele number (Crow & Kimura 1970; Caballero & Toro 2000). Common procedure employed in hatcheries is to mix ova from several females with sperm from several males in a single batch for supportive breeding purposes. This procedure induces sperm competition leading to reduction in the effective sample size of breeders and to increase in genetic variance in the population (Withler 1988; Withler & Beacham 1994). Although sperm potency leads to increased individual fertilization success, it may also lead to decreased genetic variation within populations when sperm traits are heritable due to increased variance in fertilization success between competing males and consequently decreased effective population size. Therefore equalizing milt volume in hatcheries should not necessarily reduce loss of genetic variation even at expense of favouring younger males in competition (Vladić et al 2010), because this practice reduces opportunity for natural selection to cleanse locally maladapted genetic contribution (Wedenkind et al 2007). Effective number of breeders is defined as (Ridley 1993):

$$Ne = \frac{4NmNf}{Nm + Nf}$$

(5)

where Ne is the effective number of breeders contributing to the gene pool in the following generation, Nm and Nf are numbers of male and female breeders, respectively. Ne is the effective size of population with Mendelian segregation of genes (ideal population: equal sex ratio, constant population size and random probability of survival to adulthood) derived from the probability that two alleles at a locus are derived from the same grandparent. Means and variances are derived from the numbers of offspring surviving to maturity rather than from individual reproductive success (Campton 2004). Numbers of offspring surviving to maturity are affected by life history patterns of fish under consideration, as it was shown that salmon precocious parr males produce ejaculates of greater quality (Vladić & Järvi 2001: Vladić et al 2010), in accordance with predicted inverse relationship between fish age and gamete quality (Fleming 1996). Therefore, maximizing number of founders is expected to maximize genetic diversity in the population.

To summarize, genetic effects of captive breeding in hatcheries for supplementation of wild fish after destruction of natural spawning paths due to building of artificial dams for powerplant energy production may create fish that are reproductively inferior in the wild (Fleming et al 2000; Araki et al 2007). Nevertheless, negative effects of fish supplementation might be dependent on species and on the fish strain, since detrimental effects of the seven generation rearing on reproductive performance were not found in the brown trout from the river Dalälven stock (Dannewitz et al 2004).

## SPERM COMPETITION AND MATERNAL EFFECTS

Rather than paternal, maternal genes were found to influence sperm phenotype (Froman & Kirby 2005), including sperm length in Callosobrushus maculates (Gay et al 2009). In addition to effects on sperm size, maternal genes strongly influence sperm motility, notably through predominantly (see Ankel-Simons & Cummins 1996; Rand 2001) maternally inherited haploid mtDNA (Ruiz-Pesini et al 2000; Froman & Kirby 2005). Selective forces are expected to differ between internally and externally fertilizing species due to discrete differences in the physical environment in which sperm compete to fertilize eggs. In externally fertilizing species where fertilization occurs instantaneously and sperm are relatively short-lived, sperm velocity may be naturally selected trait that confers advantage in competition for fertilizations. Nevertheless, in cases when insemination and fertilization are temporally separated as is the case for most internally fertilizing species, different sperm traits might be selected for (Parker 1993). In such cases, sperm longevity should be selected for (Taborsky 1998). Therefore, we expect that selective forces shape different ejaculate traits, which are advantageous depending of the mode of reproduction. In addition, sperm quality and quantity may trade off due to constraints imposed by conflicting life-history demands of simultaneous investment in body growth and gamete quality (Stearns 1992). However, if sperm quality and morphology are polygenic and/or unlinked traits, we should not expect to detect simple correlation between gamete quality and morphology. Sexual selection selects for the two male traits: intrinsic genetic quality (indirect mechanism) and paternal care (direct mechanism). There is no paternal care in salmonid fishes. Body size is the male trait directly preferred by females (Andersson & Simmons 2006); the mechanism suggested by which a cryptic female choice for male's genetic quality is exerted in externally fertilising fish is through ovarian fluid that facilitates sperm function in the Arctic charr (Turner and Montgomerie 2002) and in the Atlantic cod (Litvak and Trippel 1998). The effect of ovarian fluid on sperm performance depends on the physiological compatibility between male and female partners (Rosengrave et al 2008). Mate choice is believed to optimize variation on Major Histocompatibility Complex (MHC) genes that confer resistance to disease. However, in the brown trout, female eggs were fertilized preferentially by males with intermediate molecular divergence in MHC genes, as males with great amino acid divergence on the MHC loci might have lower adaptation to locally adapted pathogens (Forsberg et al 2007). In the Atlantic salmon, variation in the MHC I class gene was biased toward similar genotypes, suggesting thus suppression of hybridization and outbreeding depression as the possible mechanism in salmonid mate choice (Yeates et al 2009). Thus, local adaptation may be disrupted by interbreeding between wild and escaped farmed fish in small isolated wild populations (reviewed by Hutchings and Fraser 2006).

## CONCLUSIONS AND FUTURE PERSPECTIVES

Salmonid reproductive strategies are determined by intrasexual competition between males for fertilization and by female body condition. Human interruption in natural spawning that is practiced in hatcheries during compensatory breeding programmes can potentially diminish stock genetic diversity. Therefore it is a task of utmost importance to understand processes that intervene with the evolutionary mechanisms that maintain alternative reproductive phenotypes in the remaining natural salmonid populations. Some questions are unanswered still:

1.  How to maintain current levels of aquacultural production without simultaneously reducing existing wild fish stocks genetic diversity due to interbreeding of escaped with wild fish?

2.  Perform hatchery precociously mature parr males "the best of a bad job" strategy or produce ejaculates of greater quality than their wild counterparts?

3.  What costs, if any, do wild females pay if mated with precociously mature parr males originating from hatcheries?

## ACKNOWLEDGMENTS

I thank the editors of this publication for inviting me to provide this work. Professor Sören Nylin is acknowledged for commenting the manuscript. This paper is dedicated to the memory of professor Björn Afzelius.

## REFERENCES

1.  Alm, G. (1959) Connection between maturity, size and age in fishes. Rep. Inst. Freshwater Res. Drottningholm 40: 5- 145.

2.  Adams, C.E. & Thorpe J.E. (1989) Photoperiod and temperature effects on early development and reproductive investment in Atlantic salmon (Salmo salar L.). Aquaculture, 79: 403-409.

3.  Afzelius, B. (1959) Electron microscopy of the sperm tail. Results obtained with a new fixative. J. Biophys. Biochem. Cytol., 5: 269-278.

4.  Andersson, M. & Simmons, L.W. (2006) Sexual selection and mate choice. Trends Ecol. Evol., 21: 296-302.

5.  Ankel-Simons, F. & Cummins, JM (1996) Misconceptions about mitochondria and mammalian fertilization: Implications for theories on human evolution. Proc. Natl. Acad. Sci., 93: 13859- 13863.

6.  Anonymous (2009) Fisheries, Sustainability and Development. (eds Wramner, P., Ackefors, H., Cullberg M). Royal Swedish Academy of Agriculture and Forestry (KSLA), Halmstad,

7.  Araki, H., Cooper, B. & M.S. Blouin (2007) Genetic effects of captive breeding cause a rapid, cumulative fitness decline in the wild. Science, 318: 100- 103.

8.  Armstrong, J.D. & West, C.L. (1994) Relative ventricular weight of wild Atlantic salmon parr in relation to sex, gonad maturation and migratory activity. J. Fish Biol., 44: 453- 457.

9.  Atkinson, D.E. (1968) The energy charge of the adenylate pool as a regulatory parameter. Interaction with feedback modifiers. Biochemistry, 7: 4030- 4034.

10. Baccetti, B. (1985) plasticity of the sperm cell. in Biology of Fertilization (eds. Metz, C. B. & Monroy, A.) pp. 3- 58. Acad. Press, New York.

11. Bagliniere, J L & Maisse, G (1999) Biology And Ecology Of The Brown Sea Trout. Springer, London Ltd.

12. Ball, M.A. & Parker, G.A. (1996) Sperm competition games: External fertilization and "adaptive" infertility. J. theor. Biol., 180: 141- 150.

13. Ball, M.A. & Parker, G.A. (1997) Sperm competition games: inter- and intra-species results of a continuous external fetilization model. J. theor. Biol., 186: 459- 466.

14. Belding, D.L. (1934) The cause of the high mortality in the Atlantic salmon after spawning. Trans. Am. Fish. Soc., 64: 219- 224.

15. Billard, R. (1983) Ultrastructure of trout spermatozoa: changes after dilution and deepfreezing. Cell Tiss. Res., 228: 205-218.

16. Brommer, J.E. (2000) The evolution of fitness in life-history theory. Biol. Rev., 75: 377-404.

17. Burness, G., Casselman, S.J., Schulte-Hostedde, A.I., Moyes, C.D. & R. Montgomerie (2004) Sperm swimming speed and energetics vary with sperm competition risk in bluegill (Lepomis macrochirus). Behav. Ecol. Sociobiol., 56: 65- 70.

18. Caballero, A., & Toro, M. A. (2000). Systems of mating to reduce inbreeding in selected populations. Interrelations between effective population size and other pedigree tools for the management of conserved populations. Genet. Res., Camb. 75: 331-343.

19. Campton, D.E. (2004) Sperm competition in salmon hatcheries: the need to institutionalize genetically benign protocols. Trans. Am. Fish. Soc., 133: 1277- 1289.

20. Christen, R., Gatti, J-L. & R. Billard (1987) Trout sperm motility. The transient movement of trout sperm is related to changes in the concentration of ATP following the activation of the flagelar movement. Eur. J. Biochem., 160: 667-671.

21. Cosson, J., Billard, R., Cibert, C., Dréanno, C & Suquet, M. (1999) Ionic fractors regulating the motility of fish sperm. In The Male Gamete: From Basic Science to ClinicalApplications (ed. Gagnon, C.) Cache River Press, pp. 161- 186.

22. Cornwallis, C.K. & Birkhead, T.R. (2007) Changes in sperm quality and numbers in response to experimental manipulation of male social status and female attractiveness. Amer. Natur., 170: 758- 770.

23. Crow, J. F. & Kimura, M. (1970). An Introduction to Population Genetics Theory. New York: Harper & Row.

24. Dalley, D.L., Andrews, C. W. & Green, R. H. (1983) Precocious male Atlantic salmon parr (Salmo salar) in insular Newfoundland. Can. J. Fish. Aquat. Sci. 40: 647-652.

25. Dannewitz, J., Petersson, E., Dahl, J., Prestegaard, T., Löf. A.C. & T. Järvi (2004) Reproductive success of hatchery- produced andc wild-born brown trout in experimental stream. J Appl Ecol, 41: 355-364.

26. Einum, S. & Fleming, I.A. (1999) Maternal effects of egg size in brown trout (Salmo trutta): norms of reaction to environmental quality. Proc. R. Soc Lond. B, 266: 2095- 2100.

27. Fjalestad KT, Gjedrem T & B Gjerde (1993) Genetic improvement of disease resistance in fish: an overview. Aquaculture, 111: 65-74.

28. Fleming, I.A. (1996) Reproductive strategies of Atlantic salmon: ecology and evolution. Rev. Fish Biol. Fish., 6: 379- 416.

29. Fleming, I.A., Hindar K, Mjølnerød IB, Jonsson, B, Balstad, T & A Lamberg (2000) Lifetime success and interactions of farm salmon invading a native population. Proc. R. Soc. Lond. B, 267: 1517-1523.

30. Froman, D.P. & Kirby, J.D. (2005) Sperm mobility: Phenotype in roosters (Gallus domesticus) determined by mitochondrial function. Biol. Reprod., 72: 562-567.

31. Forsberg, LA, Dannewitz, J, Petersson, E & Grahn, M (2007) Influence of genetic dissimilarity in the reproductive success and mate choice of brown trout- females fishing for optimal MHC dissimilarity. J evol Biol, 20: 1859-1869.

32. Franzén, Å (1956) On Spermiogenesis, Morphology of the Spermatozoon, and Biology of

33. Fertilization Among Invertebrates. Zoologiska Bidrag från Uppsala, 31, 355-482.

34. Gage, M.J.G., Stockley, P. & Parker, G.A. (1995) Effects of alternative male mating strategies on characteristics of sperm production in the Atlantic salmon (Salmo salar): theoretical and empirical investigations. Phil. Trans. R. Soc. Lond. B 350: 391- 399.

35. Gage, M.J.G., Macfarlane, C.P., Yeates, S., Ward, R.G., Searle, J.B. & G.A. Parker (2004) Spermatozoal traits and sperm competition in Atlantic

salmon: relative sperm velocity is the primary determinant of fertilization success. Current Biology, 14: 44- 47.

36. Garant, D., Dodson, J.J. & L. Bernatchez (2003) Differential reproductve success and heritability of alternative reproductive tactics in wild Atlantic salmon (Salmo salar L.). Evolution, 57: 1133-1141.

37. Garcia-Vasquez, E., Moran, P., Perez, J., Martinez, J.L., Izquierdo, J.I., de Gaudemar, B. & E. Beall (2002) Interspecific barriers between salmonids when hybridisation is due to sneak mating. Heredity, 89: 282-292.

38. Gay, L., Hosken DJ, Vasudev, R, Tregenza, T & PE Eady (2009) Sperm competition and maternal effects differentially influence testis and sperm size in Callosobruchus maculatus. J evol Biol, 22: 1143-1150.

39. Gibbons, I. R. (1981) Cilia and flagella of eukaryotes. Discovery in cell biology. J. Cell Biol.91: 107s- 124s.

40. Gibson, R. J. (1993) The Atlantic salmon in fresh water: spawning, rearing and production. Rev. Fish Biol. Fish. 3: 39- 73.

41. Ginsburg, A. S. (1972) Fertilization in Fishes and the Problem of Polyspermy. Akademiya Nauk SSSR, Institut Biologii Razvitiya. (ed. Detlaf, T. A. ). Translated from Russian by Israel Program for Scientific Translations, Jerusalem.

42. Glebe, B.D. & Saunders, R.L. (1986) Genetic factors in sexual maturity of cultured Atlantic salmon (Salmo salar) parr and adults reared in sea cages. Can. Spec. Publ. Fish. Aquat.Sci., 89: 24- 29.

43. Gray, J & Hancock, GJ (1955) The propulsion of sea urchin spermatozoa. J Exp Biol, 32: 802- 814.

44. Gross, M.R. (1996) Alternative reproductive strategies and tactics: diversity within sexes. Trends Ecol. Evol., 11: 92- 98.

45. Harold, F.M. (1986) The Vital Force: A Study of Bioenergetics. W. H. Freeman and Company.

46. Harrison, A & King, SM (2000) The molecular anatomy of dynein. Essays in Biochemistry, 35: 75-87.

47. Heath, D.D., Heath, J.W., Bryden, C.A., Johnson, R.M. & C.W. Fox (2003) Rapid evolution of egg size in captive salmon. Science, 299: 1738-1740

48. Humphries, S, Evans, JP & LW Simmons (2008) Sperm competition: linking form to function. BMC Evolutionary Biology, 8: 319.

49. Hutchings, J.A. (1986) Lakeward migrations by juvenile Atlantic salmon, Salmo salar. Can. J. Fish. Aquat. Sci., 43: 732- 741.

50. Hutchings, J.A. & Fraser, D.J. (2008) The nature of fisheries- and

farming-induced evolution. Molecular Ecology, 17: 294-313.

51.  Hutchings, J. A. & Myers, R. A. (1987) Escalation of an asymmetric contest: mortality resulting from mate competition in Atlantic salmon, Salmo salar. Can. J. Zool. 65: 766- 768.

52.  Hutchings, J.A. & Myers, R. A. (1988) Mating success of alternative maturation phenotypes in male Atlantic salmon, Salmo salar. Oecologia, 75: 169- 174.

53.  Jannsson, H. & Öst, T. (1997) Hybridization between Atlantic salmon(Salmo salar) and brown trout (S. tritta) in a restored section of the river Dalälven, Sweden. Can. J. Fish. Aquat. Sci., 54: 2033- 2039.

54.  Jamieson B.G.M. (1991) Fish Evolution and Systematics: Evidence from Spermatozoa. Cambridge University Press, Cambridge.

55.  Jones, J.W. (1959) The Salmon. London: Collins.

56.  Jonsson, B., Jonsson, N. & Hansen, L.P. (1990) Does juvenile experience affect migration and spawning of adult Atlantic salmon? Behav. Ecol. Sociobiol., 26: 225- 230.

57.  Jonsson, B. & Jonsson, N. (2005) Lipid energy reserves influence life-history decision of Atlantic salmon (Salmo salar) and brown trout (S. trutta) in fresh water. Ecol. Freshw. Fish, 14: 296- 301.

58.  Jonsson, B. & Jonsson, N. (2006) Cultured Atlantic salmon in nature: a review of their ecology and interaction with wild fish. ICES J Mar Sci, 63: 1162-1181.

59.  Kinosita, K. Jr., Yasuda, R. & Noji, H. (2000) F1-ATPase: a highly efficient rotary ATP machine. Essays in Biochemistry, 35: 3-18.

60.  Levitan, D.R. (2000) Sperm velocity and longevity trade off each other and influence fertilization in the sea urchin Lytechinus variegatus. Proc. R. Soc. Lond. B, 267: 531- 534.

61.  Litvak, MK & Trippel, EA (1998) Sperm motility pattern of Atlantic cod (Gadus morrhua) in relation to salinity: effects of ovarian fluid and egg presence. Can. J. Fish. Aquat. Sci.55: 1871-1877.

62.  Lundqvist, H. (1980) Influence of photoperiod on growth in Atlantic salmon parr (Salmo salar) with special reference to the effect of precocious sexual maturation. Can. J. Zool., 58: 940- 944.

63.  Lundqvist, H., Borg. B. & Berglund, I. (1989) Androgens impair seawater adaptability in smolting Baltic salmon. Can. J. Zool., 67: 1733- 1736.

64.  Maynard Smith, J. (1982) Evolution and the Theory of Games. Cambridge University Press, Cambridge

65.  McGinnity P, Prodöhl P, Ferguson A, Hynes R, Ó Maoiléidigh N, Baker

N, Cotter D, O'Hea B, Cooke D, Rogan G, Taggart J & T Cross (2003) Fitness reduction and potential extinction of wild populations of Atlantic salmon, Salmo salar, as a result of interactions with escaped farm salmon. Proc. R. Soc. Lond. B, 270: 2443–2450

66.  McNamara, J.M. & Houston, A.I. (1996) State-dependent life histories. Nature, 380: 215- 221.

67.  McNamara, JM, Trimmer, PC, Eriksson, A, Marshall, JAR & AI Houston (2011) Environmental variability can select for optimism or pessimism. Ecology Letters, 14: 58-62.

68.  Mills, D. (1971) Salmon and trout: A Resource, its Ecology, Conservation and Management. Oliver & Boyd, Edinburgh.

69.  Mills, D. (1989) Ecology and Management of Atlantic Salmon. Chapman and Hall, London.

70.  Mohri H. & Ogawa K. (1975) Tubulin and dynein in spermatozoan motility. in The Functional Anatomy of Spermatozoon. (ed. Afzelius, B. A.) pp. 161- 168. Pergamon Press Ltd.

71.  Morisawa, M. (1994) Cell signaling mechanisms for sperm motility. Zool. Sci., 11: 647- 662.

72.  Morisawa, M. & Hayashi, H. (1985) Phosphorilation of a 15 K axonenmal protein is the trigger initiating trout sperm motility. Biomed. Res., 6: 181- 184.

73.  Morisawa, M. & Okuno, M. (1982) Cyclic AMP induces maturation of trout sperm axoneme to initiate motility. Nature, 295: 703- 704.

74.  Morisawa, M. & Suzuki, K. (1980) Osmolality and potassium ion: Their roles in initiation of sperm motility in teleosts. Science, 210: 1145- 1147.

75.  Morisawa, M., Oda, S., Yoshida, M. & H. Takai (1999) Transmembrane signal transduction for the regulation of sperm motility in fishes and ascidians. In The male Gamete: FromBasic Science to Clinical Applications (ed. Gagnon, C.) Cache River Press, pp. 149- 160.

76.  Myers, R.A. (1984) Demographic consequences of precocious maturation of Atlantic salmon (Salmo salar). Can. J. Fish. Aquat. Sci., 41: 1349- 1353.

77.  Ohta, T. (1982) Linkage disequilibrium due to random genetic drift in finite subdivided populations. Proc. Natl. Acad. Sci., 79: 1940- 1944.

78.  Omoto, C.K. (1991) Mechanochemical coupling in cilia. Int. Rev. Cyt., 131: 255- 292.

79.  Omoto, C.K. & Brokaw C.J. (1982) Structure and behaviour of the sperm terminal filament. J. Cell Sci., 58: 385- 409.

80.  Ota, K, Heg, D, Hori, M & M Koda (2010) Sperm phenotypic plasticity in a cichlid: a territorial male´s counterstrategy to spawning takeover. Behav, Ecol., 21: 1293-1300.

81.  Parker, G. A. (1970) Sperm competition and its evolutionary consequences in the insects. Biol. Rev. 45: 525- 567.

82.  Parker, G. A. (1990a) Sperm competition games: raffles and roles. Proc. R.. Soc. Lond. B 242: 120- 126.

83.  Parker, G. A. (1990b) Sperm competition games: sneaks and extra-pair copulations. Proc. R. Soc. Lond. B 242: 127- 133.

84.  Parker, G. A. (1993) Sperm competition games: sperm size under adult control. Proc. Roy. Soc. Lond. B. 253: 245- 254.

85.  Parker, G.A. (1998) Sperm competition and the evolution of ejaculates: towards a theory base. in Sperm Competition and Sexual Selection (eds. T.R. Birkhead and A.P. Møller), pp. 3- 54.

86.  Petersson, E. & Järvi, T. (1997) Reproductive behaviour of sea trout (Salmo trutta)- the consequences of sea- ranching. Behaviour,134: 1- 22.

87.  Pizzari, T., Cornwallis, C.K. & D.P. Froman (2007) Social competitiveness associated with rapid fluctuations in sperm quality in male fowl. Proc. Roy.Soc.Lond.B, 274: 853- 860.

88.  Rand, DM (2001) The units of selection on mitochondrial DNA. Annu. Rev. Ecol. Syst., 32: 415-448.

89.  Reusch, T.B.H., Häberli, M.A., Aeschlimann, P.B. & Milinski, M. (2001) Female sticklebacks count alleles in a strategy of sexual selection explaining MHC polymorphism. Nature, 414: 300- 302.

90.  Retzius, G (1904) Zur Kenntnis der Spermien der Evertebraten. Biologische Untersuchungen. N.F., 11: 1-32.

91.  Ridley, M (1993) Evolution, Blackwell.

92.  Rosengrave, P., Gemmell, N.J., Metcalf, V., McBride, K. & Montgomerie, R. (2008) A mechanism for cryptic female choice in chinook salmon. Behav. Ecol., 19, 6, 1179- 1185

93.  Rosengrave, P., Montgomerie, R., Metcalf, V. & Gemmell N.J. (2009) Sperm traits in Chinook salmon depend upon activation medium: implications for studies of sperm competition in fishes. Can J Zool, 87: 920-927.

94.  Rudolfsen, G., Figenschou, L., Folstad, I., Tveuten, H. & M. Figenschou (2006) Rapid adjustments of sperm characteristics in relation to social status. Proc.Roy.Soc.Lond.B, 273: 325- 332.

95.  Rueffler, C., Van Dooren, T.J.M., Leimar, O. & PA Abrams (2006)

Disruptive selection and than what? Trends Ecol. Evol. 21: 238- 245.

96. Ruiz-Pesini, E., Lapeña, A.C., Díez-Sánchez, C., Pérez-Martos, A., Montoya, J., Alvarez, E., Díaz, E., Urriés, A., Montoro, L., López-Pérez, M.J. & J.A. Enríquez (2000) Human mtDNA haplotypes associated with high or reduced spermatozoa motility. Am. J. Hum.Genet., 67: 682-696.

97. Stearns, S.C. (1976) Life-history tactics: A review of the ideas. Quart. Rev. Biol., 51: 3- 47.

98. Stearns, S.C. (1992) The Evolution of Life Histories. Oxford University press.

99. Stockley, P., Gage, M.J.G., Parker, G.A. & A.P. Møller (1997) Sperm competition and the evolution of testis size and ejaculate characteristics. Am. Nat., 149: 933- 954.

100. Stoss, J. (1983) Fish gamete preservation and spermatozoan physiology. in Fish Physiology.

101. Vol. IX. Reproduction Part B. Behaviour and Fertility Control (eds. Hoar, W. S., Randall, D. J. and Donaldson, E. M.) pp. 305- 350. Academic Press, Inc. New York, London.

102. Taborsky, M. (1994) Sneakers, sattelites, and helpers: parasitic and cooperative behavior in fish reproduction. Adv. Study Behav. 23: 1- 100.

103. Taborsky, M. (1998) Sperm competition in fish: 'bourgois' males and parasitic spawning. Trends Ecol. Evol. 13: 222- 227.

104. Taylor, GI (1951) Analysis of the swimming of microscopic organisms. Proc R Soc Lond A 209: 447-471.

105. Thomaz, D., Beall, E & Burke T. (1997) Alternative reproductive tactics in Atlantic salmon: factors affecting mature parr success. Proc. R. Soc. Lond. B, 264: 219- 226.

106. Thorpe, J.E. (1986) Age at first maturity in Atlantic salmon, Salmo salar: Freshwater period influences and conflicts with smolting. Can. Spec. Publ. Fish. Aquat. Sci., 89: 7-14.

107. Thorpe, J.E., Mangel, M., Metcalfe, N.B. & Huntingford, F.A. (1998) Modelling the proximate basis of salmonid life-history variation, with application to Atlantic salmon, Salmo salar L. Evol. Ecol., 12: 581- 599.

108. Tomkins, J. L. & Hazel, W. (2008) The status of conditional evolutionary stable strategy. Trends Ecol. Evol. 22: 522-528.

109. Turner, E. & Montgomerie, R. (2002) Ovarian fluid enhances sperm movement in Arctic charr. J. Fish Biol., 60:1570-1579.

110. Vladić, T. (2000) The effect of water temperature on sperm motility of adult male and precocious male parr of Atlantic salmon and brown trout.

Verh. Int. Verein. Limnol. 27: 1070- 1074.

111. Vladić, T. (2001) Gonad and Ejaculate Allocation in Alternative Reproductive Tactics of Salmon and Trout with Reference to Sperm Competition (Thesis, Stockholm University, 68 pp.).

112. Vladić, T. (2006) Sperm quality and egg size in the brown trout: implications for sperm competition and cryptic male choice. Verh. Int. Verein. Limnol. 29: 1331- 1340.

113. Vladić, T. & Järvi, T. (1997) Sperm motility and fertilization time span in Atlantic salmon and brown trout- the effect of water temperature. J. Fish Biol., 50: 1088- 1093.

114. Vladić, T.V. & Järvi, T. (2001) Sperm quality in alternative reproductive tactics of Atlantic salmon: the importance of the loaded raffle. Proc. Roy. Soc. Lond., B, 268, 2375- 2381.

115. Vladić, T.V., Afzelius, B.A. & Bronnikov, G.E. (2002) Sperm quality as reflected through morphology in salmon alternative life histories. Biol. Reprod., 66, 98-105.

116. Vladić, T., Forsberg, LA & Järvi, T (2010) Sperm competition between alternative reproductive tactics of the Atlantic salmon in vitro. Aquaculture, 302: 265-269.

117. Wedenkind, C., Rudolfsen, G., Jacob, A., Urbach, D. & R. Müller (2007) The genetic consequences of hatchery-induced sperm competition in a salmonid. Biol Conserv, 137: 180-188.

118. Withler, R. E. (1988) Genetic consequences of fertilizing Chinook salmon (Oncorhynchus tschawyutscha) eggs with pooled milt. Aquaculture, 68: 15- 25.

119. Withler, R.E. & Beacham, T.D. (1994) Genetic consequences of the simultaneous or sequential addition of semen from multiple males during hatchery spawning of chinook salmon (Onchorhynchus tschawytscha). Aquaculture, 126: 11- 23.

120. Witman, G.B. (1990) Introduction to cilia and flagella. in Ciliary and Flagellar Membranes (ed. R.A. Bloodgood), Plenum Publishing Corporation, pp. 1- 30.

121. Woolley, D. (2000) The molecular motors of cilia and eukaryotic flagella. Essays in Biochemistry, 35: 103-115.

122. Woltereck, R (1909) Weitere experimentelle Untersuchungen über Artverenderung, speziell über das Wessen quantitative Artunterschiede bei Daphniden. Verhandlungen der Deutchen Zoologischen Gesellschaft, 110-172.

123. Yeates, S, Searle, J, Ward, RG & MJG Gage (2007) A two-second delay confers first male fertilization precedence within in vitro sperm competition experiments in Atlantic salmon. J Fish Biol, 70: 318-322.

124. Yeates, SE, Einum, S, Fleming, IA, Megens, H-J, Stet, RJM, Hindar, K, Holt, WV, Van Look, KJW, & MJG Gage (2009) Atlantic salmon eggs favour sperm in competition that have similar major histocompatibility alleles. Proc. R. Soc. Lond. B 276, 559-566

# Chapter 5

# ENVIRONMENTAL EFFECT OF USING POLLUTED WATER IN NEW/OLD FISH FARMS

Y. Hamed[1], Sh. Salem[2], A. Ali[3] and A. Sheshtawi[1]

[1]Civil Engineering Department, Faculty of Engineering, Port Said University, Egypt

[2]Ministry of Water Resources and Irrigation, Egypt

[3] Irrigation and Hydraulics Department, Faculty of Engineering, Ain Shams University, Egypt

## INTRODUCTION

One of the most dangerous hazards affecting the environment situation in arid and semiarid countries like Egypt is the water and soil pollution. Due to the lack of fresh water for irrigation, countries in arid and semi-arid areas are forced to use marginal waters for irrigation and raising fish. The effect of using such kind of low quality water is rather dangerous on the environmental situation. Besides, countries like Egypt are facing great problems to get rid of untreated waste water and industrial disposal in addition to drainage water. The spill of such kind of waters in drains and lakes will cause great problems to the eco-system and environment in general. What will make the problem more complicated is the use of these waters for irrigation or raising fish due to lack of fresh waters. The using of polluted water in fish farms has a very dangerous environmental effect on soil and ground water. The water level in fish farms is higher than the original land level. Consequently, water flows from fish farms to the adjacent land and cause problems if the water was polluted. Bahr El-Baqar drain is considered as one of the most polluted drains in Egypt (Abdel-Shafy & Aly 2002). It receives and carries the greatest part of wastewater (about 3 BCM/year) into Lake Manzala through a very densely populated area of the Eastern Delta passing through four highly populated Governorates. Unfortunately, at the last decades, great areas on both sides of the drain were using its polluted water for irrigation and raising fish. As a polluted drain with high risk to the surrounding environment, Bahr El Baqar has received considerable concern by many scientists. Ali et al. (1993), Abdel-

Azeem et al. (2007) studied the effect of prolonged use of drain water for irrigation on the total heavy metals content of south Port-Said city soils. They found that using such kind of water will cause high concentration of heavy metals in soil and plants roots and shots. Water quality, chemical composition, and hazardous effects on Lake Manzala water and living organisms caused by Bahr El-Baqar drain water has also been studied by several investigators like: Rashed & Holmes (1984), Khalil (1985) and Ezzat (1989). Special attention has been paid to the effect of environmental pollution from microbiological and toxicological points of view (Zaki 1994). Fish farms located on both sides of the Bahr El-Baqar drain are using the polluted water for rising fish since long time ago. Furthermore, many agricultural lands located in these areas are also using such kind of water for irrigation. The hazardous of such kind of water on environment is enormous. Not to mention to the fish itself coming from these farms. The fish production from these farms goes directly to the market. Consequently, the risk to the human health is very high.

**Photo 1:** Fish farms adjacent to Bahr El-Baqar drain in Egypt (From Hamed et al. 2011).

Hamed 2008, studied the effect of fish farms on both of soil structure and soil salinity in area near the study area of the current research and with the same type of soil (heavy clay soil). He used blue dye (food-grade Vitasyn-Blau AE 90 from Swedish Hoechst Ltd) mixed with water in order to test the soil structure properties. He found that fish farming does not contribute to decrease in the soil salinity. He concluded that increasing fish farming activities may lead to increasing soil salinity problems in agricultural lands. The results showed also that there is no evidence that soil properties are enhanced by fish farming. On the contrary, the soil nutrient state appears to be decreasing. A layer of 10 cm thickness of mud layer with cracks is formed at the surface when the farm dried (Photo 2.).

**Photo 2:** High amount of macropores and cracks at the mud layer accumulated at the surface In fish farm site (From Hamed 2008).

Shang et al. (1998) concluded that extensive fish farming (shrimp) rapidly depleted the soil organic matter content. Other studies (Beverage & Phillips, 1993; Deb, 1998; Flaherty et al., 1999) have reported that intensive and semi-intensive fish farming result in high volume of organic and inorganic effluents and toxic chemicals to the ecosystem that gives hypernutrification, eutrophication, and high soil toxicity. In Bangladesh, studies have shown that fish farming destroys the mangrove forest, increases soil acidity, salinity (farmers use sea water), and water pollution (Deb, 1998; Guimaraes, 1989; Hossain et al., 2004 Rahman, 1994). Ali (2006) examined the impact of shrimp farming on rice ecosystem in a village in southwestern Bangladesh. He reported that prolonged shrimp farming using sea water for a 5-, 10-, and 15-year period increased soil salinity, acidity, and significantly degraded the area's soil quality. It also drastically reduced the rice production and destroyed the aquatic and non-aquatic habitat inherent in the rice ecosystem. A national project called El Salam Canal has recently finished. It relies on mixing water from the Damietta Branch of the Nile River with water from two major agricultural drains to be used for irrigation of 600,000 Feddan in the western side of Suez Canal and North Sinai. A total annual water requirement of 4.45 billion m3 of mixed water is required to irrigate 600,000 Feddan as follows:

- 2.11 billionm3 fresh water from the Damietta branch,
- 0.435 billionm3 drainage water from the Elserw drain, and
- 1.905 billionm3 drainage water from the Bahr Hadous drain.

The 200000 Feddan located within the service area of El Salam Canal in western part of Suez Canal spill their drainage water into Bahr El Baqar drain. Some agricultural areas located near the drain use El-Salam Canal water for irrigation. Due to the lack of the canal water, fish farms in the area forced to use other source of water. Unfortunately, the easier source is the polluted Bahr El-Baqar drain. The objective of this chapter is to conduct an integrated environmental assessment for fish farms located adjacent to the most polluted drain in northeastern Egypt and use its water for raising fish. A comparison between the pollution level in new/old fish farms using polluted water and agricultural lands use the same kind of waters will be conducted. The comparison will include also agricultural lands using fresh water from El-Salam Canal, lands subjected to fill from the drain and moor lands which play a reference role. An investigation of the available clues of the problem include using another drain to decrease pollution will be conducted. The level of pollution in soil and water will be recorded as a result of using the polluted water from the drain for irrigation and raising fish. The problem of seepage from the drain to the adjacent fish farms or from fish farms to the adjacent lands will be investigated. In order to achieve that water and soil samples have been collected and analyzed in order to calculate the concentrations of five main heavy metals (Pb, Zn, Cd, Cu and Mn). Samples were collected from different depths ranging from 1m to 4m in 24 different locations for the study area. In order to conduct a complete integrated environmental assessment to the problem, different locations have been chosen in new/old fish farms, moor lands, fill lands and agricultural lands using polluted drain water and fresh water from the canal.

## LOCATIONS OF SAMPLES COLLECTED

The field experiments were conducted by Salem et al 2011 in year 2009 in area located at the last 20 km of the Bahr El Baqar drain before it spills its water in Manzala Lake. The study area consists of fish farms, agricultural lands, lands subjected to fill from the drain and moor lands. The area is located within the service area of the national project El Salam Canal south of Port Said city. Total of 24 boreholes were dug in 8 horizontal sections for different locations at the study area. Every horizontal section has length of 120-160 m from the Bahr El Baqar drain side and contains three boreholes. Fig 1, shows the location of the horizontal sections while Fig 2. Shows the locations of the boreholes. Four sections were taken at each side of the drain. The depth of each borehole is 2-4 m. five of boreholes were conducted in new/old fish farms. Three boreholes were conducted near the old fish farms (near the border of the farms). Six boreholes were conducted in moor lands and seven boreholes in fill land adjacent to the drain. Three boreholes were conducted in agricultural lands; two of them in agricultural land using the polluted water and one in agricultural land using fresh canal water. The agricultural land using fresh canal water has used polluted water from the drain for irrigation for 20 years and changed to use fresh water from the canal 5 years ago. The soil type in the study area is heavy clayey soil with more than 60% clay. Boreholes taken in moor lands will be used for comparison.

## SOIL/WATER SAMPLES

Two soil samples were collected from each borehole. Chemical analyses were carried out within 24 hr of sampling at the laboratory. The concentrations of six heavy metals (Pb, Zn, Cu,Mg, and Cd) were analyzed for each soil sample. Heavy metals were analyzed by the total adsorbed metals method according to USEPA (1986) using atomic spectrophotometer (model PYE UNICAM SP9, England). The fish farms and the agricultural lands in the area are supposed to use fresh water from El-Salam canal or its branches but due to water shortage especially in summer most of the lands use the drain water for irrigation or for raising fish. There is a branch drain called Sarhan drain parallel to Bahr El Baqar drain at the eastern side located 100 m far from the drain. It collects the drainage water from the area nearby and spills it again to Bahr El-Baqar drain. Boreholes were taken by rotary boring method at 8 sections, 4 sections at each side of Bahr El Baqar drain. Every section contains three bore holes. Two soil samples were taken from every borehole. One soil samples was taken above water table and the other taken under water table. The total depth of each borehole is 2-4 m. Table 1. shows locations and depths of boreholes. It shows also the type of land use of each borehole.

### Results of water samples from the drain

Table (2) shows the concentrations of heavy metals for water samples taken from Bahr El Baqar drain at the same study area one year before the study was conducted (Hamed et al 2011). The results reflect the size of pollution of the drain water along the whole year.

### Comparison due to land use

The objective of this section is to compare between new/old fish farms using polluted water and different land uses with different water quality used. The comparison will be between old fish farms, new fish farms, lands adjacent to old fish farms, agricultural lands use polluted water from the drain and agricultural lands use fresh water from Elsalam canal and fill resulted from the excavation of the drain sides and bottom and finally natural lands (moor).

**Table 1**: Locations and depths of boreholes and type of land uses (U: upper layer. L: lower layer) (From Salem et al. 2011).

| Sector | Side | Pore hole | Distance From Bahr El-Baqar drain(m) | Symbol / no. | Depth (m) | using | Water table depth |
|---|---|---|---|---|---|---|---|
| S1 | right | S1b1 | 60 | S1b1U | 1.5 | Old Fish farm(10 years old ) | 1.65 |
| | | | | S1b1L | 4 | | |
| | | S1b2 | 142 | S1b2U | 1 | land adjacent to old fish farm | 1.2 |
| | | | | S1b2L | 3 | | |
| | | S1b3 | 162 | S1b3U | 1 | New Fish farm(two years old ) | 1.3 |
| | | | | S1b3dL | 4 | | |
| S2 | right | S2b1 | 32.5 | S2b1U | 0.8 | Old Fish farm(10 years old ) | 2.2 |
| | | | | S2b2L | 2.5 | | |
| | | S2b2 | 65 | S2b2U | 0.7 | Old Fish farm(10 years old ) | 2.5 |
| | | | | S2b2L | 2.75 | | |
| | | S2b3 | 131 | S2b3U | 0.7 | Fill from Sarhan drain(5 years old ) | 1.5 |
| | | | | S2b3L | 2 | | |
| S3 | left | S3b1 | 10 | S3b1U | 0.5 | Fill from Bahr El-Baqar drain (15 years old ) | 0.7 |
| | | | | S3b2L | 2 | | |
| | | S3b2 | 27 | S3b2U | 0.5 | | 0.85 |
| | | | | S32b2L | 1.5 | moor | |
| | | S3b3 | 44 | S3b3U | 0.5 | | 1 |
| | | | | S3b3L | 1.5 | moor | |
| S4 | right | S4b1 | 35 | S4b1U | 0.5 | Fill from Sarhan drain (5 years ago) | 1 |
| | | | | S4b1L | 3 | | |
| | | S4b2 | 70 | S4b2U | 1 | | 1 |
| | | | | S4b2L | 3.5 | moor | |
| | | S4b3 | 106.5 | S4b3U | 1.5 | Cultivated land using El-salam canal water | 2.3 |
| | | | | S4b3L | 3.5 | | |

| | | | | S5b1U | 1 | Cultivated lands using Bahr El-Baqar drain water | 3.5 |
|---|---|---|---|---|---|---|---|
| S5 | left | S5b1 | 35 | S5b2L | 4 | | |
| | | S5b2 | 72 | S5b2U | 1 | land adjacent to old fish farm | 2.3 |
| | | | | S5b2L | 3 | | |
| | | S5b3 | 120 | S5b3U | 0.5 | New Fish farm(two years old ) | 0.7 |
| | | | | S5b3L | 3 | | |
| S6 | right | S6b1 | 35 | S6b1U | 0.5 | Fill from Bahr El-Baqar drain (20 years old) | 0.7 |
| | | | | S6b2L | 2 | | |
| | | S6b2 | 67 | S6b2U | 0.5 | moor | 3.6 |
| | | | | S6b2L | 3.75 | | |
| | | S6b3 | 114 | S6b3U | 0.5 | Cultivated land using Bahr El-Baqar drain | 1.75 |
| | | | | S6b3L | 2.5 | | |
| S7 | left | S7b1 | 35 | S7b1U | 1 | moor | 1 |
| | | | | S7b1L | 3 | | |
| | | S7b2 | 63 | S7b2U | 0.8 | moor | 1 |
| | | | | S7b2L | 2 | | |
| | | S7b3 | 73 | S7b3U | 0.5 | land adjacent to old fish farm | 0.7 |
| | | | | S7b3L | 2 | | |
| S8 | left | S8b1 | 40 | S8b1U | 1 | Fill from Bahr El-Baqar drain (15 years old) | 2.25 |
| | | | | S8b1L | 2.75 | | |
| | | S8b2 | 80 | S8b2U | 1 | Fill from Bahr El-Baqar drain (15 years old) | 2.75 |
| | | | | S8b2L | 3 | | |
| | | S8b3 | 115 | S8b3U | 1 | Fill from Bahr El-Baqar drain (20 years old) | 1.9 |
| | | | | S8b3L | 2.2 | | |

**Table 2:** The values of heavy metals in drain water for one year (From Hamed et al. 2011).

| Sites | Concentration in mg/liter | | | | | | | | | | | | |
|---|---|---|---|---|---|---|---|---|---|---|---|---|---|
| | Year 2008 | | | | Year 2009 | | | | | | | | |
| | Sep. | Oct. | Nov. | Dec. | Jan. | Feb. | Mar. | Apr. | Jul. | Aug. | Sep. | Oct. | Nov. |
| Cu | 0.151 | 0.022 | 0.204 | 0.245 | 0.062 | 0.005 | 0.045 | 0.106 | 0.083 | 0.014 | 0.012 | 0.008 | 0.011 |
| Pb | 0.749 | 0.235 | 0.273 | 0.195 | 0.088 | 0.030 | 0.287 | 0.041 | 0.033 | 0.053 | 0.009 | 0.031 | 0.066 |
| Zn | 0.139 | 0.549 | 2.066 | 1.438 | 0.431 | 0.333 | 0.688 | 0.031 | 0.095 | 0.199 | 0.024 | 0.028 | 0.051 |
| Cd | 0.069 | 0.061 | 0.017 | 0.062 | 0.101 | 0.199 | 0.032 | 0.128 | 0.001 | 0.025 | 0.021 | 0.016 | 0.016 |
| Mn | 2.573 | 3.935 | 0.084 | 0.065 | 0.281 | 0.56 | 0.278 | 0.519 | 0.064 | 0.010 | 0.011 | 0.125 | 0.816 |

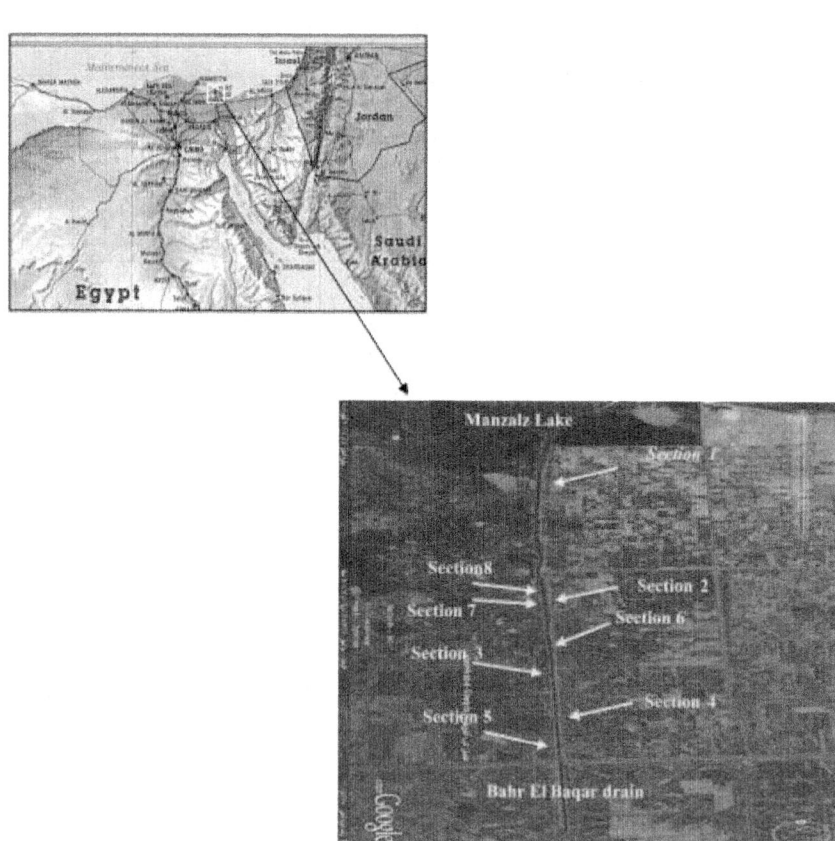

**Figure 1:** Locations of the study area (From Salem et al. 2011).

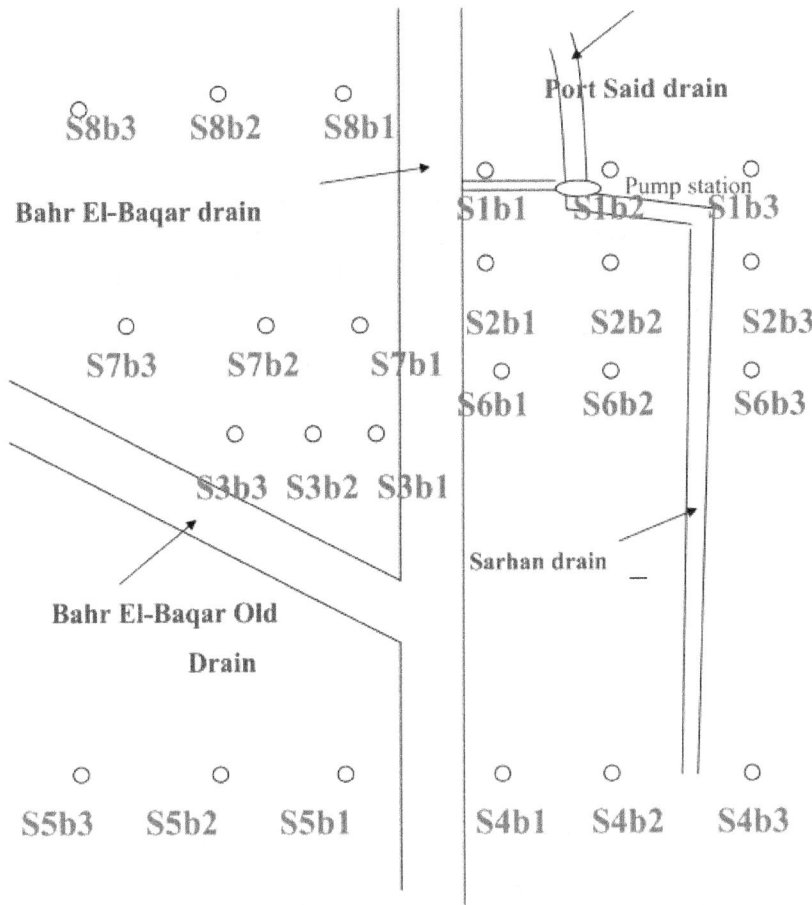

**Figure 2:** Layout of the boreholes (From Salem et al. 2011)

Fig (3) shows the difference between the concentrations of heavy metals in samples of top soil as a result of different land uses. The results showed that the old fish farms that used Bahr El-Baqar drain water have the highest percentage of heavy metals. And the second highest rate exists in the agricultural land irrigated with polluted water from Bahr El-Baqar drain for along time. And the third highest concentration of heavy metals is in the land adjacent to the fish farm that uses Bahr El-Baqar drain water. It indicates that the fish farms using polluted water significantly affect the neighboring land.

Next is land exposed to fill which exists only in top soil. In this case the concentration of heavy metals depends mainly on the age of the fill and its original location. There are two types of fill, fill comes from polluted Bahr-

ElBaqar drain and fill comes from the agricultural minor drain (Sarhan drain). For fill from Bahr El-Baqar drain, the concentration is decreased with the increase of the fill age. The lowest concentration is for the fill has age of 20-25 years. It could be attributed to the increase in pollution of the drain bottom soil by time. The reason could be the increase in population and the increase in industry spread in the recent decades which cause increase in untreated waste water and industrial disposal spilled into the drain. In spit of the new fill comes from the relatively less polluted drain, Sarhan drain, it has a high concentration ratio of heavy metals. It is known that Sarhan drain receives drainage water from fish farms and agricultural lands using polluted water from Bahr El Baqar drain for irrigation and raising fish. Next are the new fish farms that used Bahr El-Baqar drain water, and finally, the lowest heavy metal concentrations were found in both of the agricultural land uses fresh water from ElSalam Canal and the natural lands (moor lands). Fig (4) shows the difference between the concentrations of heavy metals in samples of lower soil as a result of different land uses. Nearly the same results as in top soil were obtained except for that the land adjacent to the fish farms which used Bahr El-Baqar drain water has the highest percentage of heavy metals. This is a good proof that the fish farms using polluted water significantly affect the neighboring land especially in lower soil layers. The old fish farms that used Bahr El-Baqar drain water receive the second highest rate in the concentrations of heavy metal. And the third highest concentration of heavy metals is the agricultural land irrigated with water from Bahr El-Baqar drain for along time, Next is the new fish farms that used Bahr El-Baqar drain water, and the finally is the moor lands.

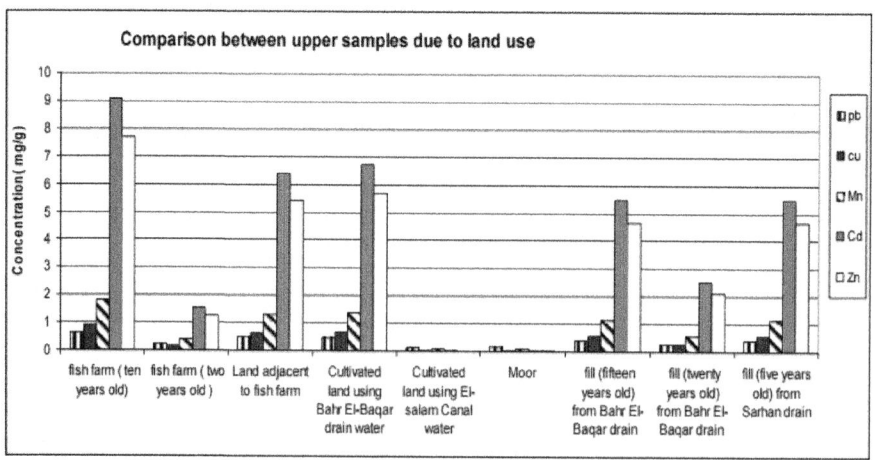

**Figure 3:** Mean heavy metals concentrations in upper soil samples for different land uses, mg/liter (From Salem et al. 2011).

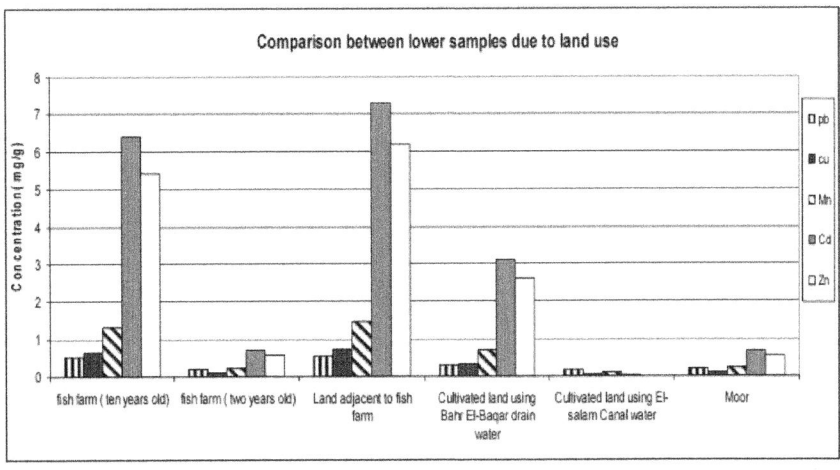

**Figure 4:** Mean heavy metals concentrations in lower soil samples for different land uses, mg/liter (From Salem et al. 2011).

## Comparison due to location from Bahr El-Baqar drain

In the current section, comparison of new/old fish farms and other land uses will be investigated in details for each section and each borehole. Moreover, effect of seepage from the polluted drain or from fish farms to the adjacent lands will be studied. Boreholes were dug in different distances and locations in lands with different land uses. The distances from the drain were kept nearly constants for different sections. Here, the effect of minor drain (Sarhan drain) parallel to Bahr El-Baqar drain and at 100 m far will be investigated. There is a question needs to be answered, will the minor drain work as a defend barrier for seepage from the polluted drain to the lands located on the minor drain side? Will it affect the fish farms using polluted water? For the coming section, only the data for the lower soil layer will be used in order to investigate the seepage from the drain and fish farms. Fig (5) shows heavy metals concentrations of boreholes in section (1). It shows concentrations in boreholes S1b1, S1b2 and S1b3 located in old fish farms, land adjacent to old fish farms and new fish farms respectively. It is clear from the figure that the seepage factor from the main drain is not dominated here. The highest heavy metal concentrations are found in bore hole S1b2 (land adjacent to old fish farm) not in S1b1 the closest borehole to the main drain. The effect of the minor drain (Sarhan drain) is not clear here since borehole S1b2 has a high ratio of heavy metals concentration. It is probably due to the influence of pump station near the section. It seems that the type of land use is the dominated factor for the pollution concentration in this section. Fig (6) shows mean heavy metals concentrations of boreholes in section (2). It shows concentrations in boreholes S2b1, S2b2 and S2b3 located in old fish farms for the first two boreholes and fill from Sarhan drain for the third one. The highest heavy metal concentrations are found in borehole S2b2. Also, in this section it is clear that the seepage factor from the main drain is not dominated. The values of concentrations in old fish farm (10 years old) 65 m far from the drain are higher than that the corresponding values in old fish farms 32.5 m away. The pollution in fill from Sarhan drain (5 years old) is lesser than that in old fish farms. It is probably due to the high pollution accumulation in old fish farms and the fill location near Sarhan drain. Fig (7) shows heavy metals concentrations in section (3). It shows concentrations in boreholes S3b1, S3b2 and S3b3 located in fill from Bahr El Baqar drain for the first borehole and moor land for the second and the third ones. The dominated factor here is the land use factor. The higher value of concentration exists in second borehole. The ratio of concentration in this section is small compared with other sections. It is probably due to moor lands (natural land) which contain less amount of pollution and the low effect of drain seepage. In fill borehole, the effect of fill is small since the soil sample was taken in 2 m depth. Fig (8) shows heavy metals concentrations in section (4). It shows concentrations in boreholes S4b1, S4b2 and S4b3 located in fill from Sarhan drain (5 years ago), moor land and agricultural lands using water from ElSalam canal respectively. The ratio of concentration in this section is small compared with other sections. The seepage factor here also is not dominated. The less concentration is found in agricultural lands using water from ElSalam canal. The location of Sarhan drain adjacent to the agricultural lands could be another factor contributing for lower pollution concentration in its soil. Fig (9) shows heavy metals concentrations in section (5). It shows concentrations in boreholes S5b1, S5b2 and S5b3 located in cultivated lands using Bahr El-baqar drain water, land adjacent to old fish farm and New Fish farm (two years old) respectively. The higher concentrations of heavy metals are located in both of cultivated land with Bahr El baqar water and land adjacent to fish farms. These results reflect the bad effect of using polluted water for irrigation and the effect of fish farms on the adjacent lands. The relatively less concentration is located in new fish farms. However, only two years of fish farming using polluted water has raised the pollution concentration many times (see Fig (3) moor land and new fish farms). Fig (10) shows heavy metals concentrations in section

(6). It shows concentrations in boreholes S6b1, S6b2 and S6b3 located in Fill from Bahr El-Baqar drain (20 years old), moor lands and cultivated land using Bahr El-Baqar drain respectively. Soil sample was taken at 2 m depth in fill borehole. Consequently, the effect of fill is small. The ratios of concentration in moor land are nearly the same as that for old fill from Bahr El Baqar drain. The unexpected results here are the lower ratio of concentration for cultivated land using Bahr El-Baqar drain although the soil sample was taken from the root zone area. It is probably due to the effect of Sarhan drain adjacent to the lands. However, the ratio of concentrations still relatively high compared with land using fresh water for irrigation. Again, seepage is not a dominated factor here. Fig (11) shows heavy metals concentrations in section (7). It shows concentrations in boreholes S7b1, S7b2 and S7b3 located in moor lands for the first two boreholes and in land adjacent to old fish farms for the third one. The effect of old fish farm on the lands nearby is quit clear here. The difference in heavy metals concentrations between land adjacent to fish farms and moor land is quite high. It reflects the damage effect of fish farms not only on its own soil but also on the soil nearby. It is probably due to the high polluted water level in fish farms which infiltrate to the adjacent lower level lands. Fig (12) shows heavy metals concentrations in section (8). It shows concentrations in boreholes S8b1, S8b2 and S8b3 located in fill from Bahr El-Baqar drain for 15, 15 and 20 years old respectively. Unlike other sections, seepage factor could be effective here. Another possible reason is the existence of some fill traces in deep layers.

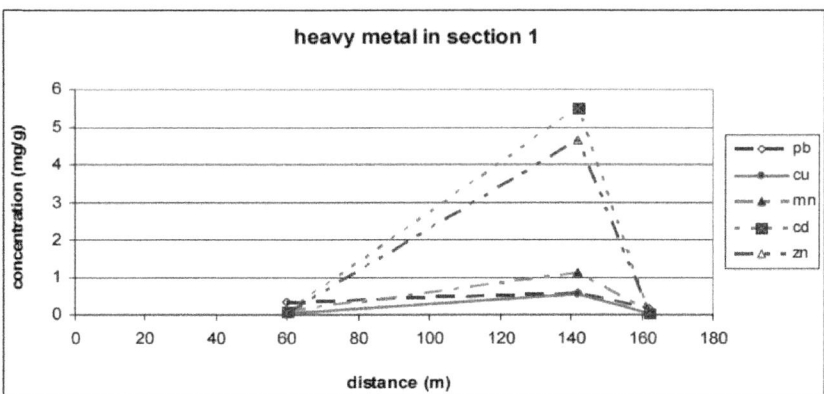

**Figure5:** Heavy metals concentrations in section (1), mg/liter (From Salem et al. 2011).

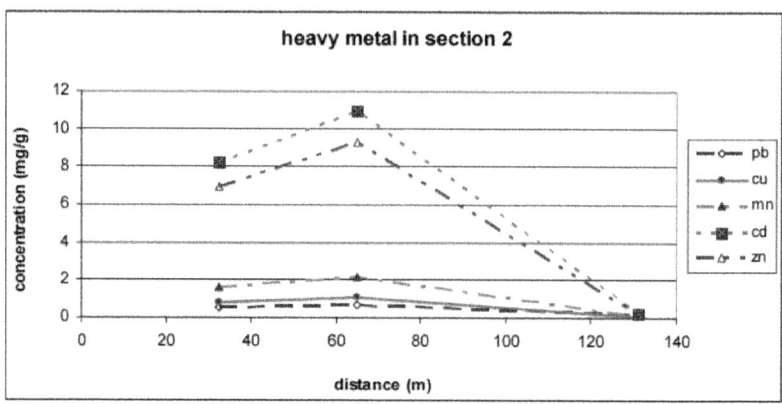

**Figure 6:** Heavy metals concentrations in section (2), mg/liter (From Salem et al. 2011).

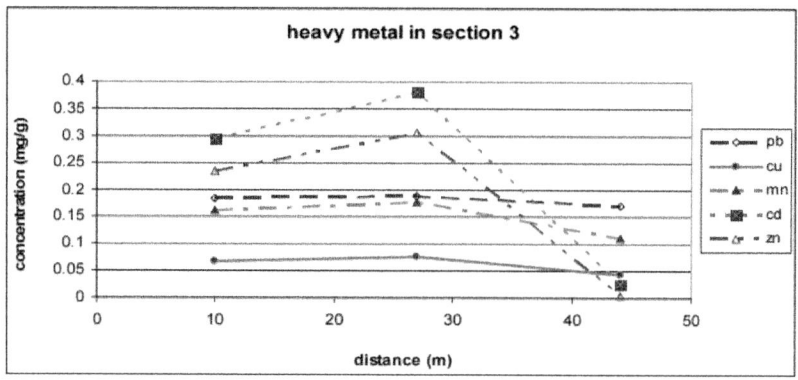

**Figure 7:** Heavy metals concentrations in section (3), mg/liter (From Salem et al. 2011).

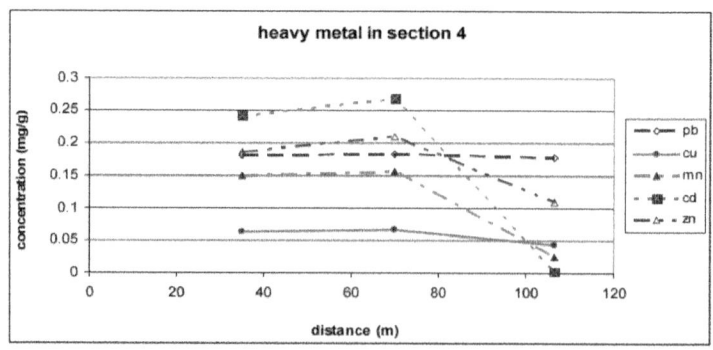

**Figure 8:** Heavy metals concentrations in section (4), mg/liter (From Salem et al. 2011).

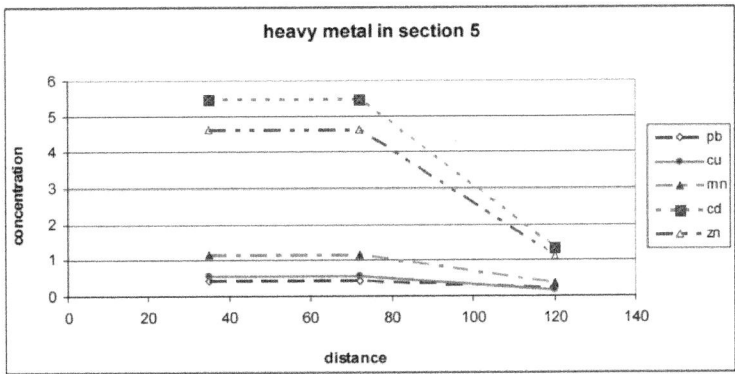

**Figure 9:** Heavy metals concentrations in section (5), mg/liter (From Salem et al. 2011).

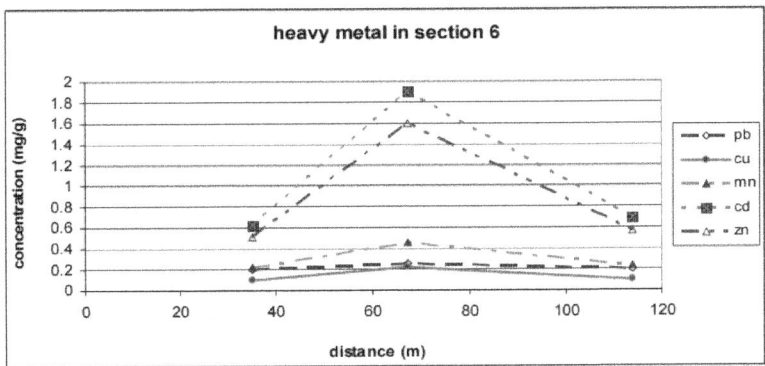

**Figure 10:** Heavy metals concentrations in section (6), mg/liter (From Salem et al. 2011).

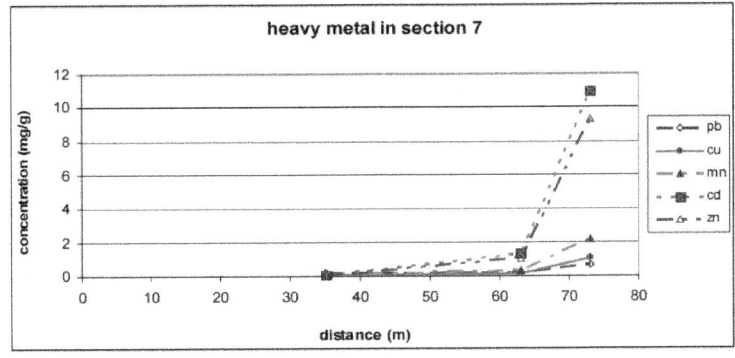

**Figure 11:** Heavy metals concentrations in section (7), mg/liter (From Salem et al. 2011).

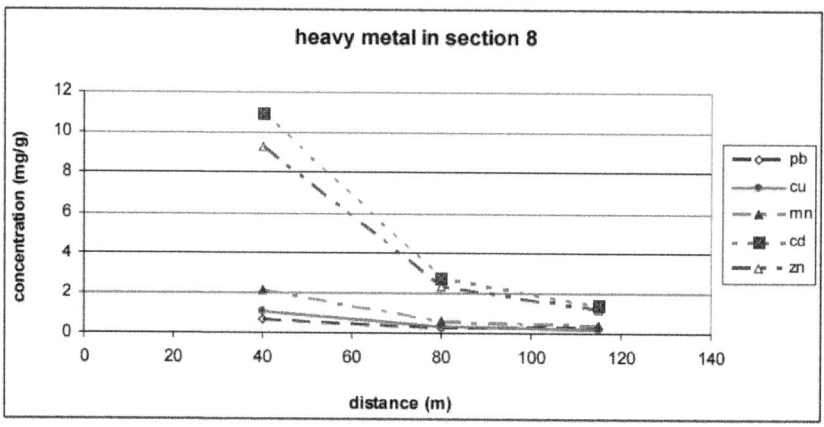

**Figure 12:** Heavy metals concentrations in section (8), mg/liter (From Salem et al. 2011).

## SUMMARY AND CONCLUSIONS

In this chapter, an integrated environmental assessment has been conducted for fish farms using polluted water in area located within the service area of the most polluted drain in Egypt (Bahr el Baqar drain). The chapter focuses on two previous studies (Salem et al. 2011), (Hamed, et al. 2011). Total of 24 boreholes in 8 different sections on both sides of the drain have been dug in order to collect soil samples for depths ranging from 1 to 4 m. Samples were sent to the laboratory in order to measure the concentrations of five heavy metals (Pb, Zn, Cd, Cu and Mn). Boreholes were excavated in different land uses and in different spaces from the polluted drain in order to conduct an integrated comparison between fish farms using polluted water and other land uses. Results showed that the most polluted areas are the old fish farms (10 years old) using polluted water from the drain for raising fish and the land adjacent to it. It reflects the harmful effect of using polluted water for long time for raising fish not only on farms soil itself but also on the soil adjacent to the fish farms. Consequently, the pollution will reach fish produced from these farms and transferred to human affecting their health. Moreover, even in new fish farms with only less than two years old, the increase in heavy metals concentrations in soil is quite high during short period of time. The level of pollution in fish farms using polluted water is much more the level of pollution in agricultural lands using polluted water from the drain in irrigation. Both of fish farms and the agricultural lands have the same age in using polluted water from the drain. Also, the chapter warns about using polluted water for irrigation in agricultural lands. It proved the bad effect of such kind of water on soil and hence in plants since the digging depth is within the root zone (1-3). This conclusion will stand against those people supporting the use of polluted drain water for irrigation or for raising fish. Fill from both of Bahr ElBaqar drain and from agricultural drain (Sarhan drain) comes after as a third higher ratio of pollution. The quality of older fill from the polluted drain is better than the recent one. It is probably due to the increase in concentration of pollution by time in drain bottom soil. The difference in concentration is too high in a relatively short period of time. This reflects the rapid deterioration of the environmental situation for Bahr El Baqr drain by time. Since fish farms and agricultural lands using polluted water are using the minor drain (Sarhan drain) for drainage, the fill coming from this drain contains high ratio of heavy metals concentration. Unfortunately, many fish farms owners use such kind of fill for constructing banks around their farms. They use the fill also for increasing the level of their farms bed soil. For agricultural lands which have used polluted water for irrigation for long time (20 years) and changed to use fresh water for relatively less period of time (5 years), the improvement of its soil quality is quite clear. The decrease in heavy metals due to use of good quality of water is rather high. It will give an optimistic view for obtaining a clue for pollution in the area. Furthermore, the existing of minor drain parallel to the major polluted drain in relatively small distance (90-100m) contributes for reducing pollution for lands located after the minor drain in most cases. Consequently, as a current solution, the local government should force fish farms owners to use fresh water for washing their lands for long period of time before they use it again for fish farms. The Chapter revealed that the overall environmental situation at the area on both sides of the drain is quite dangerous. Five dangerous heavy metals with different concentrations have been found in each soil sample on surface or deep on the ground. This pollution hazardous level has its bad effect on both of fauna and flora at the area. Finally, the Chapter revealed that fish farming with polluted water has very bad consequences to the surrounding environment, not to mention to the fish itself. Moreover, previous studies have proved that the fish farming for long time with good quality water has a bad effect on the soil structure and soil salinity. Hence, we can imagine the level of environmental deterioration as a result of using polluted water in fish farming.

## ACKNOWLEDGMENTS

The field and publishing work was financially supported by the Swedish Research Council (SIDA) through a cooperation project between Suez Canal University, Port Said branch (Port Said University now) (Egypt) and Lund University (Sweden) under the title: "Sustainable use of Cairo waste water;

environmental effects of the Bahr el Baqar Drain", in which Yasser Hamed is the principal investigator and Prof Atef Alam El-Din the University Vice President is the main supervisor.

## REFERENCES

1.   Abdel-Azeem, A. M; Abdel-Moneim, T. S.; Ibrahim, M. E.; Hassan, M. A. A. & Saleh, M. Y. (2007). Effects of Long-Term Heavy Metal Contamination on Diversity of Terricolous Fungi and Nematodes in Egypt - A Case Study. Water Air Soil Pollut. Journal, No. 186, pp.:233–254.

2.   Abdel-Shafy, H. I., & Aly, R. O. (2002). Water issue in Egypt: Resources, pollution and protection endeavors. CEJOEM, 8(1), 3–21.

3.   Ali, A.M.S., 2006. Rice to shrimp: land use/land cover changes and soil degradation in Southwestern Bangladesh. Journal of Land Use Policy 23, 421–435.

4.   Ali, O. M.; El-Sikhry, E. M., & El-Farghal, W. M. (1993). Effect of prolonged use of Bahr El Baqar drain water for irrigation on the total heavy metals content of South Port Said soils. In: Proc. 1st Conf. Egypt. Hung. Env. Egypt, pp. 53–57.

5.   Beverage, M. & Phillips, M., (1993). Environmental impact of tropical inland aquaculture. In: Pullin, R., Rosenthal, H., Maclean, J. (Eds.), Environment and Aquaculture. Center for Tropical Aquaculture Research, Manila, pp. 213–236.

6.   Deb, A.K., (1998). Fake blue revolution: environmental and socioeconomic impacts of shrimp culture in coastal Bangladesh. Ocean and Coastal Management 41, 63–88.

7.   Ezzat, A. I. (1989). Studies on phytoplankton in some polluted areas of Lake Manzala. Bulletin of the National Institute of Oceanography and Fisheries, ARE, 15(1), 1–19.

8.   Flaherty, M., Vandergeest, P. & Miller, P., (1999). Rice paddy or shrimp pond: tough decisions in rural Thailand.World Development 27 (12), 2045–2060.

9.   Guimaraes, J.P. de Compos, (1989). Shrimp culture and market incorporations: a study of shrimp culture in paddy fields in Southwest Bangladesh. Development and Change 20 (4), 333.

10.  Hamed, Y. (2008). Soil structure and salinity effects of fish farming as compared to traditional farming in northeastern Egypt. Land Use Policy Journal 25(3) pp 301- 308, July 2008.

11.  Hamed, Y., Shawky, T., Abd-Elrehim, M., ElKiki, M., Berndtsson, R. & Persson,K.,(2011) Case Study: Investigation of different potential causes of pollution in Lake Manzala northeastern of Egypt. Article in Press.

12.  Hossain, S., Alam, S.M.N., Lin, C.K., Demaine, H., Khan, Y.S.A., Das, N.G. & Roup, M.A., (2004). Integrated management approach of shrimp culture development in the coastal environment of Bangladesh. World aquaculture development in the coastal environment of Bangladesh. World Aquaculture 35 (4), 35–44.

13.  Khalil, M. T. (1985). The effect of sewage and pollutional wastes upon Bahr El-Baqar Drain and the southern area of Lake Manzala, Egypt. Egyptian Journal of Wildlife and Natural Resources, 6, 162–171.

14.  Rahman, A., 1994. The impact of chrimp culture on the coastal environment. In: Rahman, A.A., Haider, R., Huq, S., Jansen, E.G. (Eds.), Environment and Development in Bangladesh. University Press Ltd., Dhaka, pp. 490–524.

15.  Rashed, I. G., & Holmes, P. G. (1984). Chemical survey of Bahr El Bakar Drain system and its effects on Manzala Lake. In: Proceedings of the 2nd Egyptian Congress of Chemical

16.  Engineering, (pp. 1–10), Cairo, Egypt, March 18–20, 1984.

17.  Salem, Sh, Hamed,Y., Sheshtawy, A, and Ali, A(2011). Environmental assessments for areas located both sides of Bar El-Baqar polluted drain northeastern Egypt. Article in Press

18.  Shang, Y.C., Leung, P. & Ling, B.H., (1998). Comparative economics of shrimp farming in Asia. Aquaculture 164 (1–4), 183–200.

19.  U. S. Environmental Protection Agency, USEPA (1986). Test methods for evaluating solid waste: physical/chemical methods. SW-846. Washington, D. C.: USEPA, Office of Solid Waste and Emergency Response.

20.  Zaki, M. M. M. (1994). Microbiological and toxicological study of the environmental pollution of Lake Manzala (108 pp). MSc Thesis, Faculty of Science, Suez Canal University, Ismailia, Egypt

# Chapter 6

## FISHERY-INDUCED CHANGES IN THE SUBTROPICAL PACIFIC PELAGIC ECOSYSTEM SIZE STRUCTURE: OBSERVATIONS AND THEORY

Jeffrey J. Polovina, Phoebe A. Woodworth-Jefcoats

Pacific Islands Fisheries Science Center, NOAA Fisheries, Honolulu, Hawaii, United States of America

## ABSTRACT

We analyzed a 16-year (1996–2011) time series of catch and effort data for 23 species with mean weights ranging from 0.8 kg to 224 kg, recorded by observers in the Hawaii-based deep-set longline fishery. Over this time period, domestic fishing effort, as numbers of hooks set in the core Hawaii-based fishing ground, has increased fourfold. The standardized aggregated annual catch rate for 9 small (<15 kg) species increased about 25% while for 14 large species (>15 kg) it decreased about 50% over the 16-year period. A size-based ecosystem model for the subtropical Pacific captures this pattern well as a response to increased fishing effort. Further, the model projects a decline in the abundance of fishes larger than 15 kg results in an increase in abundance of animals from 0.1 to 15 kg but with minimal subsequent cascade to sizes smaller than 0.1 kg. These results suggest that size-based predation plays a key role in structuring the subtropical ecosystem. These changes in ecosystem size structure show up in the fishery in various ways. The non-commercial species lancetfish (mean weight 7 kg) has now surpassed the target species, bigeye tuna, as the species with the highest annual catch rate. Based on the increase in snake mackerel (mean weight 0.8 kg) and lancetfish catches, the discards in the fishery are estimated to have increased from 30 to 40% of the total catch

# INTRODUCTION

The North Pacific subtropical gyre is a large oceanic gyre bounded on the south by the North Equatorial Current, on the west by the Kuroshio Current, on the north by the Kuroshio Extension Current and the North Pacific Current, and on the east by the California Current [1]. Although low in primary productivity, the warm, vertically stratified oligotrophic waters of the subtropical gyre contain a highly diverse food web populated by tunas, sharks, and billfishes at the top trophic levels [2], [3]. Since the 1950s, large-scale fisheries have targeted the tunas, billfishes, and other large predators in this ecosystem. Several studies have suggested possible ecosystem impacts from fishing [3]–[5]. A comparison of catch, size, and species composition between a research longline survey in the 1950s and observer data from commercial longliners in the 1990s suggested a substantial decline in the abundance of large predators, the mean size of these predators, and some evidence of an increased abundance of formerly rare species [5]. Models of the North Pacific subtropical gyre were generated with Ecopath with Ecosim (EwE) to investigate whether the ecosystem contained any keystone species [3], [4]. The results suggested that there was not any single species group that functioned as a keystone, but that a broad reduction of apex predators as a result of fishing might result in an increase in prey in response to a decreased predation [3], [4]. In effect the fishing fleet is the keystone predator [4]. However, another modeling effort using an EwE model that incorporated some size-class structure found that while fishing decreased predator abundance there was limited evidence of trophic cascades or other ecosystem impacts based on the decline in predators [6].

A more recent analysis of catch rates for the 13 most abundant species caught in the deep-set Hawaii-based longline fishery over the past decade (1996–2006) provided evidence of a top-down response of the North Pacific subtropical ecosystem. Catch rates for apex predators such as blue shark (*Prionace glauca*), bigeye (*Thunnus obesus*) and albacore (*Thunnus alalunga*) tunas, shortbill spearfish (*Tetrapturus angustirostris*), and striped marlin (*Tetrapturus audax*) declined from 3 to 9% per year while catch rates for 4 mid-trophic species, mahimahi (*Coryphaena hippurus*), sickle pomfret (*Taractichthys steindachneri*), escolar (*Lepidocybium flavobrunneum*), and snake mackerel (*Gempylus serpens*), increased by 6 to 18% per year [7].

Ecosystem food webs and models for the central North Pacific subtropical pelagic ecosystem have traditionally been built from a species-specific perspective [2]–[4], [6]. Recent analysis of the temporal ecosystem dynamics used trophic levels derived from species-specific models [7]. However, recent applications of size-based ecosystem models across various ecosystems show they are emerging as a powerful tool, particularly in pelagic environments where predation is more strongly driven by body size than species' taxonomic identity [8]–[13]. A key advantage to size-based models is that they are based on broad ecological and physiological relationships requiring few region-specific parameter estimates apart from sea surface temperature (SST), the size structure at the base of the food web, and fishing gear selectivity.

In this paper we further examine ecosystem changes in the subtropical pelagic system from a size-based perspective and compare observations from the Hawaii-based longline fishery with simulations from a dynamic size-based ecosystem model.

## MATERIALS AND METHODS

Catch and fishing effort data from the Hawaii-based longline fishery are collected in two ways. Federally mandated logbooks are required from all fishers licensed in that fishery. The logbooks report daily records of fishing activity including location, catch by species, number of hooks per set, and other data on the fishing operation. While logbook data provides complete coverage from all vessels, it is most reliable for the landed portion of the catch while catches of discarded species including sharks and fishes with low or no economic value are often not recorded. Fishery observers are placed on a subset of all vessels to record all the species caught, fishing effort, and various operational aspects and since 2006 they have recorded the length of every third fish caught. Over the period 1996–2011, 16% of the deep-set trips had observer coverage. However, even with this relatively modest observer coverage, observer and logbook catch rates for the commercial species were highly correlated. For example, over the period 1996–2006, annual catch rates from observer and logbook data had correlations ranging from over 0.93 for albacore, striped marlin, shortbill spearfish bigeye tuna, and pomfret, between 0.80 and 0.89 for mahimahi, ono, and yellowfin tuna, 0.78 for skipjack, and 0.76 for blue shark [7].

The Hawaii-based longline fishery consists of two components: the daytime deepset fishery targeting bigeye tuna, and the nighttime shallow-set fishery targeting swordfish (*Xiphias gladius*). The deep-set fishery typically sets hooks between 100 m and 400 m with the median hook depth at about 250 m while the median depth of the deepest hook in the shallow-set fishery is 60 m [14]. Deep sets and shallow sets can be identified based on a very strong bimodal distribution of the number of hooks between floats. Shallow sets use 2–6 hooks per float while deep sets use 20–32 hooks per float [14]. For our analysis we identified deep sets as those with 10 or more hooks per float and shallow sets as those with fewer than 10 hooks per float. The shallow-set fishery occurs primarily in the winter and spring within a narrow band of 28°–32° N latitude. The shallow-set fishery was closed for several years to reduce interactions with sea turtles. This paper focuses exclusively on the deepset fishery occurring throughout the year over a broad geographic region and provides an uninterrupted observed catch and effort time series from 1996. Restricting our analysis to the deep-set fishery provides a relatively standardized depth range and method of gear deployment. This analysis was further restricted to data that were obtained from the core region of the fishing ground defined as bounded by 12°–27° N latitude. In some years, the fishery made excursions as far south as the equator and as far north as 32° N latitude; however, fishing in these areas was inconsistent over the study period.

Over the past two decades fishing effort in the Hawaii-based longline fishing ground and fishing mortality over the basin has increased about fourfold. For example, from 1996 to 2008 the number of hooks set in the Hawaii-based longline fishery increased from 10 million to 40 million. Recent stock assessments for yellowfin and bigeye tuna in the central and western Pacific estimated an increase in fishing mortality from 0.1 in 1990 to 0.3–0.4 in 2010 [15]. Unfortunately, we do not have any estimates of fishing mortality for either target species or the ecosystem that apply specifically to the central North Pacific, the area covered by the Hawaii deepset fishery.

We based this study on the catch and effort data for 23 species defined as those that have a mean catch per unit effort (CPUE) of at least 0.05 fish per 1000 hooks set over the 1996–2011 period. For those 23 species we estimated the species' mean weights from published length-weight equations by using lengths recorded by observers pooled over the 2006–2011 period (Table S1). Although some species had sufficient lengths to compute annual weights we chose to use mean sizes pooled over the entire time period for all species. The reason was our focus on changes in the ecosystem rather than within-species size structure, and many of our 23 species did not have sufficient length data for finer temporal resolution. An analysis of temporal changes in length for the most abundant tunas and billfishes in the catch is presented in Gilman *et al.* [16]. As a robust indicator of ecosystem size structure, we computed annual combined catch for small species (those with mean weights less than 15 kg) and large species (those with mean weights equal to or greater than 15 kg). The value of 15 kg was determined from the species-specific regressions from Table 1 as discussed in the results section. A generalized additive model (GAM) was fit to these two time series to estimate a standardized CPUE time series for small and large species.

Static size-based models, based on metabolic theory and empirical relationships between body size and trophic level, have been applied to investigate unexploited production and biomass of larger marine animals in the global oceans based on current environmental conditions [17]. Dynamic size-spectrum models can extend this approach by considering the time-dependent and continuous growth and mortality processes that result from size-structured feeding, representative of pelagic ecosystems [8], [12]. They can be used to predict the consequences of fishing mortality and changes in primary production as well as temperature effects on dynamic changes in the community size spectrum [13]. A key attribute of these models is that the probability of a predator of size $M$ eating an encountered prey of smaller size $m$ is given by a lognormal probability density function, with a mean value representing the preferred predator–prey mass ratio and a standard deviation that represents the breadth of the relative prey mass. A key strength of this approach is that realized predator–prey mass ratios in fish communities do not appear to vary systematically with temperature or primary production in the world's oceans [18].

**Table 1:** Change in catch rate estimated from statistically significant ($P<0.01$) linear regressions over 1996–2011, in order by by fish size. doi:10.1371/journal.pone.0062341. t001

| Species | % Annual Change in CPUE* (P-value) | Mean Weight in kg (N) |
|---|---|---|
| Blue Marlin (*Makaira nigricans*) | −5.0 (0.005) | 224.0 (1295) |
| Blue Shark (*Prionace glauca*) | −3.7 (0.004) | 106.4 (22856) |
| Striped Marlin (*Tetrapturus audax*) | −5.0 (0.004) | 93.5 (3800) |
| Shortbill Spearfish (*Tetrapturus angustirostris*) | −4.2 (0.008) | 75.7 (4078) |
| Shortfin Mako Shark (*Isurus oxyrinchus*) | 0 | 48.3 (624) |
| Swordfish (*Xiphias gladius*) | 0 | 42.0 (1509) |
| Yellowfin Tuna (*Thunnus albacares*) | 0 | 33.5 (9224) |
| Opah (*Lampris guttatus*) | −4.1 (0.008) | 30.2 (3923) |
| Bigeye thresher Shark (*Alopias superciliosus*) | 0 | 24.0 (1922) |
| Unidentified Tuna | 0 | 24.0 (49) |
| Bigeye Tuna (*Thunnus obesus*) | −2.1 (0.005) | 22.5 (41456) |
| Oceanic White-tip Shark (*Carcharinus longimanus*) | −6.9 (<0.0001) | 19.0 (277) |
| Albacore Tuna (*Thunnus alalunga*) | −6.9 (<0.0001) | 17.1 (4718) |
| Wahoo (*Acanthocybium solandri*) | 0 | 16.4 (4172) |
| Escolar (*Lepidocybium flavobrunneum*) | 12.1 (<0.0001) | 12.1 (9817) |
| Mola (*Ranzania laevis* and *Mola mola*) | 0 | 8.8 (521) |
| Skipjack Tuna (*Katsuwonus pelamis*) | 0 | 7.9 (9352) |
| Mahi Mahi (*Coryphaena hippurus*) | 0 | 7.4 (19346) |
| Lancetfish (*Alepisaurus ferox*) | 2.2 (0.026) | 7.1 (34186) |
| Great Barracuda (*Sphyraena jello*) | 0 | 5.9 (1198) |
| Pomfrets (*Taractichthys steindachneri* and *Brama japonica*) | 0 | 4.9 (14898) |
| Pelagic Stingray (*Pteroplatytrygon violacea*) | −5.4 (<0.0001) | 3.0 (4165) |
| Snake Mackerel (*Gempylus serpens*) | 15.1 (<0.0001) | 0.8 (15371) |

A size-based ecosystem model based on the pelagic component of the model detailed in Blanchard *et al.* [19] and adapted for the subtropical pelagic ecosystem [20] was used to simulate the response of the size structure to fishing pressure. Input for the model consists of small (<5 μm) and large (>5 μm) phytoplankton densities, SST, size of entry to the fishery, and gear selectivity as a function of size. We used phytoplankton densities and SST output by the NOAA Geophysical Fluid Dynamics Laboratory prototype Earth System Model 2.1 [20], [21], averaged both spatially (12°–27° N, 180°–140°W) and temporally (1996–2011) so the only variable input to the size-based model is fishing mortality.

## RESULTS

Linear regressions fit to the annual CPUE time series for each of the 23 species found 12 species had statistically significant positive or negative slopes (Table 1). Eight species 16.4 kg or larger had declining CPUE trends while the remaining 6 species 16.4 kg or larger had no significant trend (Table 1). Three species 12.1 kg or smaller had increasing CPUE trends while one, pelagic stingray, had a declining trend and the remaining 5 had no trend (Table 1). The declining species included billfishes, sharks, and tunas, with linear CPUE trend declines ranging from 2 to 7% annually over the 16-year time period. The species with increasing linear CPUE trends were escolar and two noncommercial species, snake mackerel and lancetfish, with increases of about 12, 15, and 2% annually, respectively (Table 1).

To more rigorously and robustly examine the different CPUE time trends for large and small fishes, we split the catch data into 2 groups consisting of 9 species with mean weights <15 kg, termed small fishes and 14 species with mean weights ≥15 kg, termed large fishes (Table 1). The value of 15 kg used to classify large and small fishes was based on the individual CPUE trends in Table 1 showing that 8 of the 9 species with declining CPUE trends had mean weights of 16.4 kg or larger and all of the species with declining CPUE trends had mean weights 12.1 kg or smaller. A GAM was fit to the catch in numbers per longline set for each size group using independent variables: hooks per set, set latitude, set longitude, SST at set location recorded by the observers, and year (all were significant with $p<2\times10^{-16}$). The GAM was then used to estimate annual CPUE for each size group by first using the model to estimate catch for each set, then cumulating the estimated catch for each year, and dividing annual estimated catch by annual observed effort to obtain annual estimated CPUE. The resultant estimated annual CPUE time series fits the observed annual CPUE time series quite well with correlations between the observed and estimated CPUEs of 0. 97 for large fishes and 0.86 for small fishes (Fig. 1). From the GAM, standardized annual CPUE time series for large and small fishes are computed as a function of year by replacing the set SST, latitude, and longitude by mean SST, longitude, and latitude from the 16 year period in the GAM and following the steps outlined previously to obtain annual CPUE. This standardized CPUE time series, computed with mean SST, latitude and longitude, is standardized to eliminate any trends in CPUE due to changes in these variables over the 16-year time period. The temporal trend in standardized annual CPUE for small fishes increased about 25% while it decreased by about 50% for large fishes over the time period (Fig. 2a). To check how robust the observer CPUE pattern was we computed standardized CPUEs for large and small fishes from the logbook data. The results, presented as supplemental material (Table S2 and Fig. S1) show the same pattern of about a 50% decrease in large fish CPUE and about a 33% increase in small fish CPUE (Fig. S1) very similar to that seen with the observer data, although the logbook data reports only the commercially valuable species omitting lancetfish and snake mackerel in the small size group and sharks in the large size groups.

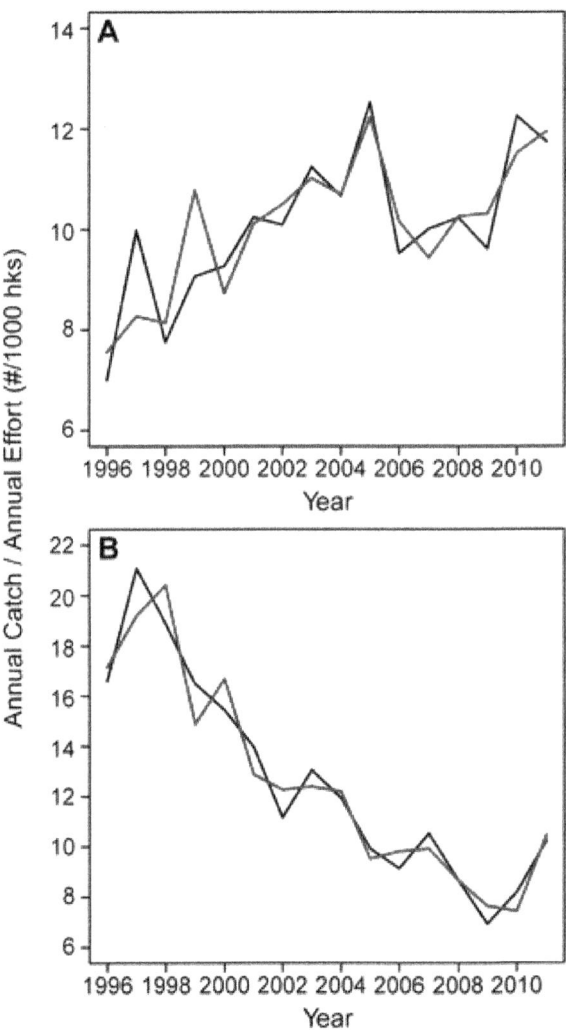

**Figure 1:** The annual observer and generalized additive model CPUE (# fish per 1000 hooks). Panels indicate (**A**) fishes <15 kg and (**B**) fishes ≥15 kg. In both panels black line represents CPUE from observer data, blue line represents CPUE estimated from the generalized additive model. doi:10.1371/journal.pone.0062341.g001

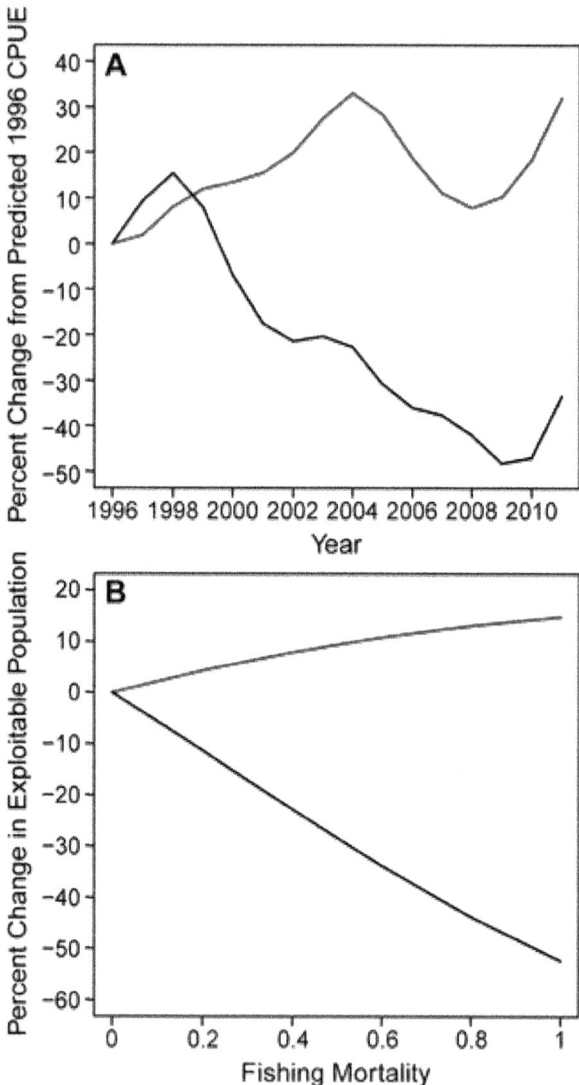

**Figure 2:** Percent change in standardized CPUE and population size for small and large fishes. Panels indicate (**A**) change in generalized additive model (GAM) standardized CPUE and (**B**) change in size-based model estimated population size for fishes <15 kg (blue) and fishes ≥15 kg (black). doi:10.1371/journal.pone.0062341.g002

The mean weights of fishes caught in the Hawaii-based deep-set fishery range from 0.8 kg for snake mackerel to 224 kg for blue marlin (Table 1). Thus for the size at entry to the fishery we use 1.0 kg. To estimate the selectivity of the gear we examine the weight-frequency distribution. The weight-frequency distribution of the catch pooled over the 16-year period shows a typical exponential frequency decline with weight above about 15 kg suggesting that fish above this size are largely fully exploited while this is not the case for smaller fishes (Fig. 3). To further define a selectivity function our pelagic size-based model was run with fishing mortality (F) for F=0.4 and F=0.6 to generate catch size distributions. We found a simple size selectivity function where fishes greater than 15 kg experience the full level of F while for fishes in the range of 1–15 kg, F is one fourth the level for the larger fishes generated catch size structures similar to that in Fig. 3.

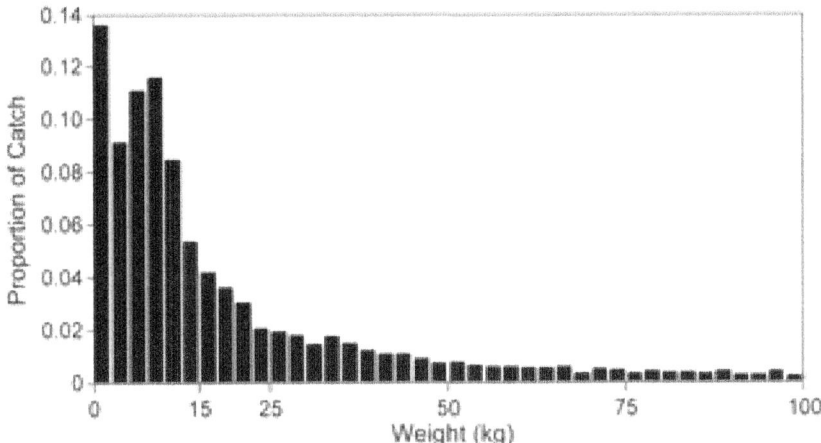

**Figure 3:** Longline catch weight-frequency distribution. The estimated longline catch weight-frequency distribution for the 23 species listed in Table 1 from the observer data 1996–2011. Distribution truncated above 100 kg. doi:10.1371/journal.pone.0062341. g003

Using this size selectivity function we compute population abundance as a function of F with the size-based model. As F increases from 0 to 1.0, the population of large fishes declines about 60%, relative to F=0, while the population of small fishes increases about 20%, relative to F=0 (Fig. 2b). The standardized fishery CPUE derived from the GAM shows the same pattern of a decline in large fish CPUE and concurrent increase in small fish CPUE (Fig. 2a). Further, both observed and model trends show the decline in large fishes is substantially greater than the increase in small fishes (Fig. 2).

Next we use the size-based model to examine the change in the entire ecosystem (fished and unfished) size structure in response to various fishing levels. We express the change in ecosystem size structure as a percent change in the size frequency distribution in the absence of fishing relative to that of a fished ecosystem for various levels of F. The change in ecosystem size structure as a function of F shows a decline in the abundance of fish's larger than 15 kg, the size of full recruitment to the fishery, and an increase in abundance of fishes within the size range 0.1–15 kg (Fig. 4). For any level of F, the magnitude of the increase in the 0.1–15 kg size range is less than the decline above 15 kg (Fig. 4). A portion of the small size class (1–15 kg) is also fished so this group is responding to both fishing and top-down impacts. For organisms weighing less than 0.1 kg, there is a very slight decrease in abundance but essentially the top-down or size-based cascade has only one cascade with declines for fishes greater than 15 kg resulting in increases for fishes between 0.1–15 kg (Fig. 4).

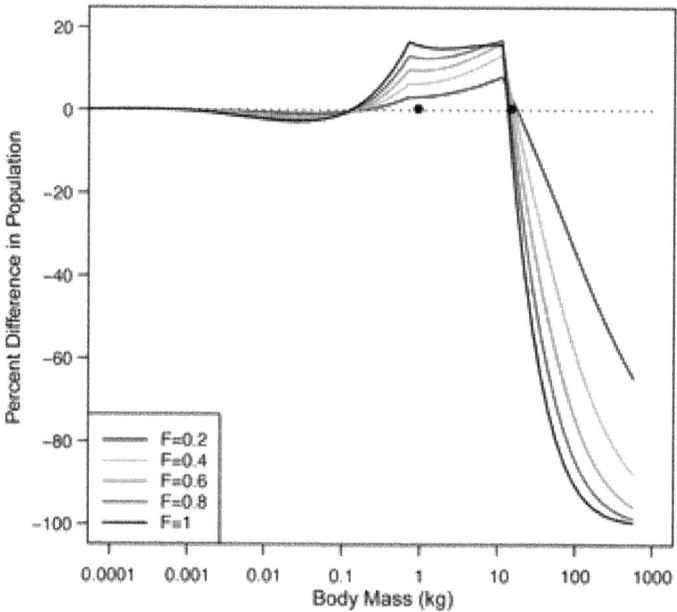

**Figure 4:** Change in fished ecosystem size structure relative to unfished ecosystem size structure. The percent change in ecosystem abundance by size between the unfished size structure and the fished size structure for F ranging from 0.2 to 1.0. The dots are located at 1 and 15 kg to indicate the size at entry to the fishery and the size of full recruitment. doi:10.1371/journal.pone.0062341.g004

Size-based indicators derived from catch data including the mean size of the catch or the proportion of large fishes have been proposed as useful indicators to monitor fishery trends and ecosystem impacts. In the presence of a size-based cascade, indicators based only on catch data will necessarily underestimate the impact from fishing on the ecosystem size structure. For example, from the observer data we can monitor the change in the size structure of the catch as the proportion of the catch greater than 15 kg. In 1996 about 70% of the catch was greater than 15 kg but this proportion has declined over time to about 45% by 2011, roughly a 25% decline (Fig. 5a). The size-based model predicts that as F increases from 0.01 to 1.0 the proportion of fishes larger than 15 kg in the catch will decline about 30% (Fig. 5b). This is similar to the decline seen in the observer data. However, we have seen from the size-based model that fishing impacts the ecosystem size structure down to 0.1 kg. Thus a more complete measure of the change in the size structure due to fishing would be to compute the ratio of the population of fish larger than 15 kg relative to all fish larger than 0.1 kg. While we can't do this with catch data, we can with the size-based model. This ratio declines, as F goes from 0.01 to 1.0, by about 55% or almost twice the decline seen in the model catch data since not only does the proportion of large fish decrease but smaller ones increase (Fig. 5b). Thus the change in ecosystem size structure due to fishing may be underestimated if computed from fisheries catch data alone.

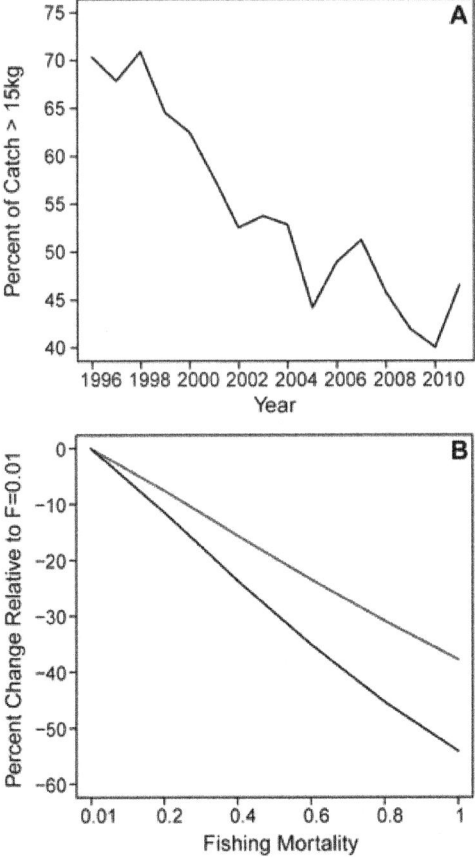

**Figure 5:** Percent change in catch and biomass. Panels indicate (**A**) the percent of the observed catch >15 kg, by year and (**B**) the percent change in the size-based model large fish catch and biomass relative to F=0.01 for F ranging from 0.2 to 1.0 for the

proportion of catch ≥15 kg (blue) and the proportion of exploitable biomass >0.1 kg that is ≥15 kg (black). doi:10.1371/journal.pone.0062341.g005

The rise in catch rates of noncommercial snake mackerel and lancetfish has an impact on the discards in the fishery. The major components of discards in the fishery are the two non-commercial fishes, lancetfish and snake mackerel, which are 100% discarded, and sharks of which about 95% are discarded based on logbook records. Pelagic stingrays are also discarded but their contribution to the total discards is minimal (Fig. 6). Using these discard proportions and the catch for these 3 species from the observer data, we estimate that in 1996 about 30% of the total catch was discarded while by 2011 this proportion had increased to nearly 40% with about one third of the total catch consisting of snake mackerel and lancetfish (Fig. 6).

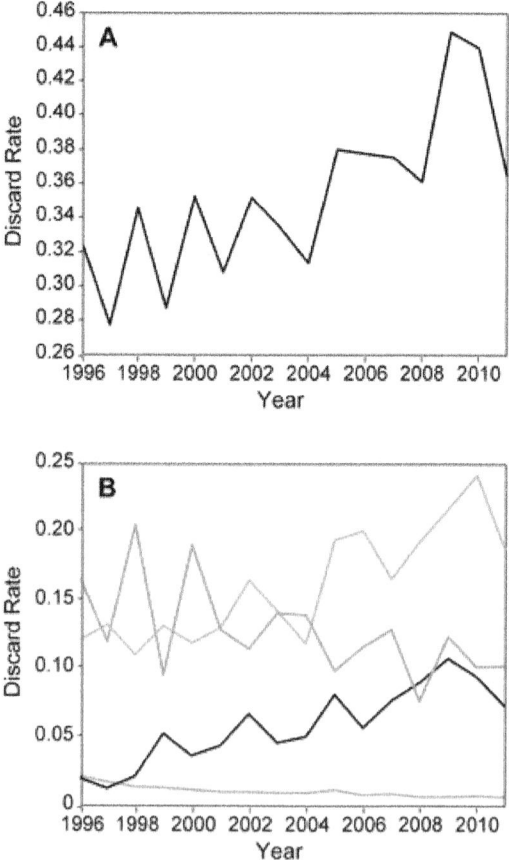

**Figure 6:** The proportion of the total observed catch estimated to be discarded. Panels indicate (**A**) the combined ratio of estimated discards consisting of the catches

of pelagic stingray, snake mackerel, lancetfish, and 95% of the shark catch to total catch and (**B**) the ratio by species of Lancetfish (blue), snake mackerel (black), pelagic stingray (grey), and 95% of the shark catch (green) to total catch. doi:10.1371/journal. pone.0062341.g006

## DISCUSSION

Evidence of fishing impacts to marine ecosystems, especially reductions in trophic structure (fishing down the food web), has been widely reported [22]. Further, decreases in the catch rates of large fishes and increases in the catch rates of small fishes based on longline catches in the central North Pacific between 1950s and 1990s has been previously reported [5]. Our results indicate this trend has continued through 2011. Further, we show that the temporal change in the species and size composition seen over the past 16 years in the Hawaii-based longline observer data is consistent with and can be explained by a size-based model. This suggests that size-based predation is the dominant mechanism in structuring the subtropical pelagic ecosystem, at least the upper trophic levels caught in the deep-set fishery. Earlier work [7] used a species-based model (EwE) to explain this temporal trend as a top-down response using estimated trophic level instead of size. Our current results and the previous ones are consistent in describing the ecosystem change as a top-down response. The fact that two different model approaches reach the same conclusion is seen as positive. We see the current results as a step forward in that a conceptually (size-based predation) and operationally (requiring many fewer estimated input parameters) simpler model explains the observed ecosystem changes.

One implication of this result is that we have a model-based description of the impact of fishing on the entire ecosystem size structure. The model describes a one-step size-based cascade where a reduction of fishes above the size that is fully exploited by the fishery increases the abundance of organisms from about the size of full entry to the fishery down to about 2 orders of magnitude in size but results in little impact on smaller micronekton and plankton. The reason the top-down impact reaches sizes as small as 0.1 kg is predators have a mean prey size that is 1/100$^{th}$ their weight with a lognormal distribution around that mean. The prey of fishes above 15 kg would have a mean size 0.15 kg and larger. The reason the size-based cascade, expressed as a percent of ecosystem abundance, diminishes with declining size is the ecosystem abundance increases exponentially as size decreases. Thus predation impacts from a fixed larger size represent a smaller and smaller fraction of the prey population as size decreases. A key result of this work is the observation that the impacts to size structure extends to sizes below those caught in the fishery, and hence catch-based indicators will underestimate the impact of fishing on the ecosystem size structure. Further, unless there is a targeted sampling for the sizes below the size at entry to the fishery these changes will not be recognized. Size-based models can help to more fully represent the full ecosystem impact of fishing on size structure. One exception to the pattern of a decline in CPUE of large fishes and concurrent increase for small fishes is the pelagic stingray (mean weight 3 kg) that exhibited a 5.4% annual decline. However, an earlier study [5] found that this species increased in the central Pacific longline catches between 1950s and 1990s and attributed this change to a reduction in predation due to declining shark abundance. Fishing effort since the 1990s has continued to increase and while the rays are not retained they are often severely damaged in the release process [23]. Thus their increased bycatch mortality may exceed the decreased predation mortality resulting in a population decline. Pelagic stingrays have been characterized as one of two elasmobranchs with the lowest risk of extinction due to their resilient life history characteristics [23] but our data suggest this may not be the case.

A recent analysis of temporal trends in catch rates of tunas, sharks, and billfishes based on a GAM using observer data the Hawaii longline fishery documented strong general declining trends in standardized catch rates for bigeye, yellowfin and albacore tunas, blue and oceanic white tip sharks, shortbill spearfish and striped marlin [16]. Our results for these species, just based on annual CPUEs but from a more geographically restricted region and over a slightly different time period, also showed significant declining trends in these species with the exception of yellowfin tuna where our estimated linear trend was not statistically significant. Additionally, Gilman *et al.* [16] looked at trends in length for tunas and billfishes and found the lengths significantly increased over time due to the distributions of length classes having shifted towards larger fish. The authors suggest reasons for this shift may include operational changes in the fishery and/or increased catches of juveniles in the purse seine fishery [16]. Initially, this shift to larger fish seems contrary to our finding of an ecosystem shift to small-sized fishes but the difference is a within-species vs. between-species comparison. We used mean weights averaged over the entire time series for each of the 23 species and described the shift in the ecosystem size structure as the shift in the relative abundance of small and large species. We did not examine temporal size trends within species, as many of our 23 species did not have sufficient length data. However, looking at the modest within species length changes presented in Gilman *et al.* [16] relative to the pretty substantial changes in the proportions of large and small fishes in the catch data we conclude that the main change to ecosystem size structure comes from changes in relative abundance between large and small species and not the smaller changes in size within species. An ecosystem approach to fisheries management looks at fishery impacts to the entire ecosystem. Clearly the longline fishery is changing the subtropical ecosystem size structure. Time series of CPUEs computed separately for the pooled small and large fishes represent an informative ecosystem indicator of this trend and should be monitored and reported in any analysis of the fishery. Current reporting in the fishery shows only catch and catch rates in numbers of fishes so managers are not as likely to be aware of the greater decline in weight per effort compared to numbers per effort, and the former may be more closely related to economics of the fishery. Lastly this work shows the value of observer data, which unlike the more commonly collected logbook data, provides information on bycatch and discarded species that contributes to a more complete understanding of ecosystem dynamics.

Our size-based model does not suggest any obvious threshold in changes to an ecosystem size structure that could serve as a management target. A recent meta-analysis of global fisheries explores tradeoffs between multispecies maximum sustainable yield (MMSY) and the collapse of individual stocks [24]. Their model finds that for a wide range of exploitation rates ranging from 0.25 to 0.60 the resultant catches equal or exceed 90% of the MMSY, but with an exploitation rate of 0.60 almost half the species in the ecosystem are expected to collapse, while with an exploitation rate of 0.25 less than 10% of the species are expected to collapse[24]. Thus in a multispecies context, taking into consideration aspects of ecosystem structure and function, the exploitation rate that achieves maximum sustainable yield should be considered an upper limit rather than a management target [24]. Unfortunately our observer data represents only a small portion of the Pacific pelagic fishery so estimating MMSY and the corresponding F is problematic. This analysis needs to be conducted on the basin-scale by the appropriate regional fisheries management organizations. The sharp decline of stingrays and oceanic white-tip shark presents concern of collapse for these species. Currently management of the longline fishery is based on single species basin-wide quotas for yellowfin tuna, bigeye tuna, and striped marlin set by the Western and Central Pacific Fisheries Commission. To the extent that these quotas cap fishing effort and mortality on all species they could prevent further ecosystems impacts. Further, ways to reduce the estimated 40% discard rate in the fishery should be a management focus as well. Lastly, this work highlights the critical importance of observer data in monitoring ecosystem changes. While this paper has focused on changes in ecosystem structure it is clear that with increases in escolar and snake mackerel CPUEs of 12 and 15% per year respectively and declines in pelagic stingray and oceanic white-tip CPUEs of 5.4 and 6.9% per year respectively we are also seeing changes in the ecosystem composition with potential significant impacts on ecosystem function. Lastly, while we have seen evidence of changes at the base of the ecosystem in the subtropical Pacific over the past decade, they have been modest relative to the substantial increase in fishing effort [25], [26]. However, going forward the impact of climate change has been projected to increase its ecosystem impact and shift the subtropical ecosystem size structure toward smaller sizes even if fishing effort remains constant [20], [27]. Thus the combined impacts of increased fishing effort and future climate change are projected to be additive and accelerate a shift of ecosystem size structure to smaller sizes. The time series of CPUE for our small fishes group shows considerably more interannual variation than the large fishes group. Many of these small fishes have faster growth rates and shorter life spans than the larger fishes and hence may be more responsive to interannual environmental changes. Thus a shift to smaller fishes may result in greater interannual variation in the longline fishery CPUE.

## Supporting Information

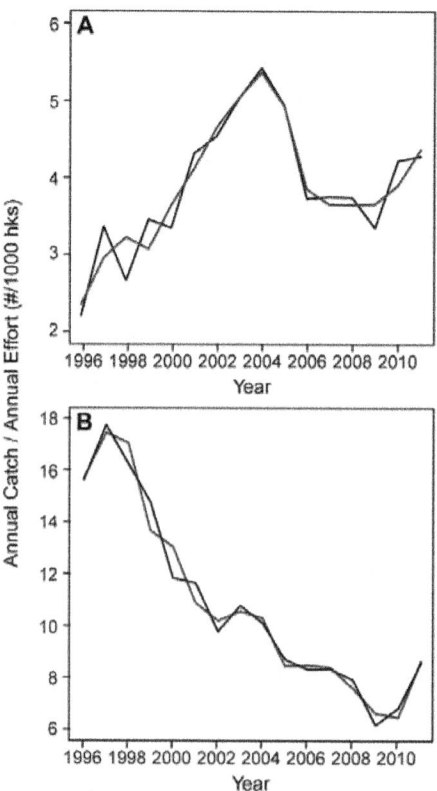

Figure S1: **The annual logbook and generalized additive model CPUE (# fish per 1000 hooks).**Panels indicate (**A**) fishes <15 kg and (**B**) fishes ≥15 kg. In both panels black line represents CPUE from logbook data, blue line represents CPUE estimated from the generalized additive model. doi:10.1371/journal.pone.0062341.s001 (TIF)

Table S1: Mean species weight, length, and length-weight conversion factors. From left, columns indicate species, mean species weight as determined form length-weight conversions, mean length from those recorded by observers measuring every third fish from 2006–2011, $a$, $b$, and reference listing length-weight conversion factors. To convert length ($L$) in cm to weight ($W$) in g, the equation $aL^b = W$ was used. doi:10.1371/journal.pone.0062341.s002 (DOCX)

**Table S1. Mean species weight, length, and length-weight conversion factors.**

| Species | Mean Weight (kg) | Mean Length (cm) | a | b | Reference[a] |
|---|---|---|---|---|---|
| Blue Marlin (*Makaira nigricans*) | 224.0 | 166.3 | 0.0411 | 3.0363 | 1, 2 |
| Blue Shark (*Prionace glauca*) | 106.4 | 171.7 | 0.0025 | 3.3835 | 2 |
| Striped Marlin (*Tetrapturus audax*) | 93.5 | 132.8 | 0.0057 | 3.3760 | 2 |
| Shortbill Spearfish (*Tetrapturus angustirostris*) | 75.7 | 134.8 | 0.0005 | 3.8340 | 2 |
| Shortfin Mako Shark (*Isurus oxyrinchus*) | 48.3 | 177.3 | 0.0167 | 2.8470 | 2 |
| Swordfish (*Xiphias gladius*) | 42.0 | 105.5 | 0.0078 | 3.2100 | 2 |
| Yellowfin Tuna (*Thunnus albacares*) | 33.5 | 109.2 | 0.0129 | 3.1125 | 2, 3 |
| Opah (*Lampris guttatus*) | 30.2 | 101.6 | 0.0281 | 3.0000 | 2 |
| Bigeye thresher Shark (*Alopias superciliosus*) | 24.0 | 166.5 | 0.0238 | 2.6750 | 2 |
| Unidentified Tuna | 24.0 | 107.2 | 0.0155 | 3.0203 | [b] |
| Bigeye Tuna (*Thunnus obesus*) | 22.5 | 113.7 | 0.0155 | 2.9725 | 2, 3 |
| Oceanic White-tip Shark (*Carcharinus longimanus*) | 19.0 | 141.8 | 0.0254 | 2.6910 | 2 |
| Albacore Tuna (*Thunnus alalunga*) | 17.1 | 104.7 | 0.0407 | 2.7800 | 2, 3 |
| Wahoo (*Acanthocybium solandri*) | 16.4 | 127.0 | 0.0024 | 3.2422 | 2, 4 |
| Escolar (*Lepidocybium flavobrunneum*) | 12.1 | 75.1 | 0.0353 | 2.9053 | 2, 5 |
| Mola (*Ranzania laevis* and *Mola mola*) | 8.8 | 51.4 | 0.0454 | 3.0500 | 2 |
| Skipjack Tuna (*Katsuwonus pelamis*) | 7.9 | 70.9 | 0.0085 | 3.2160 | 2 |
| Mahi Mahi (*Coryphaena hippurus*) | 7.4 | 87.3 | 0.0172 | 2.8868 | 2, 4 |
| Lancetfish (*Alepisaurus ferox*) | 7.1 | 107.7 | 0.0047 | 3.0000 | 2 |
| Great Barracuda (*Sphyraena jello*) | 5.9 | 99.9 | 0.0098 | 2.8750 | 2 |
| Pomfrets (*Taractichthys steindachneri* and *Brama japonica*) | 4.9 | 56.7 | 0.0160 | 3.0940 | 2 |
| Pelagic Stingray (*Pteroplatytrygon violacea*) | 3.0 | [c] | - | - | 6 |
| Snake Mackerel (*Gempylus serpens*) | 0.8 | 101.7 | 0.0007 | 3.0000 | 2 |

From left, columns indicate species, mean species weight as determined from length-weight conversions, mean length from those recorded by observers measuring every third fish from 2006 – 2011, $a$, $b$, and reference listing length-weight conversion factors. To convert length ($L$) in cm to weight ($W$) in g, the equation $aL^b = W$ was used.

[a]When multiple references are listed, the $a$ and $b$ values are an average of those listed in the literature.

[b]The $a$ and $b$ values for unidentified tuna are an average of those of all tuna listed in this table.

[c]Pelagic stingray lengths were not recorded.

**Table S2:** Change in logbook catch rate estimate from statistically significant ($P<0.01$) linear regressions over 1996–2011, ordered by fish size. From left, columns indicate species, annual percent change in CPUE based on linear regression ($P$-values for significant trends in parentheses, insignificant fits denoted by a 0% change), and mean species weight as determined form length-weight conversion. doi:10.1371/journal.pone.0062341.s003 (DOCX)

| Species | % Annual Change in CPUE[a] (P-value) | Mean Weight in kg[b] |
|---|---|---|
| Blue Marlin (*Makaira nigricans*) | -4.2 (0.0002) | 224.0 |
| Blue Shark (*Prionace glauca*) | -3.4 (<0.0001) | 106.4 |
| Striped Marlin (*Tetrapturus audax*) | -5.0 (0.0007) | 93.5 |
| Shortbill Spearfish (*Tetrapturus angustirostris*) | -3.5 (0.007) | 75.7 |
| Shortfin Mako Shark (*Isurus oxyrinchus*) | 0 | 48.3 |
| Swordfish (*Xiphias gladius*) | 0 | 42.0 |
| Yellowfin Tuna (*Thunnus albacares*) | 0 | 33.5 |
| Opah (*Lampris guttatus*) | -4.2 (0.002) | 30.2 |
| Bigeye Thresher Shark (*Alopias superciliosus*) | 0 | 24.0 |
| Unidentified Tuna | -[c] | 24.0 |
| Bigeye Tuna (*Thunnus obesus*) | -2.3 (0.001) | 22.5 |
| Oceanic White-tip Shark (*Carcharinus longimanus*) | -[c] | 19.0 |
| Albacore Tuna (*Thunnus alalunga*) | -7.2 (<0.0001) | 17.1 |
| Wahoo (*Acanthocybium solandri*) | 0 | 16.4 |
| Escolar (*Lepidocybium flavobrunneum*) | 88.6 (<0.0001) | 12.1 |
| Mola (*Ranzania laevis* and *Mola mola*) | -[c] | 8.8 |
| Skipjack Tuna (*Katsuwonus pelamis*) | 0 | 7.9 |
| Mahi Mahi (*Coryphaena hippurus*) | 0 | 7.4 |
| Lancetfish (*Alepisaurus ferox*) | -[c] | 7.1 |
| Great Barracuda (*Sphyraena jello*) | -[c] | 5.9 |
| Pomfrets (*Taractichthys steindachneri* and *Brama japonica*) | 0 | 4.9 |
| Pelagic Stingray (*Pteroplatytrygon violacea*) | -[c] | 3.0 |
| Snake Mackerel (*Gempylus serpens*) | -[c] | 0.8 |

From left, columns indicate species, annual percent change in CPUE based on linear regression ($P$-values for significant trends in parentheses, insignificant fits denoted by a 0% change), and mean species weight as determined from length-weight conversion.
[a]from linear fit.
[b]as determined from observer recorded lengths and length-weight conversions.
[c]not included in logbook records.

## ACKNOWLEDGMENTS

We thank Dr. Julia Blanchard who provided code and assistance to greatly facilitate our application of the size-based model. We also thank all the scientific observers employed thorough the Pacific Islands Region for their collection of observer data from the longline fishery.

## AUTHOR CONTRIBUTIONS

Analyzed the data: PWJ JJP. Wrote the paper: JJP.

## REFERENCES

1.    Pickard GL, Emery WJ (1990) Descriptive physical oceanography: an introduction, 5th ed. Oxford: Pergamon Press. 336 p.

2.    Seki MP, Polovina JJ (2001) Ocean gyre ecosystems. In: Steele JH et al., editors. The Encyclopedia of Ocean Sciences, vol. 4. San Diego, CA: Academic Press. 1959–1964.

3.  Kitchell JF, Essington TE, Boggs CH, Schindler DE, Walters CJ (2002) The role of sharks and longline fisheries in a pelagic ecosystem of the central Pacific. Ecosystems 5: 202–216. doi: 10.1007/s10021-001-0065-5

4.  Kitchell JF, Boggs C, He X, Walters CJ (1999) Keystone predators in the Central Pacific. In: Ecosystem approaches to fisheries management. Univ. Alaska Sea Grant. AL-SG-99-01: 665–683. doi: 10.4027/eafm.1999.47

5.  Ward P, Myers RA (2005) Shifts in open-ocean fish communities coinciding with the commencement of commercial fishing. Ecology 86: 835–847. doi: 10.1890/03-0746

6.  Cox SP, Essington TE, Kitchell JF, Martell SJD, Walters CJ, et al. (2002) Reconstructing ecosystem dynamics in the central Pacific Ocean, 1952–1998. II. A preliminary assessment of the trophic impacts of fishing and effects on tuna dynamics. Canadian Journal of Fisheries and Aquatic Sciences 59: 1736–1747. doi: 10.1139/f02-138

7.  Polovina JJ, Abecassis M, Howell EA, Woodworth P (2009) Increases in the relative abundance of mid-trophic level fishes concurrent with declines in apex predators in the central North Pacific subtropical gyre, 1996–2006. Fishery Bulletin 107: 523–531.

8.  Benoit E, Rochet M (2004) A continuous model of biomass size spectra governed by predation and the effects of fishing on them. Journal of Theoretical Biology 226: 9–21. doi: 10.1016/s0022-5193(03)00290-x

9.  Blanchard JL, Jennings S, Law R, Castle MD, McCloghrie P, et al. (2009) How does abundance scale with body size in coupled size-structured food webs? Journal of Animal Ecology 78: 270–280. doi: 10.1111/j.1365-2656.2008.01466.x

10. Blanchard JL, Law R, Castle MD, Jennings S (2011) Coupled energy pathways and the resilience of size-structured food webs. Theoretical Ecology 4: 289–300. doi: 10.1007/s12080-010-0078-9

11. Jennings S, Pinnegar JK, Polunin NVC, Boon T (2001) Weak cross-species relationships between body size and trophic level belie powerful size-based trophic structuring in fish communities. Journal of Animal Ecology 70: 934–944. doi: 10.1046/j.0021-8790.2001.00552.x

12. Law R, Plank MJ, James A, Blanchard JL (2009) Size-spectra dynamics from stochastic predation and growth of individuals. Ecology 90: 802–11. doi: 10.1890/07-1900.1

13. Maury O, Shin Y, Faugeras B, Benari T, Marsac F (2007) Modeling environmental effects on the size-structured energy flow through marine

ecosystems. Part 2: Simulations. Progress in Oceanography 74: 500–514. doi: 10.1016/j.pocean.2007.05.001

14. Bigelow KA, Musyl MK, Poisson F, Kleiber P (2006) Pelagic longline gear depth and shoaling. Fisheries Research 77: 173–183. doi: 10.1016/j. fishres.2005.10.010

**15.** .Harley SJ, Williams P, Nicol S, Hampton J (2011) The western and central Pacific tuna fishery: 2010 overview and status of stocks. Tuna Fisheries Assessment Report 11. Noumea, New Caledonia: Secretariat of the Pacific Community.

16. Gilman E, Chaloupka M, Read A, Dalzell P, Holetschek J, et al. (2012) Hawaii longline tuna fishery temporal trends in standardized catch rates and length distributions and effects on pelagic and seamount ecosystems. Aquatic Conservation: Marine and Freshwater Ecosystems 22: 446–488. doi: 10.1002/aqc.2237

17. Jennings S, Mélin F, Blanchard JL, Forster RM, Dulvy NK, et al. (2008) Global-scale predictions of community and ecosystem properties from simple ecological theory. Proceedings of the Royal Society of London B: Biological Sciences 275: 1375–1383 doi:10.1098/rspb.2008.0192.

18. Barnes C, Maxwell D, Reuman DC, Jennings S (2010) Global patterns in predator–prey size relationships reveal size dependency of trophic transfer efficiency. Ecology 91: 222–232 doi:10.1890/08-2061.1.

19. Blanchard JL, Jennings S, Holmes R, Harle J, Merino G, et al. (2012) Potential consequence of climate change on primary production and fish production in 28 large marine ecosystems. Philosophical Transactions of the Royal Society of London B: Biological Sciences 367: 2979–2989 doi: 10.1098/rstb.2012.0231.

20. Woodworth-Jefcoats PA, Polovina JJ, Dunne JP, Blanchard JL (2012) Ecosystem size structure response to 21st century climate projection: large fish abundance decreases in the central North Pacific and increases in the California Current. Global Change Biology 19(3): 724–733. doi: 10.1111/gcb.12076

21. Delworth TL, Broccoli AJ, Rosati A, Stouffer RJ, Balaji V, et al. (2006) GFDL's CM2 global coupled climate models. Part I: formulation and simulation characteristics. Journal of Climate 19: 643–374. doi: 10.1175/ jcli3629.1

22. Pauly D, Palomares ML (2005) Fishing down the food web: It is far more pervasive than we thought. Bulletin of Marine Science 76: 197–211.

**23.** .Dulvy NK, Baum JK, Clarke S, Compagno LJV, Cortés E, et al. (2008) You can swim but you can't hide: the global status and conservation of oceanic pelagic sharks and rays. Aquatic Conserv: Mar. Freshw. Ecosyst. 18: 459–482. doi: 10.1002/aqc.975

24. Worm B, Hilborn R, Baum JK, Branch TA, Collie JS, et al. (2009) Rebuilding global fisheries. Science 325: 578–585. doi: 10.1126/science.1173146

25. Polovina JJ, Howell EA, Abecassis M (2008) The ocean's least productive waters are expanding. Geophysical Research Letters 35: L03618 doi:10.1029/2007GL031745.

26. Polovina JJ, Woodworth PA (2012) Declines in phytoplankton cell size in the subtropical oceans estimated from satellite remotely-sensed temperature and chlorophyll, 1998–2007. Deep-Sea Research II 77–80: 82–88. doi: 10.1016/j.dsr2.2012.04.006

27. Howell EA, Wabnitz CCC, Dunne JP, Polovina JJ (2012) Climate-induced primary productivity change and fishing impacts on the central North Pacific ecosystem and Hawaii-based pelagic longline fishery. Climatic Change. doi 10.1007/s10584-012-0597-z.

# Chapter 7

## THE IMMUNE SYSTEM DRUGS IN FISH: IMMUNE FUNCTION, IMMUNOASSAY, DRUGS

Cavit Kum and Selim Sekkin

University of Adnan Menderes, Turkey

## INTRODUCTION

Fish is a heterogeneous group of different organisms which include the agnathans (hagfishes and lampreys), condryctians (sharks and rays) and teleosteans (bony fish). Like in all vertebrates, fish have cellular and humoral immune responses, and central organs whose the main function is involved in immune defence. Fish and mammals show some similarities and some differences regarding immune function (Cabezas, 2006; Nelson, 1994; Tort et al., 2003; Zapata et al., 1996). The fish defence system is basically similar to that described in mammals. For cellular defence systems in fish, teleosts have phagocytic cells similar to macrophages, neutrophils, and natural killer (NK) cells, as well as T and B lymphocytes. Teleosts also have various humoral defence components such as complement (classical and alternative pathways), lysozyme, natural hemolysin, transferrin and C-reactive protein (CRP). The existence of cytokines (such as interferon, interleukin 2 (IL-2), macrophage activating factors (MAF)) has also been reported (Secombes et al., 1996, Sakai, 1999). On the contrary, the morphology of the immune system is quite different between fish and mammals. Most obvious is the fact that fish lack bone marrow and lymph nodes. Instead, the head kidney serves as a major lymphoid organ, in addition to the thymus and spleen (Press & Evensen, 1999). Gut associated lymphoid tissues are also known lymphoid organs, and have been shown to function in eliciting immune responses in carp (Joosten et al., 1996). Some teleosts, such as plaice, have been shown to possess a lymphatic system that is differentiated from the blood vascular system, though the existence of such a system has been challenged in other species (Hølvold, 2007). Health of fish depends on the interrelationship of some major components of the fish and the environment in which they live (Figure 1). Tolerance of these various factors is dependent on the host and in many case the husbandry practices. The environment may be the most critical component of the fish health matrix because environmental quality influences the fish's physiological well-being, species cultured, feeding regimes, rate of growth, and ability to maintain natural and acquired resistance and immunity. Overall physiological status of the fish host is determined by the husbandry practice, environmental quality, the fish's nutritional well-being and the pathogen, all of which influence the natural resistance and acquired immunity of the host. It is common knowledge that fish stressed by one of these factors are more susceptible to infection (Magnadóttir, 2010; Plumb & Hanson, 2011). (modified from Magnadóttir, 2006 and Plumb & Hanson, 2011).

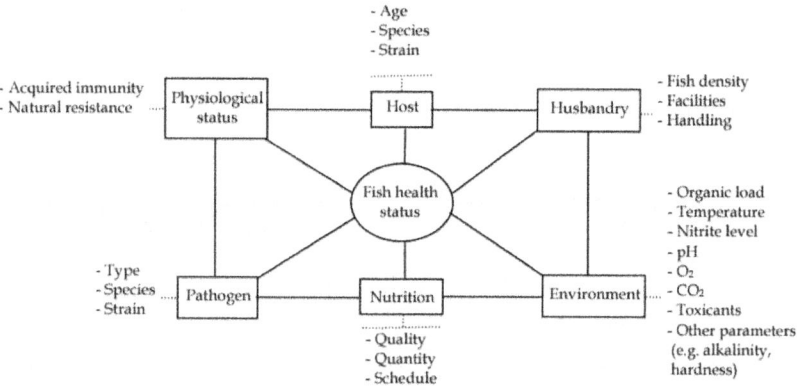

**Figure 1:** The relationship of various factors in fish health status.

In addition, in the Food and Drug Administration (FDA) and the European Union (EU) member states, although a limited number of antimicrobial agents are licensed for use in fin fish culture, various drugs such as chemotherapeutics have been used to an increasing levels treat bacterial infections in cultured fish in the last decades years. However, the incidence of drug-resistant (including multiple and cross-resistance) bacteria has become a major problem in fish culture and public health (Alderman & Hasting, 1998; Aoki, 1992; Horsberg, 2003). Vaccination is a useful prophylaxis for infectious diseases of fish and is already commercially available for bacterial infections such as vibriosis, enteric red mouth disease (ERD) and furunculosis and some viral infection such as infectious pancreatic necrosis (IPN). Vaccination may be the most effective method of controlling fish disease. Furthermore, the development of vaccines against intracellular pathogens such as Renibacterium salmoninarum has not so far been successful. Therefore, the immediate control of all fish diseases using only vaccines is impossible. Immunostimulants such as synthetic chemicals, bacterial derivatives, polysaccharides or animal and plant extracts increase resistance to infectious disease, not by enhancing specific immune responses, but by enhancing non-specific immune defence mechanisms. Although, there is no memory component and the response is likely to be of short duration. Use of these immunostimulants is an effective means of increasing the immunocompetency and disease resistance of fish. Research into fish immunostimulants is developing and many agents are currently in use in the aquaculture industry (Klesius et al., 2001; Sakai, 1999; Subasinghe, 2009). Besides, the additions of various food additives like vitamins, carotenoids, probiotics, prebiotics, synbiotics and herbal remedies to the fish feed have been tested in fish. Overall the effects have been beneficial such as reducing stress response, increasing the activity of innate parameters and improving disease resistance (Austin & Brunt, 2009; Hoffmann, 2009; Magnadóttir, 2010; Nayak, 2010).

# IMMUNE SYSTEM COMPONENTS

## Tissues and cells

Types of immune organs vary between different types of fish. In the jawless fish (hagfishes and lampreys), true lymphoid organs are absent. Instead, these fish rely on region of lymphoid tissue within other organs to produce their immune cells (Zapata et al., 1996). However, genetic differences may be small and some molecular and cellular agents similar, the anatomical and functional organisation such as the structure and form of the immune system (Press & Evensen, 1999; Randeli et al., 2008). The immune system of fish has cellular and humoral immune responses, and organs whose main function is involved in immune defence (Jimeno, 2008). Most of the generative and secondary lymphoid organs present in mammals are also found in fish, except for the lymphatic nodules and the bone marrow (Alvarez-Pellitero, 2008; Jimeno, 2008; Press & Evensen, 1999; Zapata et al., 1996). Instead, the anterior part of kidney usually called head kidney, aglomerular, assumes hemopoietic functions (Jimeno, 2008; Meseguer et al., 1995; Tort et al., 2003), and unlike higher vertebrates is the principal immune organ responsible for phagocytosis (Danneving et al., 1994; Galindo-Villegas & Hosokowa, 2004), antigen processing activity and formation of IgM and immune memory through melanomacrophagic centres (Tort et al., 2003). The most important immunecompetent organs and tissue of fish include the kidney (anterior/or head and posterior/or caudal), thymus, spleen, liver, and mucosa-associated lymphoid tissues (Figure 2) (Press & Evensen, 1999; Shoemarker et al., 2001). In fish, myelopoiesis generally occurs in the head kidney and/or spleen, whereas thymus, kidney and spleen are the major lymphoid organs (Zapata et al., 2006). Next to the thymus as the primary T cell organ head kidney is considered the primary B cell organ. Also, head kidney and spleen present macrophage aggregates, also known as melano-macrphage centres (Alvarez-Pellitero, 2008) (modified from http://www.dkimages.com/discover/previews/1171/10686362.JPG).

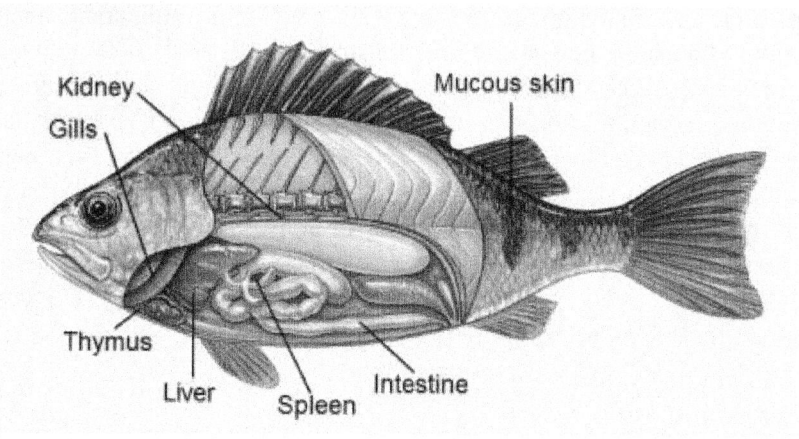

**Figure 2:** Immune structures in teleost fish.

The kidney often referred to as the head kidney tissue is important in hematopoiesis and immunity in fish. And it is predominantly a lympho-myeloid compartment (Press & Evensen, 1999). Early in development, the entire kidney is involved in production of immune cells and the early immune response. As the fish mature, blood flow through the kidney is slow, and exposure to antigens occurs. There appears to be a concentration of melanomacrophage centers are aggregates of reticular cells, macrophages, lymphocytes and plasma cells; they may be involved in antigen trapping and may play a role in immunologic memory (Galindo-Villegas & Hosokowa, 2004; Press et al., 1996; Secombes et al., 1982). The head kidney or anterior kidney (pronephros), the active immune part, is formed with two Y-arms, which penetrate underneath the gills. In addition, this structure of the kidney has a unique feature, and it is a well innervated organ, and the kidney is also an important endocrine organ, homologous to mammalian adrenal glands, releasing corticosteroids and other hormones. Thus, the kidney is a valuable organ with key regulatory functions and the central organ for immune-endocrine interactions and even neuroimmuno-endocrine connections (Press & Evensen, 1999; Tort et al., 2003). The thymus is a paired bilateral organ situated beneath the pharyngeal epitelium in the dorso-lateral region of the gill chamber. But it seems that the size of the thymus varies with seasonal changes and hormonal cycles (Galindo-Villegas & Hosokowa, 2004; Meseguer et al., 1995; Press & Evensen, 1999;

Zapata et al., 1996). The thymus appears to have no executive function. It is regarded, as a primary lymphoid organ where the pool of virgin lymphocytes in the circulation and other lymphoid organs. However, much of the data supporting this is indirect evidence obtained either by immunizing with T-dependent antigens (Ellsaesser et al., 1988) or by using monoclonal antibodies as cell surface markers (Passer et al., 1996) and functional in vitro assay. In addition, trout-labeled blood lymphocytes migrate through the thymus before reaching the spleen and kidney (Tatner & Findlay, 1991). It suggest that teleost thymus, despite its striking morphology, has the same function as in higher vertebrates, that is, it is the main source of immunocomponent T cells (Zapata et al., 1996), and research shows that the thymus is responsible for the development of T-lymphocytes, as in other jawed vertebrates (Alvarez-Pellitero, 2008; Galindo-Villegas & Hosokowa, 2004). In general, the available data support a correlation between the histological maturation of the teleost thymus, appearance of the lymphocytes in peripheral lymphoid organs, and development of the cell-mediated immune response (Zapata et al., 1996). The spleen is the major peripheral and a secondary lymphoid organ in fish which contains fewer haemopoietic and lymphoid cells than the kidney, being composed mainly of blood held in sinuses, and it is believed to be involved in immune reactivity and blood cell formation (Galindo-Villegas & Hosokowa, 2004; Manning, 1994; Zapata et al., 1996). Most fish spleen is not distinctly organized into red and white pulp, as in mammals, but white and red pulp is identifiable. It contains different sized lymphocytes, numerous developing and mature plasma cells, and macrophages in a supporting network of fibroblastic reticular cells. Lymphocyte and macrophage are present in the spleen of fish, contained in specialized capillary walls, termed ellipsoids. In addition, ellipsoids appear to have a specialised function for plasma filtration and particularly immune complex. Most macrophage is arranged in malanomacrophage centers, and it is defined that they are primarily responsible for the breakdown of erythrocytes. These centers may retain antigens as immune complexes for long periods. Although the lymphoid tissue is poorly developed in the teleost spleen, after antigenic stimulation, increased amount of lymphoid tissue does appear, and indirectly suggesting the presence of T-like and B-like cells in this group fish (Espenes et al., 1995; Galindo-Villegas & Hosokowa, 2004; Zapata et al., 1996). The spleen of teleosts has also been implicated in the clearance of blood-borne antigens and immune complexes in splenic ellipsoids and also has a role in the antigen presentation and the initiation of the adaptive immune response (Alvarez-Pellitero, 2008; Chaves-Pozo et al., 2005; Whyte, 2007). The liver is included under this chapter, because in mammals, it is responsible for production of components of the complement cascade and acute phase proteins (such as CRP), which are important in the natural resistance of the animal, defined that the liver of fish plays a similar role (Fletcher, 1981). On the contrary, research to support this claim is lacking (Galindo-Villegas & Hosokowa, 2004; Shoemarker et al., 2001). The mucosa-associated lymphoid tissues in fish are distributed around the intestine referred to as the gut, skin and gills, thus complementing the physical and chemical protection provided by the structure (Jimeno, 2008; Press & Evensen, 1999; Tort et al., 2003). Teleost lack organized mucosa-associated lymphoid tissues such as Peyer's patches of mammals, though there is evidence that skin, gills and intestine contains populations of leucocytes (Jimeno, 2008; Press & Evensen, 1999) and innate and adaptive immunity act in case of attack of microorganisms (Ellis, 2001; Schluter et al., 1999). This equipment is completed with immunocompetent cells such as leucocytes and intraepithelial plasmatic cells (Dorin et al., 1994; Moore et al., 1998; Tort et al., 2003). Recently, several additional defences have been discovered in fish mucous membranes (Bols et al., 2001), such as the production of nitric oxide by the gill as well as antibacterial peptides and proteins by skin (Campos-Perez et al., 2000; Galindo-Villegas & Hosokowa, 2004; Ebran et al, 1999; Tort et al., 2003). Not only the mucous membranes of these tissues are an important physical barrier in fish, but also contain several components with a role in the host-parasite interaction, and release antimicrobial agents or proteins. Besides that among the epidermal secretions, complement, lysozyme, lectins (or pentraxins), alkaline phosphatase and esterase, trypsin (or trypsinlike), natural antibodies or immunoglobulins are often found, although their amount and activity depend on the species, and hemolysine are among the substances present with biostatic or biocidal activities (Alexander & Ingram, 1992; Alvarez-Pellitero, 2008; Aranishi & Mano, 2000; Arason, 1996; Balfry & Higgs, 2001; Ellis, 1999; Galindo-Villegas & Hosokowa, 2004; Jones, 2001; Fast et al., 2002; Magnadóttir, 2006; Palaksha et al., 2008; Shoemarker et al., 2001; Tort et al., 2003). Most research on the presence of immunoglobulin or antibody in the mucus suggests that mucus immunoglobulin is not a result of the transduction of immunoglobulin from the serum (Shoemarker et al., 2001). Mucous or goblet cells secrete mucus, which has at least three different types of defensive roles: (1) Mucus interrupts establishment of microbes by being continually sloughed off. (2) If establishment is accomplished, mucus acts as a barrier to be crossed. (3) The mucus on skin, and presumably the other surfaces, contains a variety of humoral factors with antimicrobial properties (Galindo-Villegas & Hosokowa, 2004; Tort et al., 2003). All multicellular organisms possess a selection of cells and molecules that interact in order to ensure production from pathogens (Abbas & Lichtmann, 2006). This collection of highly specialised components makes up the immune system, and poses a physiological defence against microbe invasion (Jimeno, 2008). Fish immune cells show the same main features as those of other vertebrates, and lymphoid and myeloid cell families have been defined. Key cell types involved in non-specific cellular defence responses of teleost fish include the phagocytic cells monocytes/macrophages, non-specific cytotoxic cells (or NK cells), thrombocytes, granulocytes (or neutrophils) and lymphocytes (Table 1) (Buonocore & Scapigliati, 2009; Hamerman et al., 2005; Hølvold, 2007; Magnadóttir, 2006; Jimeno, 2008; Shoemarker et al., 2001). Epithelial and antigen presenting cell also participate in the innate defence in fish, and some teleost have been reported to have both acidophilic and basophilic granulocytes in the peripheral blood in addition to the neutrophils. Furthermore, recently it has been observed that basophilic granular cells (acidophilic/eosinophilic granule cells or mast cells) of fish Perciformes order, the largest and most evolutionarily advanced order of teleosts, are endowed with histamine (Garcia-Ayala & Chaves-Pozo, 2009; Jimeno, 2008; Magnadóttir, 2006; Murelo et al., 2007; Whyte, 2007). Mononuclear cells in fish include the macrophages (and/or tissue macrophages) and monocytes. These cells are probably the single most important cell in the immune response in fish. Not only are they important in the production of cytokines, but they also are the primary cells involved in phagocytosis and the killing of pathogens upon first recognition and subsequent infection (Buonocore & Scapigliati, 2009; Cabezas, 2006; Clem et al., 1985; Garcia-Ayala & Chaves-Pozo, 2009; Secombes et al., 2001; Shoemarker et al., 2001). Macrophages also play major roles as being the primary antigen-presenting cell in teleost, thus linking the non-specific and acquired immune response (Balfry & Higgs, 2001; Galindo-Villegas & Hosokowa, 2004; Jimeno, 2008; Shoemarker et al., 2001; Vallejo et al., 1992). Thrombocytes are thought to be a nucleated version of the mammalian platelet. These cells are involved in blood clotting and have recently been thought to have phagocytic properties (Balfry & Higgs, 2001; Secombes, 1996). (modified from Hølvold, 2007; Shoemarker et al., 2001).

**Table 1:** Non-specific immune cells in fish and their functional characteristics and mode of action

| Cellular components | Functional characteristics and mode of action |
| --- | --- |
| Monocytes/Macrophages | Phagocytosis, and phagocyte activation, cytokine production, intracellular killing, antigen processing and presentation, Secretion of growth factors and enzymes to remodel injured tissue, T-lymphocyte stimulation. |
| Granulocytes (or Neutrophils) | Phagocytosis, secretion and phagocyte activation, cytokine production, extracellular killing, inflammation. |
| Non-specific cytotoxic cells (or natural killer cells) | Recognition and target cell lysis, induce apoptosis of infected cells, Synthesize and secrete **interferon-gamma** (IFN-γ). |

Fish possess polymorph nuclear cells, or granulocytes (especially neutrophils, and eosinophils, and basophils), that contain granules, the contents of which are released upon stimulation (Balfry & Higgs, 2001). These cells are highly mobile cell, phagocytic, produce reactive oxygen species, traveling via the blood and lymphatic systems to sites of infection and injure, thereby playing a vital role in the inflammatory response. Also, neutrophils are the primary cells involved in the initial stages of inflammation in fish, between 12 to 24 hours, and the function of the granulocytes may be cytokine production to recruit immune cells to the area of damage or infection (Galindo-Villegas & Hosokowa, 2004; Manning, 1994; Shoemarker et al., 2001). However, eosinophilic granular cells found in the stratum granuloma of the gut, gills and skin, and surrounding major blood vessels, are not considered to be eosinophils but rather mast cells (Vallejo & Ellis, 1989; Reite, 1998; GalindoVillegas & Hosokowa, 2004). Cells mediating the lytic cycle to occur and destroy tumour target cells lines following receptor binding in fish have been denominated non-specific cytotoxic cells (Galindo-Villegas & Hosokowa, 2004), and are similar to (or closely related in function) the mammalian NK cells (Shoemarker et al., 2001). These cells capable of be important in protozoan parasites (Evans & Gratzek, 1989; Evans & Jaso-Friedman, 1992), and viral immunity of fish (Hogan et al., 1996), and are found in the blood, lymphoid tissue, and gut of fish (Balfry & Higgs, 2001). Lymphocytes are the cells responsible for the specificity of the specific immune response. The two different classes of lymphocytes (T and B) are the acknowledged cellular pillars of adaptive immunity, and can be distinguished by their cell surface markers and subsequent function (Balfry & Higgs, 2001; Garcia-Ayala & Chaves-Pozo, 2009; Pancer & Cooper, 2006). T lymphocytes recognize antigen that is presented by antigen-presenting cells such as macrophages, and are primarily responsible for cell-mediated immunity. These cells are also important sources of cytokines, which are particularly important in the inflammatory response (Balfry & Higgs, 2001). On the other hand, B lymphocytes are responsible for humoral immunity, and recognize antigen and produce specific antibodies to that antigen. T and B cells can be worked together and with other types of cells to mediate effective adaptive immunity (Garcia-Ayala & Chaves-Pozo, 2009; Jimeno, 2008; Miller et al., 1998; Pancer & Cooper, 2006). Interestingly, B cells from rainbow trout have high phagocytic capacity, suggesting a transitional period in B lymphocyte evolution during which a cell type important in innate immunity and phagocytosis evolved into a highly specialized component of the adaptive arm of the immune response in higher vertebrates (Jimeno, 2008; Li et al., 2006).

## Humoral molecules

The classification of humoral parameters is commonly based on their pattern recognition specificities or effector function. Most non-specific humoral molecules involved in the natural resistance of fish are presented with composition and mode of action in Table 2 (Magnadóttir, 2006; Shoemarker et al., 2001). These components are act in several ways to kill and/or prevent the growth and spread of pathogens. Other acts as agglutinins (aggregate cells) or precipitins (aggregate molecules). There are also opsonins that bind with the pathogen and, in doing so, facilities its uptake and removal by phagocytic cells. In addition, some of these substances have important role in the inflammatory immune response, such as opsonins, anaphylatoxins, neutrophil, and macrophage chemo-attractants. Briefly, these factors involve various lytic substances/or hydrolase enzymes (lyzosyme, cathepsine L and B, chitinase, chitobiase, trypsin-like), agglutinins /or precipitins (CRP, serum amyloid P (SAP), lectins, a- and natural precipitins, natural antibodies, natural hemagglutinins), enzyme inhibitors ($\alpha_2$-macroglobulin, serine-/cysteine-/ and metalproteinase inhibitors) and pathogen growth inhibitors (interferon (IFN), myxovirus (Mx)- protein, transferrin, ceruloplasmin, metallothionein). Antimicrobial peptides such as cathelicidins (CATH-1, -2), defesins (DB-1, -2, -3), hepsidins (hepsidinLEAP-1, -2), piscidins (e.g. pleurocidin, epinecidin-1, dicentracin), ribosomal proteins, histone derivates (e.g. parasin, histon H2B, SAMP H1, oncorhyncins, hipposin), which widespread in nature as defence mechanism in plant and animals are also substances that have been identified in the tissue such as mucus, liver, skin and gills of some teleost species, including halibut and flounder (Alvarez-Pellitero, 2008; Aoki et al., 2008; Aranishi & Mano, 2000; Balfry & Higgs, 2001; Buonocore & Scapigliati, 2009; Cole et al., 1997; Ellis, 1999; Ellis, 2001; Galindo-Villegas & Hosokowa, 2004; Hølvold, 2007; Lemaître et al, 1996; Magnadóttir, 2006; Rodriguez-Tovar et al., 2011; Shoemarker et al., 2001; Smith & Fernandes, 2009; Smith et al., 2000; Tort et al., 2003; Whyte, 2007; Yano, 1996).

In addition, in teleost fish, evaluating the complement system as a humoral component is an essential part of the innate immune systems, and can be activated through the two /or three pathways of complement; (1) the classical pathway such as specific immunoglobulin or IgM is triggered by binding of antibody to the cell surface but can also be activated by acute phase proteins such as ligand-bound CRP or directly by viruses, bacteria and virus-infected cells, (2) the alternative pathway such as bacteria cell wall and viral components or surface molecules of parasites is independent of antibody and activated directly by foreign microorganisms, (3) the lectin pathway is elicited by binding of a protein complex consisting mannose-binding lectins to mannans on bacterial cell surfaces. All three pathways converge to the lytic pathway, leading to opsonisation or direct killing of the microorganism (Aoki et al., 2008; Balfry & Higgs, 2001; Ellis, 1999; Ellis, 2001; Galindo-Villegas & Hosokowa, 2004; Holand & Lambris, 2002; Nakao et al., 2003; Randelli et al., 2008; Shoemarker et al., 2001; Tort et al., 2003; Whyte, 2007; Yano, 1996). (modified from Hølvold, 2007; Shoemarker et al., 2001).

**Table 2:** Non-specific humoral molecules and their composition and mode of action in fish.

| Humoral components | Composition | Mode of action |
|---|---|---|
| Antibacterial peptides (*e.g. histone H2B, cecropin P1, pleurocidin, parasin, hipposin, SAMP H1*) | Protein | Constitutive and inducible innate defence mechanism, active against bacteria, defence before development of the specific immune response in the larval fish |
| Antiproteases (*e.g. $\alpha_1$-anti-protease, $\alpha_2$-anti-plasmin, $\alpha_2$-macroglobulin*) | ---- | Restricts the ability of bacteria to invade and growth *in vivo*, active against bacteria |
| Ceruloplasmin | Protein | Copper binding |
| Complement system (*e.g. C3, C4, C5, C7, C8, C9 and their isoforms, B- and D-factors*) | Protein | Promote binding of microbes to phagocytes, promote inflammation at the of complement activation, cause osmotic lysis or apoptotic death |
| Interferons (IFNs) /Myxovirus (Mx)-proteins (*e.g. IFN-$\alpha\beta$, IFN-$\gamma$*) | Glycoprotein /or Protein | Aid in resistance to viral infection, inhibit virus replication, inducible IFN-stimulated genes |
| Lectins (*e.g. legume and cereal lectins, mannose-binding lectin, C-type lectins, intelectin, cod, ladder lectin*) | Glycoprotein and/or specific sugar binding protein | Induce precipitation and agglutination reactions, recognition, promote binding of different carbohydrates in the presence of $Ca^{+2}$ ions, active complement system, opsonin activity and phagocytosis |
| Lytic enzymes (*e.g. lysozyme, chitinase, chitobiase*) | Catalytic proteins lysozyme, complement components | Change the surface charge of microbes to facilitate phagocytosis, haemolytic and antibacterial and/or antivirucidal, antiparasitical effects, opsonic activity, inactivation of bacterial endotoxin(s) |
| Natural antibodies | ---- | Recognition and removal of senescent and apoptotic cells and other self-antigens, control and coordinate the innate and acquired immune response, activity against haptenated proteins |
| Pentaxins (*e.g. C-reactive protein, serum amyloid P*) | Protein | Opsonisation or activation of complement, promote binding of polysaccharide structures in the presence of $Ca^{+2}$ ions, induce cytokine release, coast microbes for phagocytosis by macrophage |
| Proteases (*e.g. cathepsine L and B, trypsin-like*), | ---- | Defence against bacteria, activity against *Vibrio anguillarum* |
| Transferrin/Lactoferrin | Glycoprotein | Iron binding, acts as growth inhibitors of bacteria, activates macrophage |

## Cytokines and chemokines

The initiation, maintenance, and amplification of the immune response are regulated by soluble mediators named cytokines. Cytokines are the soluble messengers of the immune system and have the capacity to regulate many different cells in an autocrine, paracrine, and endocrine fashion, and can also be immune effectors (King et al., 2001). In the last few years, much interest has been generated in the study of fish cytokines and chemokines and significant progress, and has been made in isolating these molecules from fish. In recent years, various cytokines have been described in fish, but the major drawback in identifying fish cytokines is the low sequence identity compared to their mammalian counterparts. The low sequence identities also limit the detection of proteins of fish cytokines by using the antibodies of human cytokines (Plouffe et al., 2006). Most of these have been identified in biological assays on the basis of their functional similarity to mammalian cytokine activities. Some have also been detected through their cross-reactivity with mammalian cytokines (Manning & Nakanishi, 1996).

The predominant pro-inflammatory cytokines are interleukins (ILs) (especially IL-1$\beta$ and IL- 6) and tumour necrosis factor-alfa (TNF-a) (Balfry & Higgs, 2001; Bird et al., 2005; CorripioMiyar et al., 2006; Garcia-Ayala & Chaves-Pozo, 2009; Hølvold, 2007; Jimeno, 2008; King et al., 2001; Magnadóttir, 2010; Randelli et al., 2008; Savan et al., 2005; Tort et al., 2003). These cytokines have a number of systemic effects, including body temperature elevation neutrophil mobilization, and stimulation of acute phase protein production in the liver (Balfry & Higgs, 2001; King et al., 2001; Randelli et al., 2008). Additional several cytokine /or cytokine homologues found in fish include IL-2, IL-4, IL-10, IL-11, IL-12, IL-15, IL-18, IL21, IL22, IL-26 and IFN-$_g$, (Balfry & Higgs, 2001; Bei et al., 2006; Bird et al., 2004; Buonocore & Scapigliati, 2009; Corripio-Miyar et al., 2006; Garcia-Ayala & Chaves-Pozo, 2009; Hølvold, 2007; Igawa et al., 2006; Inoue et al., 2005; Jimeno, 2008; King et al., 2001; Li et al., 2007; Magnadóttir, 2010; Randelli et al., 2008; Tort et al., 2003; Wang et al., 2005; Whyte, 2007; Yoshiura et al., 2003; Zou et al., 2004), and others cytokines in some fish species include transforming growth factor-$\beta$ family such as TGF-$\beta_1$, -$\beta_2$, -$\beta_3$, -$\beta$A, and -$\beta$B, macrophagemigration inhibition factor (MIF), macrophage-colony stimulating (M-CSF or CSF-1; such as CSF-1R or sCSF-1R), chemotactic factor (CF) and plateled activating factor (PAF). However, no antibody markers are at present available for fish TGF-$\beta$, M-CFS and PAF (Belosevic et al., 2006; Garcia-Ayala & Chaves-Pozo, 2009; Klesius et al., 2010; Manning & Nakanishi, 1996; Randelli et al., 2008; Tafalla et al., 2003). On the other hand, orthologous cytokines in teleost fish have been classed as Class I, Class II, chemokines, TNF superfamily and IL-1 family (Table 3) (Alvarez-Pellitero, 2008; Aoki et al., 2008; Lutfalla et al., 2003).

IL-1$\beta$ has been identified in 13 different species of teleost, and is produced by macrophage and also by a variety of other cells such as neutrophilic granulocytes. These ones are play a role in immune regulation through stimulation of T cells which is analogous to mammalian IL-1β. In addition, it is an important mediator of inflammation in response to infection and it has been reported in the trout to directly affect hypothalamic-pituitary-interrenal axis function, stimulating cortisol secretion. Another potentially important cytokines, TNF-a has been cloned in various fish. Besides, TNF-like protein activity has been shown to induce apoptosis, and to enhance neutrophil migration and macrophage respiratory burst activity. The number of studies in fish have provided indirect evidence suggesting that TNF-a is an important macrophage activating factor (MAF) produced by leukocytes. In some fish species, homologous MAF containing supernatants have been shown to induce a typical activated-macrophage response, evidence by increases in phagocytosis and nitric oxide production (Balfry & Higgs, 2001; Garcia-Ayala & Chaves-Pozo, 2009; Holland et al., 2002; Hølvold, 2007; Tort et al., 2003; Whyte, 2007). In addition, TNF-a has been shown increase chemotaxis of rainbow trout anterior kidney leukocytes and induces the expression of a number of genes in the immune response including IL-1β, IL-8 and cyclooxygenase-2 (COX- 2) (Zou et al., 2003). Other vital cytokines, IFNs are secreted proteins, are also pH-resistant cytokines which are produced by many cell types in response to a viral infection (within 2 days in rainbow trout injected viral haemorrhagic septicemia virus), and occurs in very young fish. In isolated Atlantic salmon macrophage stimulated with polyinosinic polycytidylic acid (poly I:C), peak IFN production occurred within 24 h and peak Mx protein production after 48 hours (Ellis, 2001; Nygaard et al., 2000). Therefore, IFN-mediated antiviral defence mechanisms are able to response during the early stages of a viral infection, which is mediated by the innate non-specific IFN responses while long-term protection is mediated by the specific immune response (Galindo-Villegas & Hosokowa, 2004; Ellis, 2001).

*: Only found in fish. (modified from Aoki et al., 2008).

**Table 3:** Cytokines of teleost fish, and their function/or structure and members.

| Class | Function/or Structure | Members |
|---|---|---|
| Cytokine class I | Involved in expansion and differentiation of cells. Have a 4-α helix bundle structure | IL-a and –b, IL-1 1-a and -b*, epo, GCSF-a and -b*, leptin, PRL, GH, M17*, M17 homologue (MSH)* |
| Cytokine class I | Involved in minimizing damage to host after insult. Contain more than 4-α helices. | IFN-α1, IFN-α2, IFN-γ, IL-10, IL-20, IL-24 |
| Chemokines | Regulate cell migration under both inflammatory and homeostasis. Small proteins with 4 conserved Cys residues. | CXC (CXCL8-like, CXC-10, -12, -13, - 14), CC (CCL19/21/25, CCL20, CCL27/28, CCL17/22, MIP, MCP) |
| TNF super family | Involved in inflammation and lymphoid organ development. Compact trimmers as membrane bound or soluble proteins. | Lymphotoxin-β, lymphotixin-β, TNF-α |
| IL-1 family | Involved in pro-inflammatory responses. Fold rich in β-strands. | IL-1α, IL-1β, IL-18 |

Chemokines are known as second-order /or chemotactic cytokines, are a superfamily of small secreted cytokines that direct migration of immune cells to sites of infection, produced by different cell types that have, among other function, chemoattractant properties stimulating the recruitment activation and adhesion of cells to sites of infection injury (Alvarez-Pellitero, 2008; Aoki et al., 2008; Ellis, 2001; Hølvold, 2007). Different chemokines have been characterized in some fish species such as rainbow trout, carp, catfish, flounder and Atlantic halibut, including members of the first two conserved cysteines in their sequence: CXC, CC, C and $CX_3C$ class /or family. Although, the CC chemokines represent the largest subfamily of chemokines, IL-8 was the first known chemokines, and other chemokines such as CXCL8 (or IL-8), $_g$IP-10, CK-1 and CK-2 belongs to the subfamily. Chemokines play a key role in the movement if immune effector cells to sites of infection and it is becoming increasingly clear that their function is also necessary to translate an innate immune response into an acquired adaptive immune (Alvarez-Pellitero, 2008; Aoki et al., 2008; Hølvold, 2007; Peatman & Liu, 2007; Whyte, 2007).

## FISH IMMUNE SYSTEM DESCRIPTION

In this section, since complexity and due this component of the immune system including innate (non-specific) and acquired (specific / or adaptive) immune systems in fish is out of the scopes of this chapter, will not be described in detail, but will be briefly mentioned herein. Hereof, components of these systems and its mode of action were given in detail at Section 2.

The classical division of the immune system is into the innate and the adaptive systems. Despite the fact that dividing immune system into the innate and the acquired immunity is a common practice, recent studies in both fish and mammalian immunology demonstrate that these are combined systems rather than independent systems. Thus, the innate immune response is also important in activating the acquired immune response (Figure 3) (Fearon & Locksley, 1996; Jimeno, 2008; Medzhitov, 2007; Shoemarker et al., 2001).

AIR: Acquired immune response. (Modified from Shoemarker et al., 2001).

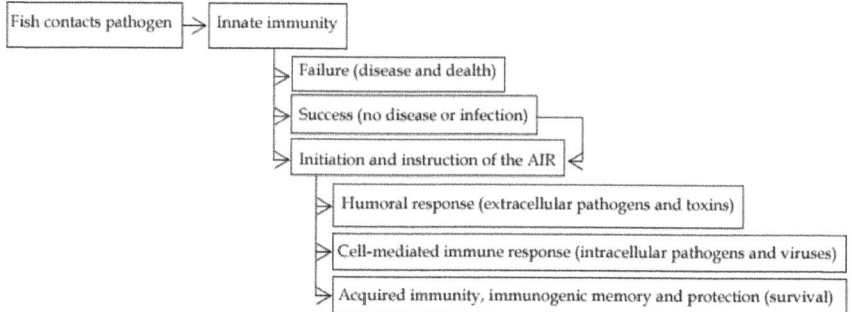

**Figure 3:** Schematic representation of the response of a fish following an encounter with a pathogen

## Innate (non-specific) immune system

The innate immune system is of prime importance in the immune defence of fish. It is commonly divided into 3 compartments: (1) physiochemical barriers and/or the epithelial and/or mucosal barrier such as scales, epithelial surface (on gills, skin and gut) with secreted mucus, (2) the humoral parameters such as cell secretions of complement, CRP, IFN, lysozyme, transferrin, lectins, antimicrobial peptides, and (3) the cellular components such as non-specific cytotoxic cells (or NK cells), monocytes/macrophages, thrombocytes, granulocytes (or neutrophils), lymphocytes (see Section 2) (Buonocore & Scapigliati, 2009; Jansson, 2002; Magnadóttir, 2010; Rodriguez-Tovar et al., 2011). The general term for these innate parameters is pattern recognition proteins or receptors. These parameters recognize pathogen associated molecular patterns (PAMPs) associated with microbes and also inhered danger signals from malignant tissue or apoptotic cells. Typical PAMP are polysaccharides and glycoproteins like bacterial lipopolysaccharide, fragellins, teichoic acid and peptidoglycans, bacterial CpG and virus associated double-stranded RNA (AlvarezPellitero, 2008; Cabezas, 2006; Ellis, 2001; Hølvold, 2007; Jimeno, 2008; Magnadóttir, 2010; Medzhitov & Janeway, 2002; Whyte, 2007). However, under normal conditions the fish maintains a healthy state by defending itself against the potential invaders by a complex system of innate defence mechanisms. These mechanisms are both constitutive and responsive and provide protection by preventing the attachment, invasion or multiplication of microbes on or in the tissue. Immune systems effecting drugs such as immunostimulants, probiotics, prebiotics and synbiotics should act through the enhancement of the innate immune response (Austin & Brunt, 2009; Galindo-Villegas & Hosokowa, 2004; Hoffmann, 2009; Magnadóttir, 2006; Nayak, 2010). The production or expression of both humoral and cellular innate parameters is commonly amplified or up-regulated during immune response, but there is believed to be no memory. This mean that a second encounter with the same pathogen will not result in enhance response as is seen in acquired immune response (Magnadóttir, 2010).

## Acquired (specific) Immune system

If a pathogen evades or overwhelms the innate defence mechanism of the lost, causing the foreign antigen to persist beyond the first several days of infection, an acquired immune system components is initiated. In addition, the antigen-specific lymphocytes of acquired immune response are capable of swift clonal expansion and of a more rapid and effective immune response on subsequent exposures to the pathogen (King et al., 2001). However, activation of the acquired immune system is relatively slow, requiring specific receptor selection, cellular proliferation and protein synthesis but it is long lasting (Magnadóttir, 2010). In contrast to the innate immune systems components, the acquired immune system produces effector cells (T- and B-lymphocytes) and molecules (immunoglobulins (Igs)/or specific antibodies), which are highly specific to the antigen of the invading microbe. The B-cells, similar to the B1-subset of mammalian B-cells, are involved in the humoral response while the T-cells are responsible for the cell-mediated response (GalindoVillegas & Hosokowa, 2004; Jansson, 2002; King et al., 2001; Magnadóttir, 2010). Furthermore, the other key elements in the evolution of the acquired immune system are the appearance of the thymus, the recombination activation gene (RAG; especially RAG 1 and 2 genes) enzymes, which through gene rearrangement generate the great diversity of the Ig superfamily (T- and B-cell receptors) and major histocompatibility complex (MHC). On the other hand, the key humoral parameter of the acquired system is the Igs (antibodies), expressed either as B-lymphocytes receptor or secreted in plasma. The trigger for activation of the acquired immune system, the activation and proliferation of lymphocytes, take place in organized lymphoid tissue. Following activation by a specific antigen, either in soluble form or in association with the MHC marker on antigen presenting cells, the B-cells proliferate and differentiate into long lasting memory cells and plasma cells, which secrete the specific antibody. Also, T-cells, using a specific receptor, recognise pathogen only in association with the MHC marker on antigen presenting cells (Alvarez-Pellitero, 2008; Buonocore & Scapigliati, 2009; Galindo-Villegas & Hosokowa, 2004; Jansson, 2002; King et al., 2001; Magnadóttir, 2010; Rodriguez-Tovar et al., 2011). Effectively only one functional Ig class, a tetrameric IgM, is demonstrated in teleost fish, and these molecule is also made up of eight heavy (mu)-/and light (lambda)- chains. This is in contrast to the pentameric Ig classes and subclasses mammals on the basis of heavy chain molecular weight and on their surface and secrete-antibodies only of the Ig class. Other Ig-like molecules have been described in some fish species, which may increase the diversity of the B-cell recognition capacity (Lorenzen, 1993; Magnadóttir, 2010; Randelli et al., 2008; Shoemarker et al., 2001; Wilson et al., 1997). Resistance to and recovery from first infection are a results of complex interactions between innate and acquired defence mechanism (Lorenzen, 1993). A summary of innate and acquired immune systems in fish is shown in Figure 4 (Jimeno, 2008).

Th1: T-helper 1, Th2: T-helper 2. (Jimeno, 2008).

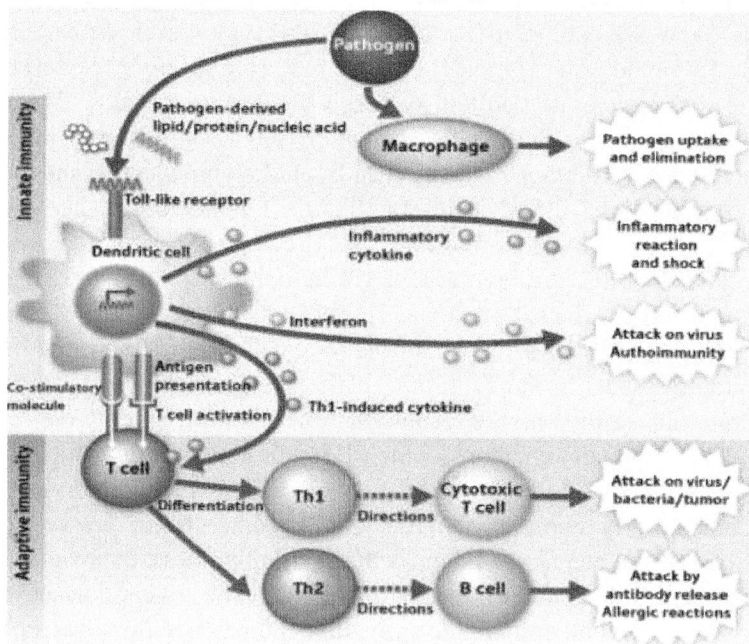

**Figure 4:** Cross-talk between innate and acquired immune systems.

Briefly, the immune reaction in fish is influenced by endogen rhythms and environmental parameters, of which temperature is by far the most important. Another important factor is nutrition, which may be subject to enormous variation within and between wild populations (Lorenzen, 1993). The immunosuppressive effects of population and stress resulting in higher disease susceptibility are well known. Choosing a universal trait or an innate component that could act as a biomarker for adverse conditions in aquaculture is however problematic. This is because of the variable effects on innate an acquired parameters depending on the type and duration of adverse conditions and on the fish species (Magnadóttir, 2006; Ortuño et al., 2001). The innate and acquired immune systems are given activity/or factor, cells involved and cellular markers in Table 4 (Jansson, 2002; Randelli et al., 2008)

APR: Acute phase response, CD: Cell-differentiation cluster; COX-2: Cyclooxygenase 2, CRP: Creactive protein, iNOS,: Inducible nitric oxide synthase, IFN: Interferon, Ig: Immunoglobulin, IL: Interleukin, LtB: Lymphotoxin B, M: Macrophage, MHC: Major histocompatibility complex, NCCRP-1: Non-specific cytotoxic cells receptor protein-1, NBT: Nitroblue tetrazolium, PLA2: Phospholypase A2, RAGs: Recombinase-activating genes, ROS: Reactive oxygen species, SAP: Serum amyloid P, TcR: T-cell receptor, Th1: T-helper 1, Th2: T-helper 2, TLRs: Toll-like receptors, TNF: Tumor necrosis factor, mAb: Monoclonal antibodies, pAb: Polyclonal antibodies. (modified from Randelli et al., 2008).

**Table 4:** The innate and acquired immune systems activity and/or factors and cellular markers.

| Activity/Factor | Cell involved | cDNA sequence coding for | Cellular marker |
|---|---|---|---|
| *Innate immunity* | | | |
| Phagocytosis | Mononuclear phagocytes B-cells | - | mAb to MΦ, and IgM, neutrophils, pAb to granulocytes, granulin |
| ROS species | Mononuclear phagocytes | iNOS | NBT, no antibodies |
| Complement, APR | Hepatocytes | C3, C4, C5, C7, C8, CRP, SAP | pAb to C3 |
| Antibacterial | Various types | Families of peptides | None |
| Antiviral | Leucocytes, fibroblasts | IFN-1, IFN, Mx-protein | None |
| Enzymes | Various types | Lysozyme, caspases, proteases | None |
| Inflammation, cytokines, monokines | Leucocytes | TNF-α, COX-2, PLA2, TLRs, ILs (1, 6, 12, 14, 16, 17, 18, 20, 21, 22), >16 chemokines | pAb for IL-1, pAb and mAb for TNF-α |
| Non-specific killing | Leucocytes | NCCRP-1 | mAb to 5C6 |
| *Acquired immunity* | | | |
| Memory, specific antibody, | B-cells | IgM, IgD, IgT, RAGs | mAb to IgM, B-cells |
| Memory, cellular recognition, | T-cells | TcR-α, -, β -γ, -δ, CD3, RAGs | DLT15, WCL38 |
| Specific killing | T-cells | CD8-α, CD8-β, MHC$_I$ | None |
| Helper activity | T-cells | CD4, MHC$_{II}$ | None |
| | Th1 / or Th2 | IFN-γ, IL-2 / or IL-4, IL-10 | None |
| | Leucocytes | ILs (7, 15, 21, 22, 26), LtB | None |

# IMMUNOASSAY

Diagnostics is the determination of the cause of a disease or clinical pathology. The techniques used range from gross observation to highly technical biomolecular-based tools. Pathogen screening is another health management technique, which focuses on detection of pathogens in sub-clinical, or apparently healthy, hosts. Schematic representation of the diagnosis using a stepwise clinical approach is presented in Figure 5 (King et al., 2001; Subasinghe, 2009).

In recent years, fish immunological research has been mainly focused on two aspects: (1) Firstly, comparative and development studies have contributed to a better understanding of the characterize, the structural and functional evolution of the immune system mechanisms and pathways from invertebrate, through fish to mammals, (2) The second aspect, and one that has received the major funding, is the requirement of the fish farming industries, and also has understated how the fish immune system responds the foreign agents. The word-wide growth in aquaculture in the past 2-3 decades has demanded the development of a comprehensive knowledge of the immune system of the commercially important fish species, and also has understated how the fish immune system responds the foreign agents. The purpose has been twofold: to secure to optimum activity of the natural immune defence of the fish through cultural conditions and the choice of fish stock (or by breeding to produce stock of fish with superior disease resistance), and also to develop and improve prophylactic measure such as vaccination, immunostimulants and probiotics (Alvarez-Pellitero, 2008; Galindo-Villegas & Hosokowa, 2004; Ellis, 2001; Magnadóttir, 2010).

CBC: Complete blood count, Ig: Immunoglobulin, IL: Interleukin. (modified from King et al., 2001).

A variety of technologies have already made an impact in reducing disease risk and many novel methods will contribute in the future (Adams & Thompson, 2006; Adams & Thompson, 2008). Improved nutrition, use of probiotics, improved disease resistance, quality control of water, seed and feed, use of immunostimulants, rapid detection of pathogens and the use of affordable vaccines have all assisted in health control in aquaculture. The success of vaccination in reducing the risk of furunculosis in salmon is an excellent example of technology having made a significant impact. This is turn led to a reduction of the use of antibiotics that has been sustained, and productivity has increased as a result of vaccination (Gudding et al., 1999; Adams et al., 2008).

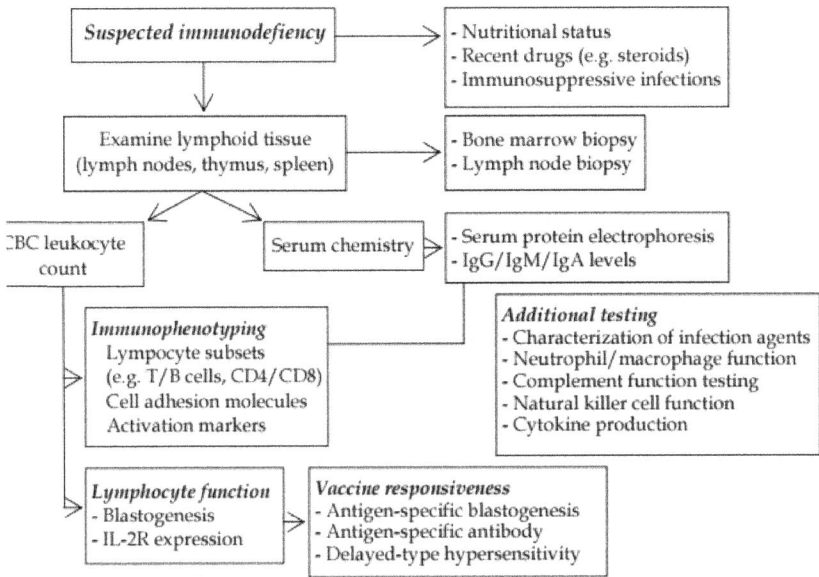

**Figure 5:** Schematic representation clinical evaluation of the immune system.

Many of the assays for detecting the changes in the protective mechanism of the fish due to immunomodulations are divided from those used in fish disease diagnostics and immunization programs. Although, most used tests in the last decades, most used assays for fish immunomodulation diagnosis are given as list in Table 5 (Adams & Thompson, 2008; Anderson, 1996; Brown-Treves, 2000, Jeney & Anderson, 1993; King et al., 2001; Lorenzen, 1993; Roque et al., 2009; Plumb & Hanson, 2011; Subasinghe, 2009). A large number of methods have been developed for immunodiagnostics and these are used routinely in many laboratories for the detection of fish and shellfish pathogens. These tools include both immunoassay and DNA-based diagnostic methods such as enzyme-linked immunosorbent assay (ELISA), radioimmunoassay (RIA), polymerase chain reaction (PCR), quantitative (or real-time)-PCR (QPCR), reverse transcriptase-PCR (RT-PCR), fluorescent antibody assays (FAT), indirect-IFAT, quantitative-FAT (QFAT), immunohistochemistry (IHC), in situ hybridization (ISH) and blot (dot-blot/dip-stick/western blot) amplification techniques (Adams et al., 2008; King et al., 2001; Plumb & Hanson, 2011; Roque et al., 2009; Subasinghe, 2009). However, with the development of Rapid Kits (immunochromatography/lateral flow) which are simple to use, sensitive and inexpensive (Adams & Thompson, 2008). (modified from Anderson, 1996).

**Table 5:** Hematological, innate and acquired immune response assays.

| Hematological/physiological assays-blood samples | Specific immune response assays |
|---|---|
| *Hematocrit*: Percent of red blood cell pack | *Scale rejection*: Transplantation indicator |
| *Leukocrit*: Percent of white blood cell pack | *Delayed hypersensitivity*: Allergenic reactions |
| *Cell counts and differentials*: Numbers of cells and types | *Trypan blue*: Killer cell activity |
| *Lysozyme levels*: Enzyme level in blood | *Chromium release*: Killer cell activity |
| *Serum immunoglobulin level*: Specific and nonspecific antibody | *Melanomacrophage centers*: Antigen processing cells, Antigen accumulation: Concentration in spleen or kidney areas, Cell aggregates: |
| *Serum protein level*: Total protein in serum | Increase in numbers of melanomacrophage cells |
| **Innate defensive mechanism or acquired immune response assays** | *Passive hemolytic plaque assay (Jerne assay)*: Antibody- producing cells |
| (These assays can be used for either response) | *Assays measuring serum antibody levels* |
| *Phagocytosis*: Percents and indexes; engulfment by phagocytic cells/or phagocytic activity: By incubating blood with a killed bacterial culture | **Immunolectrophoresis /or immunoassay and DNA-based diagnosis** |
| *Bactericidal activity*: By incubating macrophages with a live bacterial culture | *CF*: Complement fixation |
| *Rosette-forming cells*: Adherence of particles around lymphocpes | *DBH or WB*: Dot blot hybridization or Western blot |
| *Glass or plastic adherence*: Stickiness of phagocytic cells | *ELISA*: Enzyme-linked immunosorbent assay |
| *Pinocytosis*: Engulfment of fluids by phagocytic cells | *FAT*: Fluorescent antibody assays (or technique) |
| *Neutrophil activation*: Myelo-peroxidase production and NBT dye reduction by oxidative burst e.g. oxidative radicals, Chemilurninescence: light detection from oxidative burst | *IFAT*: Indirect-FAT |
|  | *QFAT*: Quantitative-FAT |
|  | *ISH*: In situ hybridization |
|  | *LAMP*: Loop-mediated isothermal amplification |
|  | *PCR*: Polymerase chain reaction |
|  | *QPCR*: Quantitative (or real-time)-PCR |
| *Blastogenesis*: Mitosis of lymphocytes cells; | *rcb-PCR*: Reserve cross blot-PCR |
| **Agglutination / or Hemagglutination** | *RT-PCR*: Reverse transcriptase-PCR |
| *Preciptinin (Ouchterlony gel)*: Measures soluble antigens in gels | *RIA*: Radioimmunoassay |
| *Immunoelectrophoresis*: For defining blood or antigenic components | *SNT or VNT*: Serum- /or virus- neutralization test |
|  | *Lateral-flow immunoassays* |
|  | Multiplex assays (e.g.*Protein array system, Micro-arrays*) |

These molecular-based techniques (immunoassay and nucleic acid assay) provide quick results, adaptable to field situation, with high sensitivity and specificity, at relativity low cost, and can be easily applied to a large number of samples, and are also particularly valuable for infections which are difficult to detect such as sub-clinical infections using standard histology and tissue-culture procedures such as histopathology, bacteriology, virology, parasitology and mycology. They can be used for non-lethal sampling, and are valuable for monitoring challenge experiments under controlled laboratory conditions. Further development of this technology is likely to enhance more rapid detection and diagnosis of disease, which is crucial for early and effective control emergent disease situations (Adams & Thompson, 2008; Subasinghe, 2009). Although, modern immunoassays are very sensitive, sometimes their result may not be easy to analyse. This is partly because the blood chemistry and/or immune parameters of fish is highly depended on environmental conditions, nutrition, and other factors such as degree of antigen purity, genetic make-up, maternal effects, age and sexual maturation. There are also differences in sensitivities and specificities for each method and in the type of samples that can be used such as formalin fixed, fresh, tissue, blood, water. Further limitations of some immunoassays are that they can be lengthy assay to perform, required cell culture expertise, specific reagent and equipment, and requiring up to 7 to 14 days before they can be evaluated. In addition, non-specific reactions in immunoassays may vary by an order of magnitude between fish caught at the same time and palace, and may eventually obscure specific antibody activity (Table 6) (Adams et al., 2008; Adams & Thompson, 2008; King et al., 2001; Lorenzen, 1993; Magnadóttir, 2010; Vatsos et al., 2003). Any antibody-based test is only as good as the antibody used in it, and a standard protocol and reliable source of standard specific antibody is crucial. Antibody probes can be produced in a number of ways, including polyclonal antibodies (prepared in animal species, and can also be very useful tools for the detection of pathogens), monoclonal antibodies (prepared using hybridoma technology), phage display antibodies or antibody fragments. However, serum contains many different types of antibodies and mixed populations of antibodies can create problems in some immunological techniques (Adams, 2004; Adams & Thompson, 2006; Adams et al., 2008), some of which are now commercial available. Although some antibody-based methods can be very sensitive and carrier status can be detected, such technology can be limited in sensitivity when environmental samples are used, such as water samples, and molecular methods are ideal in this situation. (Adams et al., 2008; Zhang et al., 2004; Zhang et al., 2006) Molecular technologies are also widely used for the detection of fish pathogens (Adams & Thompson, 2006; Cunningham, 2004; Wilson & Carson, 2003). They have been successfully utilized for the detection and identification of low levels of aquatic pathogens. In addition, molecular methods can be used for the identification to pathogens to species level and in epidemiology for the identification of individual strains and differentiating closely related strains. The DNA-based methods such as PCR are extremely sensitive. However, false positive and false negative results can cause problems due to contamination or inhibition. Real-time PCR (closed tube to reduce contamination) and Nucleic Acid Sequence Based Amplification (NASBA) are alternatives that reduce this risk and offer high sample throughput. Some of the most common PCR-based technologies used for the detection of pathogens are nested PCR, random amplification of polymorphic-DNA (RAPD), reverse transcriptase-PCR (RT-PCR), reverse cross blot-PCR (rcb-PCR) and RT-PCR enzyme hybridisation assay. In situ hybridisation is also widely used in the detection of shrimp

ELISA: Enzyme-linked immunosorbent assay, FAT: Fluorescent antibody assays (or technique), IFAT: Indirect FAT, LAMP: Loop-mediated isothermal amplification, PCR: Polymerase chain reaction, RTPCR: Reverse transcriptase-PCR, TEM: Transmission electron microscopy (modified from Adams & Thompson, 2008).

**Table 6:** Used methods, advantages and disadvantages to diagnose fish disease.

| Method | Advantage | Disadvantage |
|---|---|---|
| *Conventional methods* | | |
| Culture | Useful because the pathogen is isolated and the etiological agent can be confirmed | Labour intensive, can be expensive, not always possible to confirm identity of etiological agent |
| Histopathology | Useful for assisting in the diagnosis of disease, particularly where the causative agents of new diseases have not yet been identified | Labour intensive; skilled personnel required, not always possible to identify agent |
| Microscopy | It is an important tool in many of the methods shown in this Table. Many different types of microscopes are now available | Can be labour intensive; skilled personnel required; can be expensive if using confocal microscope or TEM. Not always possible to identify agent |
| Biochemical analysis | Useful for identifying bacteria with characteristic biochemical profiles; commercial kits available for this purpose | Can be labour intensive; skilled personnel required. Not always possible to identify agent |
| *Molecular methods* | | |
| PCR | Very sensitive, can be automated to analyse large sample numbers | Only detects presence of DNA of pathogen, not the whole organism. False positive and negative results can occur |
| Nested-PCR | Extremely sensitive method, more sensitive and specific than one-round PCR | Takes longer than the one-round PCR. False positive and negative results can occur |
| RT-PCR | Can detect live pathogens (e.g. detects RNA) | Care needed to ensure RNA is not degraded |
| Random amplified polymorphic DNA | Useful method for determining the identity of microorganisms at a strain level, assessing the genetic relationship of samples or analysing mixed pathogen populations in samples | Can be labour intensive. Skilled personnel required |
| Reverse cross blot-PCR | Useful for distinguishing closely related species | Expensive. Labour intensive. Skilled personnel required |
| RT-PCR enzyme hybridisation assay | Can detect live pathogens. Large sample numbers can be analysed | Labour intensive. Skilled personnel required |
| *In situ* hybridisation | Detects DNA or RNA of pathogen, therefore there is no need for antibodies to detect protein | Labour intensive. Skilled personnel required. Expensive, sometimes difficult to see pathology in tissue sections after procedure |
| LAMP | Fast, with results obtained in a couple of hours. Suitable for field application. Does not require skilled operator. Results easy to interpret. Sensitive | Complex to set up initially |
| Quantitative-PCR | Allows quantification of DNA that can be related to pathogen level in infected tissue. Extremely sensitive | Labour intensive. Requires specialised equipment. Skilled personnel required. Expensive |
| *Immunological methods* | | |
| Agglutination | Simple method, no requirement for specialised equipment | Not very sensitive in comparison to other immunological methods |
| ELISA-detection of pathogen | Versatile method that can be used to identify pathogens or antibodies depending on how assay is set up. Microassay– therefore small amounts of reagent needed. Quantitative; can be automated to analyse large sample numbers. Sensitive | Standardised reagents and specialised equipment needed. Need careful selection of controls and a skilled operator |

| | | |
|---|---|---|
| Immuno-histochemistry | An extension of histopathology–the pathology can be observed around the infected tissue as the slide is counterstained. Can be amplified to increase sensitivity | Need formalin-fixed, wax embedded tissue sections, therefore procedure is labour intensive. Need a skilled operator to analyse results |
| Western blot | Particularly useful for serology to identify pathogen-specific proteins | Standardised reagents and specialised equipment needed. Need careful selection of controls and a skilled operator |
| Dot blot | Versatile method which can be used to identify pathogens or antibodies depending on how assay is set up. Microassay–therefore only small amounts of reagent needed. Protein not denatured in process unlike Western blotting | Standardised reagents need to be available to perform analysis. Need a skilled operator |
| FAT/IFAT | Fast method if performed directly on infected tissue smears, takes longer if fixed tissue sections are used (e.g need to process infected tissue). Sensitive. Useful for detection of viruses | Need a skilled operator to analyse results, auto-fluorescence on tissue sections can interfere with interpretation of results. Requires specialised equipment |
| Serology-ELISA detection of fish antibodies | Non-destructive sampling method, uses ELISA format therefore can screen large numbers of samples | Indirectly detects the presence of the pathogen. Most suitable for viral infections as antibodies against Gram-negative bacteria may cross-react in assay. In order to perform the assay a specific anti-fish species antibody is required. Needs careful interpretation |
| Rapid kits | Fast (results obtained in minutes), inexpensive, suitable for field application. Easy to interpret results. Sensitive | Designed to be used with fresh tissue. Using frozen or fixed tissue may affect sensitivity of results |
| *Multiplex methods* | | |
| Protein array system (Luminex) | Versatile method that can be used to identify pathogens or antibodies depending on how assay is set up. Can detect proteins or DNA. Microassay–therefore only small amounts of reagent needed. Quantitative. Can measure several pathogens or analytes simultaneously. Sensitive | Labour intensive. Needs a skilled operator. Expensive. Standardised reagents need to be available to perform analysis. Requires specialised equipment |
| Multiplex-PCR assays | Can detect more than one pathogen with the assay. Sensitive | Difficult to standardise. Expensive |
| Micro-arrays | Can detect more than one pathogen with the assay. Allows up and down regulation of genes to be examined. Very sensitive | Needs a skilled operator, very expensive, labour intensive, designated software needed to analyse results. Requires specialised equipment |

Viruses and confirmation of mollusc parasites. Colony hybridisation has also been used successfully for the rapid identification of Vibrio anguillarum in fish (Powell & Loutit, 2004), and has the advantage of detecting both pathogenic and environmental strains (Adams et al., 2008). Serology is an alternative approach to pathogen detection, and can also be applied to the detection of pathogen-specific antibodies in fish. The ELISA is well suited to large scale screening and this can be performed in any species of fish when an anti-fish species antibody is available (Adams et al., 2008). A number of new technologies are being developed for the rapid detection of pathogens and monitoring host responses. These include immunochromatography, such as lateral flow technology, and multiplex testing using the Bio-Plex Protein Array System or microarray technologies (Adams and Thompson, 2006). Lateral Flow is simple methodology enabling accurate (high sensitivity, specificity), simple, easy to use (2 steps, no instrument required) testing that is also economic (time/labor saving). The Protein Array System (Luminex) theoretically offers simultaneous quantitative analysis of up to 100 different biomolecules from a single drop of sample in an integrated, 96-well formatted system, mainly focusing on the detection of cytokines. Therefore, it can be used in molecular and immunodiagnostics to detect pathogens directly from tissue samples or culture, or it can be used in serology to measure fish antibodies (Adams et al., 2008; Adams & Thompson, 2008; Dupont, 2005; Giavedoni, 2005).

## IMMUNOSUPPRESSION

Aquatic environment of fish is in close contact with numerous pollutants. Aquatic pollutants such as heavy metals, aromatic hydrocarbons, pesticides and mycotoxins modulate the immune system of fish, thus increasing the host susceptibility to infectious pathogens. Pollutants in the water which may be particulate or soluble can also be natural source such as metals showing the seasonal increase in lakes as well as drugs used in the prevention or treatment of disease such as cortico-steroid hormones, used drugs in terrestrial animal health in aquaculture such as florfenicol, oxolinic acid, and oxytetracycline (Table 7). Immunosuppressive effects of these compounds may occur at high concentrations or long-term exposures (Anderson, 1996; Bols et al., 2001; Brown-Treves, 2000; Duffy et al., 2002; El-Gohary et al., 2005; Enis-Yonar et al., 2011; Kusher & Crim, 1991; Lumlertdacha & Lovell, 1995; Lundén & Bylund, 2002; Lundén et al., 1998; Lundén et al., 1999; Manning, 2001; Manning, 2010)

DDT: Dichloro-diphenyl-trichloroethane, PAHs: Polynuclear aromatic hydrocarbons, PCBs: Polychlorinated biphenyls, TCDD: 2,3,7,8-tetrachlorodibenzo-p-dioxin. IHNV: Infectious hematopoietic necrosis virus. (modified from Anderson, 1996).

**Table 7:** Nonspecific defense mechanisms and specific immune response assays /or parameters in fish effected by presence of some immunosuppressive compounds.

| Substances | Parameters | Fish species |
|---|---|---|
| *Metals and organometallies* | | |
| Aluminum | Reduced chemiluminescence | Rainbow trout |
| Arsenic | Phagocytosis elevated or lowered | Rainbow trout |
| Cadmium | Elevated serum antibody | Rainbow trout |
| | Chemiluminescence reduced | Rainbow trout |
| | Lymphocyte number and mitogenic response reduced | Goldfish |
| | Antibody-binding lymphocyte reduced | Bluegill |
| Chromium | Serum antibody reduced | Brown trout, carp |
| Copper | Chemiluminescence reduced | Rainbow trout |
| | Susceptibility to IHNV increased | Rainbow trout |
| | Leukocyte respiratory burst activity inhibed | Rainbow trout |
| | Serum antibody reduced | Brown trout |
| | Antibody-producing cells reduced | Rainbow trout |
| | Susceptibility to *Vibrio anguillarum* increased | Eel |
| Lead | Serum antibody reduced | Brown trout |
| Mercury | Lymphocyte numbers reduced | Barb |
| Nikel | Serum antibody reduced | Brown trout |
| Zinc | Serum antibody reduced | Brown trout |
| | Phagocytosis decreased | Rainbow trout |
| *Aromatic hydrocarbones* | | |
| Benzidine | Non-specific agglutination rise | Estuarine fish |
| PAHs | Macrophage activity reduced | Spot, Hogchoker |
| | Melanomacrophage numbers reduced | Flounder |
| PCBs | | |
| Benzo[a]pyrene | Phagocytic capacity reduced | Rainbow trout |
| PCB 126 | Antibody-producing cells reduced | Medaka |
| | Non-specific cytotoxic cell activity reduced | Catfish |
| Aroclor 1254 | Antibody-producing cells reduced | Coho salmon |
| Aroclor 1232 | Susceptibility to disease increased | Channel catfish |
| Aroclor 254/1260 | Susceptibility to disease increased | Rainbow trout |
| Phenols | Antibody- producing cells reduced | Rainbow trout |
| Hydroquinone | Non-specific cytotoxic cell activity reduced | Carp |
| TCDD | Mitogenic response partially suppressed Susceptibility to IHNV | Rainbow trout |
| *Pesticides* | | |
| Bayluscide | Serum African antibody reduced | Catfish |
| Dichlorvos | Lysozyme activity reduced | Carp |
| DDT | Antibody-producing cell, serum antibody reduced | Goldfish |
| Endrin | Phagocytic, antibody-producing cell activities reduced | Rainbow trout |
| Malathion | Lymphocyte number reduced | Channel catfish |
| Metrifonate | Phagocytic, neutrophilic and lysozyme activity reduced , antibody-producing cell reduced, | Cichlid fish |
| Methyl bromide | Thymic necrosis | Medaka |
| Tributyltin | Chemiluminescence reduced | Oyster, Hogchoker |
| Trichlorophon | Phagocytic, neutrophilic, lysozyme activity reduced | Carp |
| *Mycotoxins* | | |
| Aflatoxin-B₁ | B-cell memory loss, neutrophilic activity reduced | Rainbow trout |
| Fumonisin-B₁ | Antibody-producing cells reduced | Catfish |
| *Antibiotics* | | |
| Florfenicol | Chemiluminescence reduced Phagocytic cells counts reduced after 5-6 weeks | Rainbow trout |
| Oxolinic acid | Antibody-producing cells reduced | Rainbow trout |
| Oxytetracycline | Mitogenic response reduced, | Carp |
| | Antibody-producing cells reduced, phagocytic activity reduced | Rainbow trout |
| *Other compounds* | | |
| Cortisol/Kenalog-40 | Antibody-producing cells reduced | Rainbow trout |
| Hydrocortisone | Phagocytic activity reduced | Striped bass |

# IMMUNOMODULATION

Immunomodulators present in the diets stimulate the innate immune systems, while antigenic substance such as bacterins and vaccines initiate the more prolonged process of antibody production and acquired immune systems. Prophylactic and therapeutics administration of immunomodulators will need to be adapted to each cultured fish species in anticipation of recognize pathogens, under known environmental conditions (Gannam & Schrock, 2001). Prophylactic and therapeutic compounds and/or drugs against infections are rarely successful or limited effects; currently there are no approved some drugs for the control and treatment fish disease in the aquaculture industry. For example, several substances, such as fumagilin and albendazole have been used in fish with potential value in controlling microsporidian infections. However, other drugs, like sulphaquinoxaline, amprolium and metronidazole have been ineffective to control the disease (Berker & Speare, 2007; Dykova, 2006; Rodriguez-Tovar et al., 2011). Most of similar drugs have ambiguous result and it is has been reported that high concentrations and prolonged treatment of infections with some drugs might cause side-effects. More promising results have been achieved by using immune-prophylactic control components such as probiotics (e.g. basillus P64, yeasts and lactic acid bacteria), prebiotics (e.g. fructo- galacto-, transgalactooligosaccharide), vaccination (e.g. Vibrio spp., Yersinia ruckerii) and immunostimulants (e.g. β-glucan, chitosan and levamisole) (Austin & Brunt, 2009; Hoffmann, 2009; Magnadóttir, 2010; Nayak, 2010; Rodriguez-Tovar et al., 2011). On the other hand, in recent years, organically produced aquatic products are increasingly available to consumers and, in particular, sea bass and sea bream from certificated fish farms (Perdikaris & Paschos, 2010). The initial legislative framework for organic aquaculture in the European Union (EU) was the Directives (EC) 834/07 and (EC) 889/08 (EU, 2007; 2008).

# IMMUNOSTIMULANTS

Various chemotherapeutic compounds have been extensively used to treat bacterial infections in cultured for about the last 20-30 years. However, the incidence of drug-resistant bacteria has become a major problem in fish culture (see Chapter 11: Section 5.2). Although, vaccination is a useful prophylaxis for infectious disease, and is also already commercially available for bacterial infections such as vibriosis, redmouth disease and for viral infections such as infectious pancreatic necrosis, the development of vaccines against intracellular pathogens such as Renibacterium salmoninarum has not so for been unmitigated successful. Therefore, the immediate control of all fish disease using only vaccines is impossible. Even thought, use of immunostimulants, in addition to chemotherapeutic drugs and vaccines, has been widely accepted by the aquaculture industry, many question about the efficacy of immunostimulants from users still continue such as whether this components can protect against infections disease (Table 8). Also, the biological activities of the immunostimulants may be so multiple and potent that some of them may be more harmful than beneficial (Dalmo, 2002; Sakai, 1999).

By definition, an immunostimulant is a naturally occurring compound that molecules that modulates the immune system by increase the host's resistance against disease that in most circumstances are caused by pathogens (Bricknell & Dalmo, 2005). However, synthetic chemicals such as isoprinosine, bestatin, levamisole, muramyl dipeptide and FK-565 wellknown as lactoyl tetrapeptide are known to possess immunostimulatory properties. It is important to note the use of the term "modulate", as a substance with the potential immunostimulatory properties may lead to a down regulation of the immune response if administered in excess amounts or long-term usage. Hence, administration of an immunestimulant prior to an infection may elevate the defence barriers of the animal and thus provide protection against an otherwise severe or lethal infection. Also, immunostimulants enhance individual components of innate immune response, but this does not always translate into increased survival. An important point to have in mind is that not by enhancing acquired immune response. Therefore, there is no memory component and the response is likely to be of short duration (Gannam & Schrock, 2001; Hølvold, 2007; Maqsood et al., 2011; Raa, 2000; Sakai, 1999).

**Table 8**: A comparison of characteristics of chemotherapeutics, vaccines and immuno-stimulants (Sakai, 1999).

|  | Chemotherapeutics | Vaccines | Immunostimulants |
|---|---|---|---|
| When | Therapeutically | Prophylactically | Prophylactically |
| Efficacy | Excellent | Excellent | Good |
| Spectrum of activity | Middle | Limited | Wide |
| Duration | Short | Long | Short |

A division of immunostimulants depended on which effects they include such as antibacterial, -viral, –fungal and –parasitic effects may be helpful but hard to accomplish. Some immunostimulants may induce both antibacterial and antiparasitic effects, whereas other may help the organism to fight virus and fungus. Generally, immunostimulants used in fish and shrimp in many countries can be divided into two main groups as biological substances and synthetic chemicals depending on their sources (Table 9) (Anas et al., 2005; Brown-Treves, 2000; Dügenci et al., 2003; Galindo-Villegas & Hosokowa, 2004; Gannam & Schrock, 2001; Gildberg et al., 1996; Glina et al., 2009;

Jiye et al., 2009; Lauridsen & Buchmann, 2010; Maqsood et al., 2011; Noga, 2010; Paulsen et al., 2003; Perera & Pathiratne, 2008; Petersen et al., 2004; Raa, 2000; Sakai, 1999). But, some immunostimulants may be included in different subgroups by some researchers, such as schizophyllan and scleroglucan. These substances may be included in bacterial derivatives-subgroups as various β-glucan products from Schizopyllum commune and S. glucanicum, respectively, or may be included in polysaccharides-subgroups as polysaccharides containing sugars.

## Dose, timing, administration-route and -period of immunostimulants

The effect of timing the administration on immunostimulant function is a very important issue. Usually, the most effective timing of antibiotics is upon the occurrence of disease, and they cannot often be used prophylactically due to risk of fostering the development of drugresistant bacteria. Researchers proposed that immunostimulants may improve health and performance of fish and shrimp in aquaculture, if used prior to: (1) before the outbreak of disease to reduce disease-related losses, (2) situations known to result in stress and impaired general performance of animals such as handling, change of temperature and environment, weaning of larvae to artificial feeds, (3) expected increased exposure to pathogenic microorganisms and parasites such as spring and autumn blooms in the marine environment, high stocking density, (4) developmental phases when animals are particularly susceptible to infectious agents such as the larvae phase of shrimp and marine fish, smoltification in salmon, sexual maturation (Raa, 2000; Sakai, 1999). In addition, the effects of immunostimulants may also be different dependent on the administration route, the dose used, the duration of the treatment and growth period. Immunostimulants does not show a linear dose/effect relationship; instead they most often show a distinct maximum at a certain intermediate concentration and even a complete absence of effect or toxicity, at high concentration. The explanations for these phenomena are still speculative and include competition for receptors (analogous to substrate inhibition of enzyme), over stimulation resulting in exhaustion and fatigue of the immune system (Bright-Singh & Philip, 2002). (modified from Galindo-Villegas & Hosokowa, 2004; Sakai, 1999).

**Table 9:** Groups, substances and examples of immunostimulants evaluated in many countries that have been tried to increase disease protection in fish species and/or shrimps.

| Groups | Substances | Compounds |
|---|---|---|
| Biological substances | Animal compounds | EF-203 (Chicken egg), Ete (Tunicate, *Ecteinascidia turbinata*), Hde (Abalone, *Haliotis discus hannai*), cod milt, firefly squid and acid-peptide fractions (fish protein hydrolysate) |
| | Plant extracts | Glycyrrhizin (Licorice, saponin in *Glycyrrhiza glabra*), quil-A saponin, ergosan (*Laminaria digitata*), C-UP III (a Chinese herb mix), laminaran (Seaweed), spirulina (*Spirulina plantensis*) *Quillaja saponica* (Soap tree), leaf extract (*Ocimum sanctum*), scutellaria extract (*Scutellaria baicalensis*), astragalus extract (*Astragalus membranaceus*), ganoderma extract (*Ganoderma lucidum*), lonicera extract (*Lonicera japonica*), phyllanthus extract (*Phyllanthus emblica*), azadirachta extract (*Azadirachta indica*), solanum extract (*Solanum trilobatum*), mistletoe (*Viscum album*), nettle (*Urtica dioica*), ginger (*Zingiber officinale*) and chevimmun (*Echinacea anguistifolia-Baptista tinctoria-Eupatorium perfoliatum*) |
| | Bacterial and yeast derivatives | β-glucans (from bacteria and mycelial fungi; MacroGard, VitaStim, SSG, Eco-Activa, Betafectin, Vetregard, Dinamune, Aquatim, AquaStim, Curdlan, Krestin), ascogen (Aquagen), peptidoglycan (*Brevibacterium lactofermentum*; *Vibrio sp.*), pDNA (*Escherichia coli*), lipopolysaccharide, fragellins (recombinant-*Borrelia*), *Vibrio anguillarum* cells, *Clostridium butyricum* cells, *Achromobacter stenohalis* cells and streptococcal components (*Bordetella pertuosis*, *Brucella abortus*, *Bacillus subtilis*, *Klebsiella pneumonia*) |
| | Cytokines | Interferon, interleukin-2, tumor necrosis factor |
| | Hormones | Growth hormone, prolactin, melanin stimulating hormone, β-endorphin and melanin concentrating hormone |
| | Nutritional factors | Vitamin-A, -C, -E, carbohydrate (Acemannan), soybean protein, trace elements (zinc, iron, copper, selenium) and nucleotides |
| | Polysaccharides | Chitin, chitosan, lentinan, schizophyllan, sclerotium, scleroglucan, protein-bound polysaccharide (PS-K), oligosaccharide and polyglucose |
| | Others | Lactoferrin |
| Synthetic chemicals | | Avridine, bestatin, DW-2929, ISK, KLP-602, FK-156 (lactoyl tetrapeptide), FK-565, fluro-quindone, Freund's complete adjuvants, imiquimod, isoprinosine, levamisole, muramyl dipeptide and sodium alginate |

It is reported that oral administration of an immunostimulant such as lipopolysaccharide is increased larval growth. This may be important in the intensive production of fish larvae and juveniles. In spite of advantages and limitations, the basic methodologies adopted are injection, immersion and oral (Table 10). Injection and immersion methods are suitable only for intensive aquaculture and both require the fish to be handled or at least confined in a small space during the procedures (Dalmo, 2002; Guttvik et al. 2002; Raa, 2000). By injections of immunostimulants enhances the function of leucocytes and protection against pathogens. However, this method is labour intensive, relatively timeconsuming and becomes impractical when fish weight less than 15 gram. By immersion, efficacies had been confirmed by several researchers (Anderson et al., 1995; Baba et al., 1993; Jeney & Anderson, 1993; Perera &

Pathiratne, 2008), although, since dilution, exposure time and levels efficacy are not well defined, caution must be taken in account by applying this methods. Oral administration is only method economically suited to extensive aquaculture, is non-stressful and allows mass administration regardless of fish size, but of course may be administration only in artificial diet (Galindo-Villegas & Hosokowa, 2004; Noga, 2010). (Galindo-Villegas & Hosokowa, 2004).

**Table 10:** Administration methods, advantages and limitations of immunostimulants in aquaculture.

| Route | Dose | Exposure time | Advantages | Limitations |
|-------|------|---------------|------------|-------------|
| Injection | Variable | 1 or 2 doses | Allows use of adjuvants, Most potent immunization route, most cost effective method for large fish | Only for intensive aquaculture, fish must be >10~15 g, stressful (anesthesia, handling), labour hard |
| Immersion | 2-10 mg/L | 10 min to hours | Allows mass immuno-stimulation of small (<5 g) fish, most cost effective method for small fish, | Only for intensive aquaculture, dip rise handling stress, potency not as high as injection route |
| Oral | 0.01–4% | Some days or longer | Only not-stressful method, Allows mass immuno-stimulation of fish any size, no extra labour cost | Poor potency, requires large amounts of immunostimulation to achieve protection, suitable only for fish fed artificial diet |

The effects of immunostimulants were dose and/or application time, route, and period related. For example, low-dose glucan content being beneficial whereas high-dose glucan content had limited effects (Ai et al., 2007). Peptidoglycan is not influence the high-dose (0.1%) in shrimp diets, and not effect after 60 days of oral administration in rainbow trout growth. On the other hand, Ete exerted a protective effect in eels injected intra-peritoneal 2 days after challenge with A. hydrophila. However, the protection was not seen when Ete was administered intra-peritoneal 2 days before or concurrently with the bacteria. The adjuvant effects of glucan against A. salmonicida vaccine oral delivery (7 days administration) and immersion (15 min). No adjuvant effects were seen with the immersion treatment, although the fish administered glucan orally showed enhanced vaccine effects (Sakai, 1999). The number of NBT-positive cell in catfish increased following oral administration of glucan and oligosaccharide over 30 days, but not over 45 days (Yoshida et al., 1995). The effects of longterm oral administration of immunostimulants are still unclear. However, the dilution, the effective administration period and the levels of efficacy require more complete investigation for each immunostimulants.

# In vivo and in vitro effects of immunostimulants

The benefit of immunostimulants is considerable. They have the potential to elevate the innate defence mechanisms of fish prior to exposure to a pathogen, or improve survival following exposure to a specific pathogen when treated with an immunostimulant. There are two main procedures for evaluating the efficacy of an immunostimulant; (1) in vivo such as protection test against fish pathogen, (2) in vitro such as the measurements of the efficiency of cellular and humoral immune mechanism (Bricknell & Dalmo, 2005; Maqsood et al., 2011). In vivo evaluation should be based at least on the following parameters: phagocytosis, antibody production, free radical production, lysozyme activity, natural cytotoxic activity, complement activity, mitogen activity, macrophage activating factor (MAF), nitroblue tetrazolium reaction (NBT), etc. The evaluation of an immunostimulant by the in vitro methods which test the effects of that substance on the immune system is to be preferred in preliminary studies (Table 11 and Figure 6) (Aly & Mohamed, 2010; Anas et al., 2005; Barman et al., 2011; Brown-Treves, 2000; Dügenci et al., 2003; Galina et al., 2009; Galindo-Villegas & Hosokowa, 2004, Gannam & Schrock, 2001; Magnadóttir, 2010; Maqsood et al., 2011; Noga, 2010; Paulsen et al., 2003; Peddie et al., 2002; Raa, 2000; Sakai, 1999; Yin et al., 2009). Nevertheless, if possible in vitro test should be performed together wit in vivo experiments in order to elucidate the basic mechanisms responsible for the protection (Bricknell & Dalmo, 2005; Brown, 2006; Brown-Treves, 2000; Dalmo, 2002; Sakai, 1999). Used aquaculture potential immunostimulants with in vivo and/or in vitro effects and administration route are given at Table 12, as well as with doses at Table 13.

MAF: Macrophage activating factor, NBT: Nitroblue tetrazolium reaction. (modified from Bricknell & Dalmo, 2005).

**Table 11:** The in vivo and in vitro responses seen in fish treated with immunostimulants.

| *In vivo* effects | *In vitro* effects |
|---|---|
| Increased survival after challenges with bacteria, antiparasitic effects including reduced settlement of sea lice, improved resistance to viral infection and increased interferon levels | Increased macrophage activity including: - Phagocytosis, free radical production, enzyme activity, migration activity, production of cytokines, nitric oxide production, bacterial killing, antibody production, respiratory burst, MAF, NBT |
| Growth enhancement | Increased cytotoxicity |
| Increased antibody production following vaccination | Increased lysozyme activity |
| Increased lysozyme levels | Increased cytokine induction Increased oxygen radical induction Increased cell proliferation |

Villegas et al., 2006; Gildberg et al., 1996; Ispir & Dorucu, 2005; Kunttu et al., 2009; Lauridsen & Buchmann, 2010; Ortuño et al., 2002; Peddie et al., 2002; Perera & Pathiratne, 2008; Sakai et al., 1995; Seker et al., 2011; Soltani et al., 2010; Yin et al., 2009; Zhao et al., 2010).

**Table 12:** Potential immunostimulants, in vivo and/or in vitro effects of this immuno-stimulants with administration route.

| Immunostimulant | Species | Route | *In vivo* or *in vitro* effects | Resistance |
|---|---|---|---|---|
| **Synthetic chemicals** | | | | |
| FK-565 | Trout | ip | phagocytosis ↑ | *A. salmonicida* ↑ |
| | | *in vitro* | antibody ↑ | |
| Freund's adjuvants | Trout | ip | - | *V. anguillarum* ↑, *Y. ruckeri* ↑ |
| | Yellowtail | ip | - | *P. piscicida* → |
| Levamisole | Carp | ip/oral | phagocytosis ↑ / NBT → | - |
| | | im | phagocytosis ↑, CL ↑ | *A. hydrophila* ↑ |
| | Trout | *in vitro* | phagocytosis ↑, NBT → | - |
| | | ip | phagocytosis ↑, CL ↑, complement ↑ | *V. anguillarum* ↑ |
| | | im | - | *A. salmonicida* ↑ |
| Muramyl dipeptide | Trout | ip | phagocytosis ↑, CL ↑ | *V. anguillarum* ↑ |
| **Bacterial and yeast derivatives** | | | | |
| *Achromobacter stenohalis* cells | Char | ip | CL ↑, complement ↑ | *A. salmonicida* ↑ |
| *Clostridium butyricum* cells | Trout | oral | phagocytosis ↑, NBT ↑ | *V. anguillarum* ↑ |
| Glucan | Trout | im / ip | phagocytosis ↑ / NBT ↑ | - |
| Lipopolysaccharide | Plaice | ip | macrophage migration ↑ | - |
| | Red sea bream | | phagocytosis ↑ | |
| | Goldfish | | macrophage activating factor ↑ | |
| | Salmon | *in vitro* | phagocytosis ↑, NBT ↑ | |
| | Catfish | | IL-1 ↑ | |
| Peptidoglucan | Trout | oral | - | *V. anguillarum* ↑ |
| | Shrimp | | - | YHB ↑ |
| | Yellow tail | | phagocytosis ↑ | *E. seriolicida* ↑ |
| | J. flounder | | lysozyme ↑, phagocytosis ↑ | *E. tarta* ↑ |
| Vibrio bacteria | Trout | im | - | *A. salmonicida* ↑, *E. seriolicida* ↑ |
| | Prawn/shrimp | ip,im,oral | - | *Vibrio sp.* ↑ |
| VitaStim | Coho | ip, oral | - | *A. salmonicida* ↑ |
| | Chinook | oral / im | - | *A. salmonicida* ↑/→ |
| | Catfish | oral | antibody ↑ | *E. ictaluri* → |
| Yeast glucan | Salmon | oral | - | *V. anguillarum* ↑, *A. salmonicida* ↑ |
| | | ip | complement ↑, lysozyme ↑, antibody ↑, NBT ↑, killing → | *A. salmonicida* →, *Y. ruckeri* ↑ |
| | Catfish | ip | phagocytosis ↑, NBT ↑, antibody ↑, killing ↑ | *E. ictaluri* ↑ |
| | | oral | CL ↑, migration ↑ | *E. ictaluri* → |
| | Shrimp | im,*in vitro* | phenoloxidase ↑, lysozyme →, NBT ↑, CL ↑ | - |
| | Trout | ip | lysozyme ↑, killing ↑, $O_2^-$ ↑, NBT → | *V. anguillarum* → |
| | Turbot | oral | lysozyme ↑, complement →, CL ↑ | *V. anguillarum* ↑ |
| Yeast glucan + Vit-C | Trout | oral | lysozyme →, complement ↑, CL ↑ | - |

| Animal and plant extracts | | | | |
|---|---|---|---|---|
| Acid-peptide fraction | Salmon | *in vitro* | NBT ↑ | - |
| EF-203 | Trout | oral | phagocytosis ↑, NBT ↑, antibody → | *R. salmoninarum* ↑ |
| | | | phagocytosis ↑, CL ↑ | *Streptecoccus sp.* ↑ |
| Ete (Tunicate) | Eel | ip | phagocytosis ↑ | *A. hydrophila* ↑ |
| | Catfish | ip | phagocytosis ↑, antibody → | *E. ictaluri* ↓ |
| Firefly squid | Trout | ip | NBT mitogen Con A, LPS killing | - |
| Glycyrrhizin | Yellowtail | oral | complement ↑ | *E. seriolicida* ↑ |
| | Trout | *in vitro* | mitogen (Con-A, LPS) ↑ macrophage activating factor ↑, O2⁻ ↑, | - |
| Hde (Abalone) | Trout | ip | phagocytosis ↑, CL↑, NK ↑ | *V. anguillarum* ↑ |
| Quil-A saponin | Yellowtail | oral | leucocyte migration ↑ | - |
| | Trout | im | serum bactericidal activity ↑ | - |
| Polysaccharides | | | | |
| Chitin | Trout | ip | phagocytosis ↑, lysozyme → | *V. anguillarum* ↑ |
| Chitosan | Trout | oral/ip,im | phagocytosis ↑, NBT ↑/NBT ↑, killing ↑ | *A. salmonicida* ↑/ - |
| Lentinan + Schizophyllan | Carp | ip | phagocytosis ↑ | *E. tarda* ↑ |
| Oligosaccharide | Catfish | oral | NBT ↑ | - |
| Polyglucose | Salmon | *in vitro* | NBT ↑, pinocytosis ↑, acid phosphatase ↑ | - |
| PS-K | Tilapia | oral | phagocytosis ↑ | *E. tarta* ↑ |
| Schizophyllan | Prawn | oral | phagocytosis ↑ | *Vibrio sp.* ↑ |
| Schizophyllan + Scleroglucan | Yellow tail | ip | complement ↑ | *E. seriolicida* ↑ |
| | | | phagocytic index ↑, lysozyme ↑ | *P. piscicida* → |
| Scleroglucan | Carp | ip | - | *A. hydrophila* ↑ |
| Hormones, Cytokines and Others | | | | |
| Growth hormone | Trout | ip | phagocytosis ↑, mitogen ↑, CL↑, NK ↑ | *V. anguillarum* ↑ |
| Interferon | Flatfish | *in vitro* | - | HRV ↑ |
| Lactoferrin | Trout | oral | phagocytosis ↑ | *V. anguillarum* ↑ |
| | | | CL ↑ | *Streptococcus sp.* → |
| | | *in vitro* | CL ↑, NBT ↑ | - |
| | Red sea bream | oral | lectin ↑, lysozyme → | *Cryptocaryon* |
| Prolactin | Trout | *in vitro* | NBT ↑ | - |

Increase, ⁻: Decrease, ®: No change, Brook: Brook trout, Chinook: Chinook salmon, CL: Chemiluminescent response, Coho: Coho salmon, Con-A: Concanavalin-A, IL-1: Interleukin-1 production, im: immersion, ip: Intraperitoneal injection, J. flounder: Japanese flounder, Killing: Bactericidal activity of macrophage, LPS: Lipopolysaccharide, NBT: Nitroblue tetrazolium reaction, NK: Natural killer cell, O2⁻: Production of superoxide anion, Prawn: Kuruma prawn, PS-K: Protein-bound polysaccharide, Salmon: Atlantic salmon, Shrimp: Black tiger shrimp, Trout: Rainbow trout, Turbot: Scophthalmus maximus, Vit-C: Vitamin-C, A. hydrophila: Aeromonas hydrophila, A. salmonicida: Aeromonas salmonicida, E. ictaluri: Edwardsiella ictaluri, E. seriolicida: Enterococcus seriolicida, E. tarda: Edwardsiella tarda, HRV: Hirame rhabdo virus, P. piscicida: Pasteurella piscicida, R. salmoninarum: Renibacterium salmoninarum, V. anguillarum: Vibrio anguillarum, Y. ruckeri:

Yersinia ruckeri, YHB: Yellow-head baculo virus. (modified from Gannam & Schrock, 2001; Sakai, 1999).

Recognition of β-glucans on fungal particles induces several dectin-1-mediated cellular responses, which might contribute to anti-fungal immunity in vivo. These include fungal uptake and killing and the production of pro-inflammatory cytokines and chemokines, such as tumour-necrosis factor (TNF) and CXC-chemokine ligand 2, in collaboration with the Toll-like receptors (TLRs), which is likely to lead to cellular recruitment and activation. Dectin-1-mediated recognition also stimulates the production of interleukin-12 (IL-12), which might result in a protective T-helper 1 (TH1)-cell response and the production of interferon-g (IFN-g), thereby activating the fungicidal activities of phagocytes. In dendritic cells, β-glucan recognition by dectin-1 can also induce the production of IL-10 and IL-2), which could potentially contribute to the development of regulatory T cells, thereby limiting inflammatory pathology and promoting fungal persistence and long-term immunity, as proposed previously. IL-10 would also inhibit the production of pro-inflammatory cytokines and chemokines. Fungi might also mask their β-glucan, by conversion from yeast to hyphal forms. This could result in the induction of non-protective TH2-cell immune responses, mediated by IL-4; this could be the result of preventing recognition by dectin-1 although the pathways leading to this response are unknown. Although dectin-1 is described here as having a central role in the generation of protective immune responses, it should be noted that many other opsonic and non-opsonic receptors such as the mannose receptor, complement receptor 3, dendritic-cell-specific ICAM3-grabbing non-integrin and TLRs also contribute to this process. (modified from Brown, 2006).

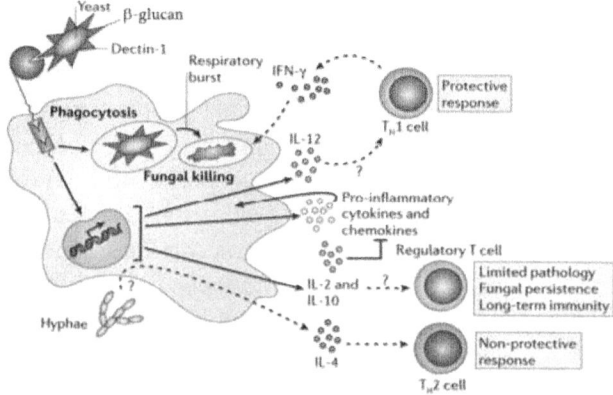

**Figure 6:** Recognition of β-glucans on fungal particles induces several dectin-1-mediated cellular responses, which might contribute to anti-fungal immunity in vivo.

**Table 13**: Doses and effects of immunostimulants as nutritional factors and nucleotides in some fish species

| Immunostimulant | Species | Dose - Route | *In vivo* or *in vitro* effects |
|---|---|---|---|
| β-glucan | Atlantic salmon | 15 mg/kg, inj | ROS ↑, lysozyme ↑ |
| | | 150 mg/kg, oral | acid phosphatase↑ |
| | | 1 ml/fish, inj | lysozyme ↑, complement ↑, antibody ↑ |
| | Coho salmon | 5 and 15 mg/kg, inj | → |
| | Japanese flounder | 3 g/kg, oral | NBT ↑ |
| | Sea bass | 2% wet bw, oral | humoral activation ↑ |
| | Dab | 0.5 µg/kg, iv | ROS ↑ |
| | Dentex | 0.5%, oral | → |
| | Turbot | 0.5 – 500 µg/kg, iv | ROS ↑ |
| | | 2 g/kg | leukocyte number ↑ |
| | Yellow tail | 2 – 10 mg/ml, inj | phagocytic activity ↑ |
| | Rainbow trout | 88 and 350 µg/g, oral | lysozyme ↑, ROS ↑, complement bacteriolytic activity ↑ |
| β-glucan + LPS | Atlantic salmon | 1 – 250 + 10 µg/ml | lysozyme ↑ |
| β-glucan + Mannose | Japanese flounder | 1%, oral | NBT ↑, lysozyme ↑ |
| | Pink snapper | 0.1 – 1% w/w, oral | ROS ↑, macrophage activation ↑ |
| β-glucan + FKC + Quillaja saponica | Pink snapper | 122 + 34 + 5 mg/kg | agglutination titers ↑ |
| CFA | Sockey salmon | 5 mg/kg, inj | antibody ↑ |
| Chitosan | M. rosenbergi larva | 0.25 – 1% v/v, 150 ml *in vitro* | Antibacterial activity ↑ |
| Chitin | Gilhead sea bream | 0.1 ml/fish, inj | → |
| | | 1 mg/fih, inj | humoral and cellular activation ↑ |
| | | 25 – 100 mg/kg | NCCs ↑, ROS ↑, phagocytic activity ↑ |
| EF-203 | Rainbow trout | 30 µg/kg, oral | NBT ↑, phagocytic activity ↑ |
| Ergosan | Rainbow trout | 1 mg/fish | complement ↑ |
| Fungi | Sockey salmon | 10 g/kg, oral | → |
| Glycyrrhizin | Yellow tail | 0 – 50 mg/kg, oral | complement ↑ |
| Laminaran | Blue gourami | 20 mg/kg, inj | CL ↑ |
| Levamisole | Atlantic salmon | 2.5 mg/L, bath | ROS ↑, phagocytic activity ↑, lysozyme ↑ |
| | Coho salmon | 5 mg/kg, inj | → |
| | Sockey salmon | 125 – 500 µg/ml, oral | phagocytosis ↑, complement ↑, ROS ↑, lymphokine ↑ |
| | | 0.5 – 500 µg/ml, iv | ROS ↑ |
| | | 75 – 300 mg/kg, oral | NCCs ↑ |
| | Japanese flounder | 125 – 500 mg/kg, oral | phagocytic activity ↑, NBT ↑, lysozyme ↑ |
| | Rainbow trout | 10 and 50 µg/ml, *in vitro* | phagocytic activity ↑, NBT ↑ |
| | | 5 mg/kg, inj | NBT ↑, lysozyme ↑, phagocytic activity ↑, killing ↑ |
| LPS | Red sea bream | 1 ml/fish, inj | phagocytic activity ↑ |

| MCFA | Coho salmon | 5 mg/kg, inj | → |
|---|---|---|---|
| Microsporidian | Flounder | 106 spores, inj | antibody ↑ |
| Myxsosporean | Sea bass | multiple, iv | ROS ↑ |
| Peptidoglycan | Japanese flounder | 1.5 – 4.5 g/kg, oral | phagocytosis ↑, complement ↑, MAF ↑, ROS ↑ |
| | Yellow tail | 0.2 mg/kg, oral | phagocytic activity ↑ |
| Quillaja | Yellow tail | 0.5 – 50 mg/kg | chemotaxis ↑ |
| Yeast | Gilhead sea bream | 1 – 10 g/kg | cellular response ↑ |
| | Sea cucumber | 5%, oral | phagocytic activity ↑, phagocytic index ↑, lysozyme ↑ |
| Wy-18, 251 | Coho salmon | 10 mg/kg, inj | → |

Increase, ⁻: Decrease, ®: No change, µg: Microgram, inj: Injection, iv: Intra venous, bw: Body weight, CFA: Complete Freund's adjuvant, CL: Chemiluminescent response, FKC: Formalin-killed Edwardsiella tarda cells, L: Litre, LPS: Lipopolysaccharide, Killing: Bactericidal activity of neutrophil and monocytes, MAF: Macrophage activating factor, M. rosenbergi: Macrobrachium rosenbergi, MCFA: Medium-chain fatty acid(s), NBT: Nitroblue tetrazolium reaction, NCCs: Non-specific cytotoxic cells, ROS: Reactive oxygen species, Turbot: Scophthalmus maximus, Wy-18, 251: an analog of levamisole at structure 3-(pchlorophenyl) thiazolol [3,2-a] benzimidazole-2-acetic acid. (modified from Galindo-Villegas & Hosokowa 2004).

Improvements in the health status of fish can certainly be achieved by balancing the diets with regard to nutritional factors (see Table 9), in particular lipids such as fatty acids, essential oils, and anti-oxidative vitamins or minerals such as vitamin-C, -E and selenium, and also minerals iron and fluoride, but this is primarily a result of an input of substrates and cofactors in a complex metabolic system. These compounds were identified as micronutrients that could affect disease resistance. This is unlike immunostimulants, which interact directly with the cells of the immune system and make them more active, because they enhance in immune system by providing substrate and co-factors necessary for the immune system to work properly. Nevertheless, some nutritional factors are so intimately interwoven with the biochemical processes of the immune system that significant health benefits can be obtained by adjusting the concentration of such factors beyond the concentration range sufficient to avoid deficiency symptoms or below a certain concentration range (Balfry & Higgs, 2001; Gannam & Schrock, 2001; Lim et al., 2001a, 2001b; Raa, 2000; Soltani et al., 2010). The modulatory effects of dietary nutritional factors on macrophage-, haemolytic-, lysozyme- and complement activation, lymphocyte proliferation, macrophage phagocytic response as well as oxidative burst, pinocytosis and bactericidal activity have been reported in aquaculture (Balfry & Higgs, 2001; Galindo-Villegas & Hosokowa, 2004; Gannam & Schrock, 2001, Sakai et al., 1999).

Other a type of nutritional factors as immunostimulants in aquaculture, nucleotides have essential physiological and biochemical functions including encoding and deciphering genetic information, mediating energy metabolism and cell signaling as well as serving as components of co-enzymes, allosteric effectors and cellular agonists. Also, these compounds have traditionally been considered to be non-essential nutrients. Nucleotides consist of a purine or a pyrimidine base, a ribose or 2'-deoxyribose sugar and one or more phosphate groups. The term nucleotide in this context refers not only to a specific form of the compounds but also to all forms that contain purine or pyrimidine bases.

1: Canadian Biosystem Inc. Calgary-Canada, 2: Chemoforma Augst-Swithzerland, 3: Amano Siyaku Co-op Tokyo, a: Before vaccination, b: Post-vaccination, bw: Body weight, d: Day, w: Week, : Increase, ‾: Decrease, ®: No change, Chinook: Chinook salmon, CL: Chemiluminescent response, J. flounder: Japanese flounder, Killing: Bactericidal activity of macrophage, NCCs: Non-specific cytotoxic cells, NBT: Nitroblue tetrazolium reaction, NT: Nucletoide, ROS: Reactive oxygen species, Trout: Rainbow trout, Turbot: Scophthalmus maximus, Vit-C: Vitamin-C, Vit-E: Vitamin-E, A. hydrophila: Aeromonas hydrophila, A. salmonicida: Aeromonas salmonicida, E. tarda: Edwardsiella tarda, V. anguillarum: Vibrio anguillarum. (modified from Galindo-Villegas & Hosokowa, 2004; Li & Gatlin, 2006; Sakai et al., 1999).

In recent years, world-wide heightened attention on nucleotide supplementation for fishes was aroused by the reports of some researches, indicating that dietary supplementation of nucleotides enhanced resistance of salmonids to viral, bacterial and parasitic infections as well as improved efficacy of vaccination and osmoregulation capacity (Burrells et al., 2001a, 2001b; Grimble & Westwood, 2000; Li & Gatlin, 2006). The modulatory effects of dietary nucleotides on lymphocyte maturation, activation and proliferation, macrophage phagocytosis, immunoglobulin responses as well as genetic expression of certain cytokines have been reported in humans and animals including some fish species such as hybrid tilapia, rainbow trout, Coho salmon, Atlantic salmon and common carp (Gil, 2002; Li & Gatlin, 2006). To date, research pertaining to nucleotide nutrition in fishes has shown rather consistent and encouraging beneficial results in fish health management, although most of the suggested explanations remain hypothetical and systematic research on fishes is far from complete. Because increasing concerns of antibiotic use have resulted in a ban on subtherapeutic antibiotic usage in some countries, research on immune nutrition for aquatic animals is becoming increasingly important. Also, research on nucleotide nutrition in fish is needed to provide insights concerning interactions between nutrition and physiological responses as well as provide practical solutions to reduce basic risks from infectious diseases for the aquaculture industry (Burrells et al., 2001a, 2001b; Li & Gatlin, 2006). In aquaculture, used immunostimulants as nutritional factors and nucleotides with dose, administration route and effects are given at Table 14 (Galindo-Villegas & Hosokowa, 2004; Li & Gatlin, 2006; Sakai et al., 1999)

**Table** 14. Doses and effects of immunostimulants as nutritional factors and nucleotides in fish species.

| Immunostimulant | Species | Dose - Period | *In vivo* or *in vitro* effects / Resistance |
|---|---|---|---|
| *Nutritional factors* | | | |
| Vitamin-C | Catfish | 150 mg/kg, 14 w | *E. tarta* ↑ |
| | | 1000 mg/kg, 7 w | neutrophil → (phagocytosis) |
| | | 3000 mg/kg | complement ↑, phagocytic index ↑ antibody → macrophage killing → *E. tarta* ↑ |
| | | 4000 mg/kg, 9 w | complement →, antibody →, *E. tarta* ↑ |
| | Red sea bream | 1000 mg/kg | complement → |
| | | 10 000 mg/kg | phagocytic activity ↑ |
| | Atlantic salmon | 2750 mg/kg, 26 w | complement ↑, NBT →, phagocytosis →, MAF ↑, *A. salmonicida* ↑ |
| | | 2980 mg/kg | antibody ↑ |
| | | 3170 mg/kg, 16 w | NBT ↑, killing ↑, migration ↑, antibody ↑, *A. salmonicida* ↑ |
| | | 4000 mg/kg, 52 d | *V. salmonicida* ↑ |
| | | 5000 mg/kg | antibody → |
| | Coho salmon | 3000 mg/kg | phagocytosis ↑, ROS ↑, complement ↑ |
| | J. flounder | 6100 mg/kg | NBT ↑ |
| | Trout | 244 mg/kg, 16 w | proliferation ↑, NBT →, MAF ↑ |
| | | 550 mg/kg, 10 d | IHNV ↑ |
| | | 2000 mg/kg, 4 w/12 w | *I. mutifiliis* ↑/*V. anguillarum* ↑ |
| | | 2000 mg/kg, 127 d | phagocytic index ↑, lysozyme ↑→ |
| | Turbot | 300 – 200 mg/kg | phagocytic activity ↑, lysozyme ↑ |
| | Sockey salmon | > requirement | → |
| | Yellow tail | 122 – 6100 mg/kg | phagocytic activity ↑, lysozyme ↑ |
| Vit-C + Vit-E | Gilhead sea bream | 2900 + 1200 mg/kg | lysozyme ↑, NCCs ↑ |
| | | Many concentration, *in vitro* | migration ↑, phagocytic activity ↑, ROS (mix) ↑ |
| Vit-C + Yeast glucan | Trout | oral | lysozyme →, complement ↑, CL ↑ |
| Vitamin-E | Catfish | 2500 mg/kg, 180 d | phagocytic index ↑, antibody → |
| | Chinook | 300 mg/kg and > requirement | → |
| | J. flounder | 600 mg/kg | phagocytic activity ↑, lysozyme ↑ |
| | Atlantic salmon | low levels | IHNV ↓ |
| | | > requirement | → |
| | | 800 mg/kg, 20 w | NBT →, complement ↑, lysozyme ↓, *A. salmonicida* ↑ |
| | Trout | 500 mg/kg, 12 w | phagocytosis ↑ |
| | | low levels | antibody ↓ |
| | Turbot | 500 mg/kg | phagocytic activity ↑ |
| Vit-E + Selenium | Catfish | 240 + 0.8 mg/kg, 120 d | NBT ↑ |
| Vitamin-A (retinol) | Atlantic salmon | oral | anti-protease activity ↑, migration ↑ |
| | Gilhead sea bream | 50 -300 mg/kg | ROS ↑ |
| α-tocopherol | Gilhead sea bream | 600 – 1800 mg/kg | complement ↑ |
| α-tocopherol acetate | Yellow tail | 119 – 5950 mg/kg | phagocytic activity ↑, lysozyme ↑ |

| Arginine | J. flounder | 150 mg/kg | | NBT ↑, lysozyme ↑ |
|---|---|---|---|---|
| Ascorbate 2-monophospate | Atlantic salmon | 20 – 1000 mg/kg | | → |
| Ascorbil 2-sulfate | Atlantic salmon | 4770 mg/kg | | → |
| | | 82, 44, 3170 mg/kg, 23 w | | antibody ↑ |
| | | 1000 mg/kg | | ROS ↑, lymphocyte number ↑ |
| | | 2750 mg/kg | | complement ↑ |
| | | 4000 mg/kg | | lysozyme ↑ |
| Axtahantin | J. flounder | 100 mg/kg | | chemotaxis ↑, NBT ↑ |
| Essential oil | Common carp | 30, 60, 120 ppm diet, 1% bw, 8 d | | antibody ↑, bactericidal activity ↑ |
| Protein hydrosilate | Atlantic salmon | 1 – 25 in vitro | | ROS ↑ |
| Soybean protein | Trout | oral | | phagocytosis ↑, NBT ↑, killing ↑ |
| **Nucleotides** | | | | |
| Ascogen P [1] | Hybrid striped bass | 5 g/kg, fixed ration approaching satiation daily | | neutrophil oxidative radical ↑, survival after challenge with Streptoccus iniae ↑ |
| Ascogen S [2] | Hybrid tilapia | 2 and 5 g/kg, 16 w | | growth ↑, survival ↑ |
| | | 5 g/kg, 120 d | | antibody after vaccination ↑ lymphocyte mitogenic response ↑ |
| | Trout | 0.62, 2.5 and 5 g/kg, diet at 1% bw/d, 37 d | | growth ↑ |
| Optimun [2] | Trout | 2 g/kg, containing 0.03% nucleotide | 2% bw/d, 3 w | survival after challenge with, V. anguillarum ↑ |
| | | | 1% bw/d, 2 w | survival after challenge with infectious salmon anaemia virus ↑ |
| | Coho salmon | | 2% bw/d, 3 w | survival after challenge with Piscirickettsia salmonis ↑ |
| | Atlantic salmon | | 2% bw/d, 3 w | sea lice infection ↓ |
| | | | 1.5% bw/d, 3 w [a] and 5 w [b] | antibody ↑, mortality ↓ |
| | | | 1.5% bw/d, 8 w | plasma chloride ↓, growth ↑ |
| | | | 10 w | intestinal fold ↑ |
| | Turbot | | to hand saniation daily | Altered immunogene expression in various tissues |
| Ribonuclease-digested yeast RNA [3] | Common carp | 15 mg/fish, by intubation, 3 d | | phagocytosis ↑, complement ↑, lysozyme ↑, respiratory burst ↑, A. hydrophila infection ↓ |

# Risks and benefits using immunostimulants

Immunostimulants are more widely applied both within the aquaculture sector and in traditional animal husbandry. There are many examples of successful use of immunostimulants to improve fish welfare, and also in vivo or in vitro effects of immune system (see Table 12, Table 13 and Table 14). One of the earliest applications of immunostimulants in fish was the use of glucans in salmon diets. These diets were considered to be effective in managing disease outbreaks after stressful events such as grading and there was believed to be some benefit in reducing sea lice settlement; allowing the stock to go longer between anti-sea lice treatments. Certainly, the use of in-diet immunomodulators has become widely accepted in aquaculture with commercially available diets supplemented with nucleotides which have been demonstrated to reduce sea lice settlement and provide better protection against A. salmonicida and V. anguillarum infection (Bricknell & Dalmo, 2005; Burrells et al., 2001a;

2001b). Immunostimulants can provide particular benefits when used in order to: (1) reduce mortality due to opportunistic pathogens, (2) prevent virus disease such as Vitamin-C on infectious hematopoietic necrosis (IHN) virus and yeast glucan on yellow-head baculovirus, (3) enhance disease resistance of farmed fish and shrimp, (4) reduce mortality of juvenile fish especially in fry and larval fish, (5) enhance the efficacy of antimicrobial as adjoint substances, if used in combination with curative antimicrobial drugs at an early phase of disease development, or prior to anticipated disease outbreak, (6) enhance the resistance to parasites or microsporidias, such as Vitamin-C on Ichthyophthirius multifiliis, lactoferrin on Cryptocaryon irritans, or glucans and chitin on Loma salmonea, (7) enhance the efficacy of vaccines, (8) improve fish welfare against stress (e.g grading, sea transfer, vaccination and environmental change), such as glucans may be helped reduce the negative effects of stress on the innate immune response, soybean lecithin may be provided higher tolerance for increased water temperature, and vitamin-E may be protected the complement system against stress-related reduction of activity, (9) promoting a greater and more effective sustained immune response to those infectious agents producing subclinical disease without risks of toxicity, carcinogenicity or tissue residues, (10) maintaining immune surveillance at heightened level to ensure early recognition and elimination of neoplastic changes in tissues, and (11) selectively stimulating the relevant components of the immune system or non-specific immune mechanism that preferentially confer protection against micro-organisms, such as via interferon release, especially for those infectious agents for which no vaccines currently exists (Ai et al., 2007; Bricknell & Dalmo, 2005; Cerezuela et al., 2009; Gannam & Schrock, 2001; Maqsood et al., 2011; Raa, 2000; Rodriguez-Tovar et al., 2011; Yin et al., 2009). Naturally, there is a risk that use of immunostimulants in aquaculture may cause unforeseen problems. Continual feeding of immunostimulants has generally been abandoned, in adult fish, in favor of pulse feeding. There are two possible outcomes of continuous feeding of an immunostimulant; (1) although, it is a very rare occurrence, the immunostimulant up-regulates the immune system to heightened levels and this is maintained until the immunostimulant is withdrawn, (2) the most obvious contra-indication as it would be in larval fish, continual exposure to an immunostimulant can induce tolerance. This is caused by the immune system of the host becoming de-sensitized to the immunostimulant and the immunostimulant response is lost, or in extreme circumstances the continued expose to an immunostimulant causes the immune response to become suppressed, giving a lower level of innate defences whilst exposure to that particular immunostimulant is maintained (Bricknell & Dalmo, 2005; Sakai et al., 1999). Besides, no research has yet been performed concerning the influence of immunostimulants at some stage such as maturation and spawning of fish. The immune systems become suppressed by sex hormones, testosterone and estradiol-17β, at these stages. Although the use of immunostimulants could cause recovery of the immune systems suppressed by sex hormones, they may disturb sexual maturation and other essential functions associated with spawning, or may include sterility through polyploidy (Cuesta et al., 2007; Magnadóttir, 2010; Piferrer et al., 2009). On the other hand, the mere deleterious side-effects of immunostimulants have not been completely investigated.

## CONCLUSIONS

Important progress has been made in recent years in our knowledge of the immunological control of fish diseases which has benefitted the growing aquaculture industry worldwide and also provided better understanding of some basic immunological phenomena. There are mainly three methods for control of fish disease: vaccination, chemotherapeutics and immunostimulants. In addition, researches in recent years about probiotics, prebiotics, and synbiotics also exhibited positive health effects in fish species. Immunostimulants and vaccines are used together to prevent infectious diseases. Immunostimulants may be used for treatment of some infectious diseases; they may not as effective as many chemotherapeutics. Antibiotic-resistant bacteria threaten treatment of fish disease using chemotherapeutics. Immunostimulants may compensate these limitations of chemotherapeutics. Immunostimulants are thought to be safer than chemotherapeutics and their range of efficacy is wider than vaccination. The combination of vaccination and immunostimulant administration may also increase the potency of vaccines. In addition continued pressure on the use of antimicrobials associated with food residue and environmental issue will encourage the use of immunostimulants. However, cautions have to be taken regarding issues such as tolerance, non-wanted side effects such as immunosuppression using too high doses of immunostimulants or non-desirable effects caused by a prolonged use of such compounds. Actual knowledge of potential immunostimulants is still obscure in several aspects, especially in those related to pathways and mechanisms in which such substances can reach their specific cell targets.

## REFERENCES

1.    Abbas, A. K. & Lichtmann, A. H. (2006). Basic Immunology: Functions and disorders of the immune system, 2nd Edition, W.B. Saunders Company, ISBN 1416029745, Philadelphia, USA

2.  Adams, A. & Thompson, K.D. (1990). Development of an ELISA for the detection of Aeromonas salmonicida in fish tissue. Journal of Aquatic Animal Health, Vol. 2, pp. 281- 288

3.  Adams, A. & Thompson, K.D. (2008). Recent applications of biotechnology to novel diagnostics for aquatic animals. Revue Scientifique et Technique–Office International des Epizooties, Vol. 27, No. 1, pp. 197-209

4.  Adams, A. (2004). Immunodiagnostics in aquaculture. Bulletin of the European Association of Fish Pathologists, Vol.24, pp. 33-37

5.  Adams, A.; Aoki, T.; Berthe, C.J.; Grisez, L. & Karunasagar, I. (2008). Recent technological advancements on aquatic animal health and their contributions toward reducing disease risks–a review, In: Diseases in Asian Aquaculture VI. Fish Health Section, M.G. Bondad-Reantaso, C.V Mohan, M. Crumlish & R.P. Subasinghe (Eds.), 71-88. Asian Fisheries Society, Manila, Philippines

6.  Ai, Q.; Mai, K.; Zhang, L.; Tan, B.; Zhang, W.; Xu, W. & Li, H. (2007). Effects of dietary beta-1, 3 glucan on innate immune response of large yellow croaker, Pseudosciaena crocea. Fish and Shellfish Immunology, Vol. 22, pp. 394-402

7.  Alderman, D.J. & Hastings, T. S. (2003). Antibiotic use in aquaculture: development of antibiotic resistance-potential for consumer health risks. International Journal of Food Science and Technology, Vol. 33, No. 2, pp. 139–155

8.  Alexander, J.B. & Ingram, G.A. (1992). Noncellular nonspecific defence mechanisms of fish. Annual Review of Fish Diseases, Vol. 2, pp. 249-279

9.  Alvarez-Pellitero, P. (2008). Fish immunity and parasite infections: from innate immunity to immunoprophylactic prospects. Veterinary Immunology and Immunopathology, Vol. 126, pp. 171–198

10. Aly, S.M & Mohamed, M.F. (2010). Echinacea purpurea and Allium sativum as immunostimulants in fish culture using Nile tilapia (Oreochromis niloticus). Journal of Animal Physiology and Animal Nutrition, Vol. 94, No. 5, pp. e31-e39

11. Anas, A.; Paul, S.; Jayaprakash, N.S.; Philip R.; Bright-Singh I. S. (2005). Antimicrobial activity of chitosan against vibrios from freshwater prawn Macrobrachium rosenbergii larval rearing systems. Disease of Aquatic Organisms, Vol. 67, pp. 177-179

12. Anderson, D.P.; Siwicki, A.K. & Rumsey, G. L. (1995). Injection or immersion delivery of selected immunostimulants to trout demonstrate enhancement of nonspecific defense mechanisms and protective

immunity. In: Diseases in Asian Aquaculture, Vol. 11, Fish Health Section, M. Shariff, R.P. Subasinghe & J.R. Arthur, (Eds.), 413- 426, Asian Fisheries Society, Manila, Philippines

13. Anderson, D.P. (1996). Environmental factors in fish health: Immunological aspects. In: The Fish Immune System: Organism, Pathogen, and Environment, G. Iwama & T. Nakanishi (Eds.), 289-310, Academic Press, ISBN 0-12-350439-2, San Diego, California, USA

14. Aoki, T.; Takano, T.; Santos, M.D. & Kondo, H. (2008). Molecular innate immunity in teleost fish: Review and future perspectives. In: Fisheries for Global Welfare and Environmental, Memorial book of the 5th Word Fisheries Congress, K. Tsukamoto, T. Kawamura, T. Takeuchi, T.D. Beard, Jr. & M.J. Kaiser (Eds.), 263-276, ISBN 978-4- 88704-144-8, Terrapub, Setagaya-ku, Japan

15. Aranishi, F. & Mano, N. (2000). Antibacterial cathepsins in different types of ambicoloured Japanese flounder skin. Fish and Shellfish Immunology, Vol. 10, pp. 87-89

16. Arason, G. (1996). Lectin as defence molecules in vertebrates and invertebrates. Fish and Shellfish Immunology, Vol. 6, pp. 277-289

17. Austin, B. & Brunt, J.W. (2009). The use of probiotics in aquaculture, Chapter 7. In: Aquaculture Microbiology and Biotechnology: Volume 1, D. Montet & R.C. Ray (Eds), 185-207, Science Publishers, ISBN 978-1- 57808-574-3, Enfield, New Hampshire, USA

18. Baba, T.; Watase, Y. & Yoshinaga, Y. (1993). Activation of mononuclear phagocyte function by levamisole immersion in carp. Nippon Suisan Gakkaishi, Vol. 59, pp. 301-307

19. Balfry, S.K. & Higgs, D.A. (2001). Influence of dietary lipid composition on the immune system and disease resistance of fish, Chapter 11, In: Nutrition and Fish Health, L. Chhorn & C.D. Webster (Eds.), 213-234, The Haworth Press, Inc., ISBN 1-56022-887- 3, Binghamton, New York, USA

20. Barman, D.; Kumar, V.; Roy, S.; Singh, A.S.; Majumder, D.; Kumar, A. & Singh, A.A. (1991). The role of immunostimulants in Indian aquaculture. Cited 21.05.2011. Available from http://aquafind.com/ articles/Immunostimulants-In-Aquaculture.php

21. Becker, J.A. & Speare D.J. (2007). Transmission of the microsporidian gill parasite, Loma salmonae. Animal Health Research Reviews, Vol. 8, pp. 59-68

22. Bei, J.X.; Suetake, H.; Araki, K.; Kikuchi, K.; Yoshiura, Y.; Lin, H R. & Suzuki, Y. (2006). Two interleukin (IL)-15 homologues in fish from two

distinct origins. Moleculer Immunology, Vol. 43, pp. 860-869

23. Belosevic, M.; Hanington, P.C. & Barreda, D.R. (2006). Development of goldfish macrophages in vitro. Fish and Shellfish Immunology, Vol. 20, pp. 152-171

24. Bird, S.; Zou, J.; Kono, T.; Sakai, M.; Dijkstra, J.M. & Secombes, C. (2004). Characterization and expression analysis of interleukin 2 (IL-2) and IL-21 homologues in the Japanese pufferfish, Fugu rubries, following their discovery by synteny. Immunogenetics, Vol. 56, pp. 909-923

25. Bird, S.; Zou, J.; Savan, R.; Kono, T.; Sakai, M.; Woo, J. & Scombes, C. (2005). Characterisation and expression analysis of an interleukin 6 homologue in the Japanese pufferfish, Fugu rubripes. Developmental and Comperative Immunology, Vol. 29, pp.775-789

26. Bols, N.C.; Brubacher, J.L.; Ganassin, R.C. & Lee, L.E.J. (2001). Ecotoxicology and innate immunity in fish. Developmental and Comparative Immunology, Vol. 25, pp. 853-873

27. Bounocore, F. & Scapigliati, G. (2009). Immune defence mechanism in the sea bass Dicentrarchus labrax L., Chapter 6, In: Fish Defenses, Volume 1: Immunology, G. Zaccone, J. Meseguer, A. García-Ayala, B.G. Kapoor (Eds.), 185-219, Science Publishers, ISBN 978-1-57808-327-5, Enfield, New Hampshire, USA

28. Bricknell, I. & Dalmo, R.A. (2005). The use of immunostimulants in fish larval aquaculture. Fish Shellfish Immunology, Vol. 19, pp. 457-472

29. Bright-Singh, I. S. & Philip, R. (2002). Use of immunostimulants in aquaculture management. In: Recent Advances in Diagnosis and Management of Diseases in Mariculture -Course Manuel; 1-5; 7-27 November 2002, Cochin, India

30. Brown, G.D. (2006). Dectin-1: a signaling non-TLR pattern-recognition receptor. Nature Reviews Immunology, Vol.6, pp. 33-43

31. Brown-Treves, K.M. (2000). Immuno-stimulants, Chapter 19, In: Applied Fish Pharmacology, Aquaculture Series 3, K.M. Brown-Treves (Ed.), 251-259, Kluwer Academic Publishers, ISBN 0-412-62180-0, Dordrecht, Netherlands

32. Burrells, C.; Williams, P.D. & Forno, P.F. (2001a). Dietary nucleotides: a novel supplement in fish feeds: 1. Effects on resistance to disease in salmonids. Aquaculture, Vol. 199, pp. 159–169

33. Burrells, C.; Williams, P.D.; Southage, P.J. & Wadsworth, S.L. (2001b). Dietary nucleotides: a novel supplement in fish feeds: 1. Effects on vaccination, salt water transfer, growth rate and physiology of Atlantic salmon. Aquaculture, Vol. 199, pp. 171-184

34. Cabezas L. R. (2006). Functional genomics in fish: towards understanding stress and immune responses at a molecular level. PhD Thesis, 1-223, Departament de Biologia

35. Cellular, Fisiologia i Immunologia, Facultat de Ciències, Universitat Autòmona de Barcelona, Barcolona, Spain

36. Campos-Perez, J.J.; Ward, M.; Grabowski, P.S.; Ellis, A.E. & Secombes, C.J. (2000). The gills are an important site of iNOS expression in rainbow trout Oncorhynchus mykiss after challenge with the Gram-positive pathogen Renibacterium salmoninarum. Immunology, Vol. 99, pp: 153-161

37. Cerezuela, R.; Cuesta, A.; Messeguer, J.; Ángeles Esteban, M. (2009). Effects of dietary vitamin D3 administration on innate immune parameters of seabream (Sparus aurata L.). Fish and Shellfish Immunology, Vol. 26, pp. 243-248

38. Chaves-Pozo, E.; Mulero, V.; Meseguer, J. & Ayala, A.G. (2005). Professional phagocytic granulocytes of the bony fish gilthead seabream display functional adaptation to testicular microenvironment. Journal of Leukocyte Biology, Vol. 75, pp. 345–351

39. Clem, L.W.; Sizemore R.C.; Ellsaesser, C.F. & Miller. N.W. (1985). Monocytes as accessory cells in fish immune responses. Developmental and Comparative Immunology, Vol. 9, pp. 803-809

40. Cole, A.M.; Weis, P. & Diamond, G. (1997). Isolation and characterization of pleurocidin; an antimicrobial peptide in the skin secretions of winter flounder. Journal of Biological Chemistry, Vol. 272, pp. 12008-12013

41. Corripio-Miyar, Y.; Bird, S.; Tsamopoulus, K. & Secombes, C. J. (2006). Cloning and expression analysis of two pro-inflamatory cytokines; IL-1beta and IL-8, in haddock (Melanogrammus aeglefinus). Molecular Immunology, Vol. 44, pp. 1361-1373

42. Cuesta, A.; Vargas-Chacoff, L.; García-López, A.; Arjona, F.J.; Martínez-Rodríguez G.; Meseguer, J.; Mancera, J.M. & Esteban, M.A. (2007). Effect of sex-steroid hormones, testosterone and estradiol, on humoral immune parameters of gilthead seabream. Fish and Shellfish Immunology, Vol. 23, pp. 693-700

43. Cunningham, C.O. (2004). Use of molecular diagnostic tests in disease control: Making the leap from laboratory to field application, Chapter 11, In: Molecular Aspects of Fish and Marine Biology Vol. 3: Current trends in the study of bacterial and viral fish and shrimp diseases, K.Y. Leung (Ed.), 292-312, World Scientific Publishing Co. Pte. Ltd., ISBN 981-238-749-8, London, UK

44. Dalmo R. A. (2002). Immunostimulation of fish. ICES CM 2002/R:09, Cited 05.04.2011. Available from http://www.ices.dk/products/CMdocs/2002/R/R0902.PDF

45. Dannevig, B. H.; Lauve, A.; Press, C.McL. & Landsverk, T. (1994). Receptor-mediated endocytosis and phagocytosis by rainbow trout head kidney sinusoidal cells. Fish and Shellfish Immunology, Vol. 4, pp. 3-18

46. Dickerson, H.W. & Clark, T.G. (1996). Immune response of fishes to ciliates. Annual Review of Fish Diseases, Vol. 6, pp. 107-120

47. Dorin, D.; Sire, M.F. & Vernier, J.M. (1994). Demonstration of an antibody response of the anterior kidney following intestinal administration of a soluble protein antigen in trout. Comparative Biochemistry and Physiology, Vol. 109, pp. 499-509

48. Duffy, J.E.; Carlson, E.; Li, Y.; Prophete, C. & Zelikoff, J.T. (2002). Impact of poly-chlorinated biphenyls (PCBs) on the immune function of fish: age as a variable in determining adverse outcome. Marine Environmental Research, Vol. 54, pp. 559-563

49. Dugenci, S.K.; Arda, N. & Candan, A. (2003). Some medicinal plants as immunostimulant for fish. Journal of Ethnopharmacology, Vol. 88, No. 1, pp. 99-106

50. Duncan, P.L. & Klesius, P.H. (1996). Effects of feeding spirulina on specific and nonspecific immune responses of channel catfish. Journal of Aquatic Animal Health, Vol. 8, pp. 308-313

51. Dupont, N.C.; Wang, K; Wadhwa, P.D.; Culhane, J.F. & Nelson, E.L. (2005). Validation and comparison of luminex multiplex cytokine analysis kits with ELISA:

52. Determinations of a panel of nine cytokines in clinical sample culture supernatants. Journal of Reproductive Immunology, Vol. 66, pp. 175-191

53. Dykova I. (2006). Phylum microspora. In: Fish Diseases and Disorders, Vol. 1, Protozoon and Metazoan Infections, 2nd Edition, P.T.K. Woo (Ed.), 205-229, CABI Publishing, ISBN 0851990150, Oxford, UK

54. Ebran, N.; Julien, S.; Orange, N.; Saglio, P.; Lemaître, C. & Molle, G. (1999). Pore-forming properties and antibacterial activity of proteins extracted from epidermal mucus of fish. Comparative Biochemistry and Physiology. Part A, Molecular & Integrative Physiology, Vol. 122, No. 2, pp. 181-189

55. El-Gohary, M.S.; Safinaz, G.M.; Khalil, R.H.; El-Banna, S. & Soliman, M.K. (2005). Immunosuppressive effects of metrifonate on Oreochromis Niloticus. Egyptian Journal of Aquatic Research, Vol. 31, pp. 448-458

56.  Ellis, A.E. (1999). Immunity to bacteria in fish. Fish and Shellfish Immunology, Vol. 9, pp. 291- 308

57.  Ellis, A.E. (2001). Innate host defense mechanisms of fish against viruses and bacteria. Developmental and Comparative Immunology. Vol. 25, pp. 827-839

58.  Ellsaesser, C.F.; Bly, J.E. & Clem, L.W. (1988). Phylogeny of lymphocyte heterogeneity: The thymus in channel catfish. Developmental and Comparative Immunology, Vol. 12, pp. 787-799

59.  Enis-Yonar, M.; Mise-Yonar, S. & Silici, S. (2011). Protective effect of propolis against oxidative stress and immunosuppression induced by oxytetracycline in rainbow trout (Oncorhynchus mykiss; W.). Fish and Shellfish Immunology, Vol. 31, pp. 318-325

60.  Espenes, A.; Press, C.; Danneving, B.H. & Landsverk, T. (1995). Immune-complex trapping in the splenic ellipsoids of rainbow trout (Oncorhynchus mykiss). Cell and Tissue Research, Vol. 282, pp. 41-48

61.  EU (European Union) (2007). Council Regulation (EC) No. 834/07 of 28 June 2007 on organic production and labelling of organic products and repealing Regulation (EEC) No. 2092/91. Official Journal of the European Union L 189, 20/07/2007, pp. 1-23

62.  EU (European Union) (2008). Council Regulation (EC) No. 889/08 of 5 September 2008 laying down detailed rules for the implementation of Council Regulation (EC) No. 834/2007 on organic production and labelling of organic products with regard to organic production; labelling and control. Official Journal of the European Union L 250, 18/09/2008, pp. 1-84

63.  Evans, D.L. & Gratzek, J.B. (1989). Immune defense mechanisms in fish to protozoan and helmint infections. American Zoologist (new name; Integrative and Comparative Biology (ICB)), Vol., 29, No. 2, pp. 409-418

64.  Evans, D.L. & Jaso-Friedmann, L. (1992). Nonspecific cytotoxic cells as effectors of immunity in fish. Annual Review of Fish Diseases, Vol. 2, No. 1, pp. 109-121

65.  Fast, M.D.; Sims, D.E.; Burka, J.F.; Mustafa, A. & Ross, N.W. (2002). Skin morphology and humoral non-specific defence parameters of mucus and plasma in rainbow trout; coho and Atlantic salmon. Comparative Biochemistry and Physiology, Vol. 132, No. 3, pp. 645-57

66.  Fearon, D.T. & Locksley, R. M. (1996). The instructive role of innate immunity in the acquired immune response. Science, Vol. 272, pp. 50-53

67.  Fletcher, T.C. (1981). Non-antibody molecules and the defense mechanisms of fish. In: Stress and Fish, A.D. Pickering (Ed.), 171-183,

Academic Press, ISBN 0125545509, New York, USA

68.  Galina, J; Yin, G.; Ardó, L. & Jeney, Z. (2009). The use of immunostimulating herbs in fish. An overview of research. Fish Physiology and Biochemistry, Vol. 35, No. 4, pp. 669- 676

69.  Galindo-Villegas, J. & Hosokawa H. (2004). Immunostimulants: Towards temporary prevention of diseases in marine fish. In: Avances en Nutrición Acuícola VII. Memorias del VII Simposium Internacional de Nutrición Acuícola, L. E. Cruz Suárez, D. Ricque Marie, M. G. Nieto López, D. Villarreal, U. Scholz & M. Gonzalez (Eds.), 279-319, 16-19 Noviembre 2004, Hermosillo, Sonara, México

70.  Galindo-Villegas, J.; Fukada, H; Masumoto, T. & Hosokawa, H. (2006). Effect of dietary immunostimulants on some innate immune responses and disease resistance against Edwardsiella tarda infection in Japanese flounder (Paralichthys olivaceus). Aquaculture Science, Vol. 54, No. 2, pp. 153-162

71.  Gannam, A.L. & Schrock, M.R. (2001). Immunostimulants in fish diets, Chapter 10, In: Nutrition and Fish Health, L. Chhorn & C.D. Webster (Eds.), 235-266, The Haworth Press, Inc., ISBN 1-56022-887-3, Binghamton, New York, USA

72.  García-Ayala, A. & Chaves-Pozo, E. (2009). Leukocytes and cytokines present in fish testis: A review, Chapter 2, In: Fish Defenses, Volume 1: Immunology, G. Zaccone, J. Meseguer, A. García-Ayala, B.G. Kapoor (Eds.), 37-74, Science Publishers, ISBN 978-1-57808-327-5, Enfield, New Hampshire, USA

73.  Giavedoni, L.D. (2005). Simultaneous detection of multiple cytokines and chemokines from nonhuman primates using luminex technology. Journal of Immunological Methods, Vol. 301, pp. 89-101

74.  Gil, A., 2002. Modulation of the immune response mediated by dietary nucleotides. Europen Journal of Clinical Nutrition, Vol. 56, No. Suppl. 3, pp. S1–S4.

75.  Gildberg, A.; Bogwald, J.; Johansen, A. & Stenberg, E. (1996). Isolation of acid peptide fraction from a fish protein hydrolysate with strong stimulary effect on atlantic salmon (Salmon salar) head kidney leucocytes. Comparative Biochemistry and Physiology Part B: Biochemistry and Molecular Biology, Vol., 114, No. 1, 97-101

76.  Grimble, G.K. & Westwood, O.M.R. (2000). Nucleotides. In: Nutrition and Immunology: Principles and Practice, German; J.B. & Keen; C.L. (Eds.), 135– 144, Humana Press Inc., Totowa, New Jersey, USA

77. Gudding, R.; Lillehaug, A. & Evensen, Ø. (1999). Recent developments in fish vaccinology. Veterinary Immunology and Immunopathology, Vol. 72, No. 1-2), pp. 203-212

78. Guttvik, A.; Paulsen, B.; Dalmo, R.A.; Espelid, S.; Lund, V. & Bøgwald, J. (2002). Oral administration of lipopolysaccharide to Atlantic salmon (Salmo salar L.) fry. Uptake, distribution, influence on growth and immune stimulation. Aquaculture, Vol., 212, pp. 35-53

79. Hamerman, J. A.; Ogasawara, K. & Lanier, L.L. (2005). NK cells in innate immunity. Current Opinion in Immunology, Vol. 17, pp. 29-35

80. Hoffmann, K. (2009). Stimulating immunity in fish and crustaceans: some light but more shadows. Aqua Culture Asia Pacific Magazine, Vol. 5, No. 5, pp. 22-25

81. Hogan, R.J.; Stuge, T.B.; Clem, L.W.; Miller, N.W. & Chinchar, V.G. (1996). Anti-viral cytotoxic cells in the channel catfish /Ictalurus punctatus). Developmental and Comparative Immunology, Vol. 20, pp. 115-127

82. Holland, J.W.; Pottinger, T.G. & Secombes, C.J. (2002). Recombinant interleukin-1 beta activates the hypothalamic-pituitary-interrenal axis in rainbow trout; Oncorhynchus mykiss. Journal of Endocrinology, Vol. 175, pp. 261-267

83. Holland, M.C.H. & Lambris, J.D. (2002). The complement system of teleosts. Fish and Shellfish Immunology, Vol. 12, pp. 399-420

84. Hølvold, L.B. (2007). Immunostimulants connecting innate and adaptive immunity in Atlantic salmon (Salmo salar). Master in Biology-Field of study Marine Biotechnology, 1- 69, Department of Marine Biotechnology, Norwegian College of Fishery Science, Univetsity of Tromso, Tromsø, Norway

85. Horsberg, T.E. (2003). Aquatic animal medicine. Journal of Veterinary Pharmacology and Therapeutics, Vol. 26, No. 1-2, pp. 39-42

86. Igawa, D.; Sakai, M. & Savan, R. (2006). An unexpected discovery of two interferon gamma– like genes along with interleukin (IL)-22 and -26 from teleost: IL-22 and -26 genes have been described for the first time outside mammals. Moleculer Immunology, Vol. 43, pp. 999-1009

87. Inoue, Y.; Kamuta, S.; Ito, K.; Yoshiura, Y.; Ototake, M.; Moritomo, T. & Nakanishi, T. (2005). Molculer cloning and expression analysis of rainbow trout (Oncorhynchus mykiss) interleukin-10 cDNAs. Fish and Shellfish Immunology, Vol. 18, pp. 335-344

88. Ispir, U. & Dorucu, M. (2005). A study on the effects of levamisole on the immune system of rainbow trout (Oncorhynchus mykiss; Walbaum).

Turkish Journal of Veterinary and Animal Sciences, Vol. 29, pp. 1169-1176

89. Jansson E. (2002). Bacterial Kidney Disease in salmonid fish: Development of methods to assess immune functions in salmonid fish during infection by Renibacterium salmoninarum. PhD Thesis, 1-52, Department of Pathology and Department of Fish, National Veterinary Institute, Swedish University of Agricultural Sciences, Uppsala, Sweden

90. Jeney, G. & Anderson, D.P. (1993). Enhanced immune response and protection in rainbow trout to Aeromonas salmonicida bacterin following prior immersion in immunostimulants. Fish and Shellfish Immunology, Vol. 3, pp. 51-58

91. Jimeno, C.D. (2008). A transcriptomic approach toward understanding PAMP-driven macrophage activation and dietary immunostimulant in fish. PhD Thesis, 1-222, Departament de Biologia Cellular, Fisiologia i Immunologia, Facultat de Ciències, Universitat Autòmona de Barcelona, Barcelona, Spain

92. Jiye, L.; XiuQin, S.; Fengrong, Z. & LinHua, H. (2009). Screen and effect analysis of immunostimulants for sea cucumber, Apostichopus japonicus. Chinese Journal of Oceanology and Limnology, Vol. 27, No. 1, pp. 80-84

93. Jones, S.R.M. (2001). The occurrence and mechanisms of innate immunity against parasites in fish. Developmental and Comparative Immunology, Vol. 25, pp. 841-852

94. Joosten, P.H.M.; Kruijer, W.J. & Rombout, J.H.W.M. (1996). Anal immunisation of carp and rainbow trout with different fractions of a Vibrio anguillarum bacterin. Fish and Shellfish Immunology, Vol. 6, pp. 541-551.

95. King, P.D.; Aldridge, M.B.; Kennedy-Stoskopf, S. & Stott, J.L. (2001). Immunology, Chapter 12, In: CRC Handbook of Marine Mammal Medicine, 2nd Edition, L.A. Dierauf & F.M.D. Gulland (Eds.), 237-252, CRC Press LLC, ISBN 0-8493-0839-9, Boca Raton, Florida, USA

96. Klesius, P.H.; Shoemaker, C.A.; Evans, J.J. & Lim, C. (2001). Vaccines: Prevention of Diseases in aquatic animals, Chapter 17, In: Nutrition and Fish Health, L. Chhorn & C.D. Webster (Eds.), 317-335, The Haworth Press, Inc., ISBN 1-56022-887-3 , Binghamton, New York, USA

97. Klesius, P.H.; Pridgeon, J.W. & Aksoy, M. (2010). Chemotactic factors of Flavobacterium columnare to skin mucus of healthy channel catfish (Ictalurus punctatus). FEMS Microbiology Letters, Vol. 310, pp. 145–151

98.  Kunttu, H.M.T.; Valtonen, E.T.; Suomalainen, L.R.; Vielma, J. & Jokinen, I.E. (2009). The efficacy of two immunostimulants against Flavobacterium columnare infection in juvenile rainbow trout (Oncorhynchus mykiss). Fish and Shellfish Immunology, Vol. 26, pp. 850-857

99.  Kusher, D.I. & Crim, W.C. (1991). Immunosuppression in bluegill (Lepomis macrochirus) induced by environmental exposure to cadmium. Fish and Shellfish Immunology, Vol. 1, pp. 157-161

100. Lauridsen, J.H. & Buchmann, K. (2010). Effects of short- and long-term glucan feeding of rainbow trout (Salmonidae) on the susceptibility to Ichthyophthirius multifiliis infections. Acta Ichthyologica et Piscatoria, Vol. 10, No. 1, pp. 61-66

101. Lemaître, C.; Orange, N.; Saglio, P.; Saint, N.; Gagnon, I. & Molle, G. (1996). Characterisation and ion channel activities of novel antibacterial proteins from the skin mucosa of carp (Cyprinus carpio). European Journal of Biochemistry, Vol. 240, No. 1, pp. 143-149

102. Li, J.; Bardera, D.R.; Zhang, Y.A.; Boshra, H.; Gelman, A.E.; LaPatra, S.; Tort, L. & Sunyer, J.O. (2006). B lymphocyte from early vertebrates have potent phagocytic and microbicidal abilities. Nature Immunology, Vol. 7, pp. 1116-1124

103. Li, J.H.; Shao, J.Z.; Xiang, L.X. & Wen, Y. (2007). Cloning; characterization and expression analysis of puffer fish interleukin-4 cDNA: the first evidence of Th2-type cytokine in fish. Molecular Immunology, Vol. 44, pp. 2088-2096

104. Li, P. & Gatlin, D.M. (2006). Nucleotide nutrition in fish: Current knowledge and future application. Aquaculture, Vol., 251, pp. 141-152

105. Lim, C.; Klesius, P.H. & Shoemaker, A.C. (2001a). Dietary iron and fish health, Chapter 9, In: Nutrition and Fish Health, L. Chhorn & C.D. Webster (Eds.), 189-199, The Haworth Press, Inc., ISBN 1-56022-887-3, Binghamton, New York, USA

106. Lim, C.; Klesius, P.H. & Webster, A.C. (2001b). The role of dietary phosphorus, zinc, and selenium in fish health, Chapter 10, In: Nutrition and Fish Health, L. Chhorn & C.D. Webster (Eds.), 201-212, The Haworth Press, Inc., ISBN 1-56022-887-3 , Binghamton, New York, USA

107. Lorenzen, K. (1993). Acquired immunity to infectious diseases in fish: implications for the interpretation of fish disease surveys. In: Fish: Ecotoxicology and Ecophysiology, T. Braunbeck, W. Hanke, H. Segner, pp. 183-196; ISBN 3527300104, Verlag Chemie, Weinheim, New York, USA

108. Lumlertdacha, S. & Lovell, R.T. (1995). Fumonisin-contaminated dietary corn reduced survival and antibody production by channel catfish challenged with Edwardsiella ictaluri. Journal of Aquatic Animal Health, Vol. 7, pp. 1-8

109. Lundén, T.; Miettinen, S.; Lonnstrom, L.G.; Lilius, E. M. & Bylund, G. (1998). Influence of oxytetracycline and oxolinic acid on the immune response of rainbow trout (Oncorhynchus mykiss). Fish and Shellfish Immunolog,; Vol. 8, pp. 217-230

110. Lundén, T.; Miettinen, S.; Lonnstrom, L.G.; Lilius, E. M. & Bylund, G. (1999). Effect of florfenicol on the immune response of rainbow trout (Oncorhynchus mykiss). Veterinary Immunology and Immunopathology, Vol. 67, pp. 317-325

111. Lundén, T. & Bylund, G. (2002). Effect of sulphadiazine and trimethoprim on the immune response of rainbow trout (Oncorhynchus mykiss). Veterinary Immunology and Immunopathology Vol. 85 pp. 99-108

112. Lutfalla, G.; Crollius, H.R.; Stange–Thomann, N.; Jaillon, O.; Mogensen, K. & Monneron, D. (2003). Comparative genomic analysis reveals independent expansion of lineage specific gene family in vertebrates: the class II cytokine receptors and their ligands in mammals and fish. BMC Genomics, Vol. 4, pp. 29

113. Magnadóttir, B. (2006). Innate immunity of fish (overview). Fish and Shellfish Immunology, Vol. 20, pp. 137-151

114. Magnadóttir, B. (2010). Immunological control of fish diseases. Marine Biotecnology, Vol. 12, pp. 361-379

115. Manning, B.B. (2001). Mycotoxins in fish feeds, Chapter 13, In: Nutrition and Fish Health, L. Chhorn & C.D. Webster (Eds.), 267-287, The Haworth Press, Inc., ISBN 1-56022-887- 3, Binghamton, New York, USA

116. Manning, B.B. (2010). Mycotoxins in aquaculture feed. In: Nutrition and Fish Health, L. Chhorn & C.D. Webster (Eds.), 267-287, The Haworth Press, Inc., ISBN 1-56022-887- 3 , Binghamton, New York, USA

117. Manning, M.J. 1994. Fishes. In: Immunology: A Comparative Approach, R.J. Turner (Ed.), 69- 100, John Wiley & Sons Ltd., ISBN 0471944009, Chichester, UK

118. Manning, M.J. & Nakanishi, T. (1996). The specific immune system: Cellular defenses In: The Fish Immune System: Organism, Pathogen, and Environment, G. Iwama & T. Nakanishi (Eds.), 159-205, Academic Press, ISBN 0-12-350439-2, San Diego, California, USA

119. Maqsood, S.; Singh, P.; Samoon, M.H. & Wani, G.B. (2011). Use of immunostimulants in aquaculture systems. Cited 11.07.2011. Available from http://aquafind.com/ articles/Immunostimulants-in-aquaculture.php

120. Medzhitov, R. & Janeway, C.A. Jr. (2002). Decoding the patterns of self and nonself by the innate immune system. Science, Vol. 296, pp. 298-300

121. Medzhitov, R. (2007). Recognition of microorganisms and activation of the immune response. Nature, Vol. 449, pp. 819-826

122. Meseguer, J.; López-Ruiz, A. & García-Ayala, A. (1995). Reticulo-endothelial stroma of the head-kidney from the seawater teleost gilthead seabream (Sparus aurata L): an ultrastructural and cytochemical study. Anatomical Record, Vol., 241, pp: 303-309

123. Miller, N.; Wilson, M.; Bengtén, E.; Stuge, T.; Warr, G. & Clem, W. (1998). Functional and molecular characterization of teleost leukocytes. Immunological Reviews, Vol. 166, pp. 187–197

124. Moore, J.D.; Ototake, M. & Nakanishi, T. (1998). Particulate antigen uptake during immersion immunisation of fish: The effectiveness of prolonged exposure and the roles of skin and gill. Fish and Shellfish Immunology, Vol. 8, pp. 393-407

125. Mulero, I.; Sepulcre, M.P.; Meseguer, J.; Garcia-Ayala, A. & Mulero, V. (2007). Histamine is stored in mast cells of most evolutionarily advanced fish and regulates the fish inflammatory response. The Proceeding of the National Academy of Science USA (PNAS), Vol. 104, No. 49, pp. 19434–19439

126. Nakano, M.; Mutsuro, J.; Nakahara, M.; Kato, Y. & Yano, T. (2003). Expansion of genes encoding complement components in bony fish: biological implications of the complement diversity. Developmental and Comparative Immunology, Vol. 27, pp. 764- 762

127. Nayak, S.K. (2010). Probiotics and immunity: A fish perspective. Fish and Shellfish Immunology, Vol. 29, pp. 2-14

128. Nelson, J.S. (2006). Fishes of the world, 4th Edition, John Wiley & Sons Inc. Publication ISBN 0- 471-25031-7, New York, USA

129. Noga, E. J. (2010). Fish Disease Diagnose and Treatment, 2nd Edition, Wiley-Blackwell: John Wiley & Sons Inc. Publication, ISBN 978-0-8138-0697-6, Iowa, USA

130. Nygaard, R.; Husgard, S.; Sommer, A.I.; Leong, J.A. & Robertsen, B. (2000). Induction of Mx protein by interferon and double-stranded RNA in salmonid cells. Fish and Shellfish Immunology, Vol. 10, pp. 435-450

131. Ortuño, J.; Esteban, M.A. & Messeguer, J. (2001). Effects of short-term crowding stress on gilthead seabream (Sparus aurata L.) innate immune response. Fish and Shellfish Immunology, Vol. 11, pp. 187-197

132. Ortuño, J.; Cuesta, A.; Rodríguez, A.; Esteban, M.A. & Meseguer, J. (2002). Oral administration of yeast; Saccharomyces cerevisiae; enhances the cellular innate immune response of gilthead seabream (Sparus aurata L.). Veterinary Immunology and Immunopathology, Vol. 85, pp. 41-50

133. Palaksha, K.J.; Shin, G.W.; Kim, Y.R. & Jung, T.S. (2001). Evolution of non-specific immune components from the skin mucus of olive flounder (Paralichthys olivaceus). Fish and Shellfish Immunology, Vol. 24, pp. 479-488

134. Pancer, Z. & Cooper, M.D. (2006). The evolution of adaptative immunity. Annual Review of Immunology, Vol. 24, pp. 497-518

135. Park, I.Y.; Park, G.B.; Kim, M.S. & Kim, S.C. (1998). Parasin I; an antimicrobial peptide derived from histone H2A in the catfish, Parasilurus asotus. FEBS Letters, Vol. 437, pp. 258-268

136. Passer, B.J.; Chen, C.H.; Miller, N.W. & Cooper, M.D. (1996). Identification of a T lineage antigen in the catfish. Developmental and Comparative Immunology, Vol. 20, pp. 441- 450

137. Paulsen, S.M.; Lunde, H.; Engstad, R.E. & Robertsen, B. (2003). In vivo effects of β-glucan and LPS on regulation of lysozyme activity and mRNA expression in Atlantic salmon (Salmo salar L.) Fish and Shellfish Immunology, Vol. 14, No. 1, pp. 39-54

138. Peatmen, E. & Liu, Z. (2006). CC chemokines in zebrafish: evidence for extensive intrachoromosomal gene duplications. Genomics, Vol. 88, pp. 381-385

139. Peatman, E. & Liu, Z. (2007). Evolution of CC chemokines in teleost fish: a case study in gene duplication and implications for immune diversity. Immunogenetics, Vol. 59, pp. 613-623

140. Peddie, S.; Zou, J. & Secombes, C.J. (2002). Immunostimulation in the rainbow trout (Oncorhynchus mykiss) following intraperitoneal administration of Ergosan. Veterinary Immunology and Immunopathology, Vol. 86, pp. 101-113

141. Pedersen, G.M.; Gildberg, A. & Olsen, R.L. (2004). Effects of including cationic proteins from cod milt in the feed to Atlantic cod (Gadus morhua) fry during a challenge trial with Vibrio anguillarum. Aquaculture, Vol. 233, pp. 31-43

142. Pellitero, P.A. (2008). Fish immunity and parasite infections: from innate immunity to immunoprophylactic prospects. Veterinary Immunology and Immunopathology, Vol. 126, pp. 171-198

143. Perdikaris, C. & Paschos, I. (2010). Organic aquaculture in Greece: a brief review. Reviews in Aquaculture, Vol. 2, No. 2, pp. 102–105

144. Perera, H.A.C.C & Pathiratne A. (2008). Enhancement of immune responses in Indian carp, Catla carta, following administration of levamisole by immersion. In: Disease in Asian Aquaculture VI, Fish Health Section, M.G. Bondad-Reantosa, C.V. Crumlish & R.P. Subasingle (Eds.), 129-142, Asian Fisheries Society, Manila, Philippines

145. Piferrer, F.; Beaumont, A.; Falguière, J.C.; Flajšhans, M.; Haffray, P. & Colombo, L. (2009). Polyploid fish and shellfish: production, biology and applications to aquaculture for performance improvement and genetic containment. Aquaculture, Vol. 293, pp. 125-156

146. Plouffe, D.A.; Hanington, P.C.; Walsh, J.G.; Wilson, E.C. & Belosevie, M. (2006). Comprasion of select innate immune mechanisms of fish and mammals. Xenotransplantation, Vol. 12, pp. 226-277

147. Plumb, J.A. & Hanson, L.A. (2011). Health Maintenance and Principal Microbial Diseases of Cultured Fishes, 3rd Edition, Wiley-Blackwell: John Wiley & Sons Inc. Publication, ISBN 978-0-8138-1693-7, Iowa, USA

148. Powell, J.L. & Loutit, M.W. (2004). Development of a DNA probe using differential hybridization to detect the fish pathogen Vibrio anguillarum. Microbial Ecology, Vol. 28, pp. 365-373

149. Press, C. McL,; Evensen, Ø.; Reitan, L.J. & Landsverk, T. (1996). Retention of furunculosis vaccine components in Atlantic salmon Salmon solar L., following different routes of administration. Journal of Fish Disease, Vol. 19, 215-224

150. Press, C.McL. & Evensen, Ø. (1999). The morphology of the immune system in teleost fishes. Fish and Shellfish Immunology, Vol. 9, pp. 309-318

151. Raa, J. (2000). The use of immune-stimulants in fish and shellfish feeds. In: Avances en Nutrición Acuícola V. Memorias del V Simposium Internacional de Nutrición Acuícola. L.E. Cruz-Suárez; D. Ricque-Marie, M. Tapia-Salazar, M.A. Olvera-Novoa & R. Civera-Cerecedo (Eds.). 19-22 Noviembre 2000, Mérita, Yucatán, Mexico

152. Randelli, E.; Buonocore, F. & Scapigliati, G. (2008). Cell markers and determinants in fish immunology. Fish and Shellfish Immunology, Vol. 25, pp. 326-340

153. Retie, O. (1998). Mast cells/eosinophilic granular cells of teleostean fish: A review focusing on standing properties and functional responses. Fish and Shellfish Immunology, Vol. 8, pp. 489-513

154. Rodriguez-Tovar, L.E.; Speare, D.J. & Markham, R.J. (2011). Fish microsporidia: immune response, immunomodulation and vaccination. Fish and Shellfish Immunology, Vol. 30, pp. 999-1009

155. Rombout, J.H.M.W.; Huttenhuis, H.B.T.; Picchietti, S. & Scapigliati, G. (2005). Phylogeny and ontogeny of fish leucocytes. Fish and Shellfish Immunology, Vol. 19, pp. 441-445

156. Roque, A.; Soto-Rodríguez, S.A. & Gomez-Gil, B. (2009). Bacterial fish diseases and molecular tools for bacterial fish pathogens detection. In: Aquaculture Microbiology and Biotechnology: Volume 1, D. Montet & R.C. Ray (Eds), 73-99, Science Publishers, ISBN 978-1-57808-574-3, Enfield, New Hampshire, USA

157. Sakai, M. (1999). Current research status of fish immunostimulants. Aquaculture, Vol. 172, pp. 63-92

158. Sakai, M.; Yoshida, T. & Kobayashi, M. (1995). Influence of the immunostimulant, EF203, on the immune responses of rainbow trout, Oncorhynchus mykiss, to Renibacterium salmoninarum. Aquaculture, Vol. 138, no. 1-4, pp. 61-67

159. Savan, R.; Kono, T.; Igawa, D. & Sakai, M. A. (2005). A novel tumor necrosis factor (TNF) gene present in tandem with the TNF-alpha gene on the same chromosome in teleosts. Immunogenetics, Vol. 57, pp. 140-150

160. Schluter, S.F.; Bernstein, R.M. & Marchalonis, J.J. (1999). Big Bang - emergence of the combinatorial immune system. Developmental and Comparative Immunology, Vol. 23, pp. 107-111.

161. Secombes, C.J. (1996). The nonspecific immune system: Cellular defenses. In: In: The Fish Immune System: Organism, Pathogen, and Environment, G. Iwama & T. Nakanishi (Eds.), 63-105, Academic Press, ISBN 0-12-350439-2, San Diego, California, USA

162. Secombes, C.J.; Manning, M.J. & Ellis, A.E. (1982). The effect of primary and secondary immunization on the lymphoid tissue of the carp, Cyprinus carpio L. Journal of Experimental Zoology, Vol. 220, pp. 277-287

163. Secombes, C.J.; Hardie, L.J. & Daniels. G. (1996). Cytokines in fish: An update. Fish and Shellfish Immunology, Vol. 6, pp. 291-304

164. Secombes, C.J.; Wang, T.; Hong, S.; Peddie, S.; Crampe, M.; Laing, K.J.; Cunningham, C. & Zou, J. (2001). Cytokines and innate immunity of

fish. Developmental and Comparative Immunology, Vol. 25, No. 8-9, pp. 713-723.

165. Seker, E.; Ispir, U. & Dorucu, M. (2011). Immunostimulating effect of levamisole on spleen and head-kidney leucocytes of rainbow trout (Oncorhynchus mykiss; Walbaum 1792). Kafkas Universitesi Veteriner Fakultesi Dergisi, Vol. 17, no. 2, pp. 239-242

166. Shoemaker, C.A.; Klesius, H.P. & Lim C. (2001). Immunity and disease resistance in fish, Chapter 7, In: Nutrition and Fish Health, L. Chhorn & C.D. Webster (Eds.), 149-162, The Haworth Press, Inc., ISBN 1-56022-887-3 , Binghamton, New York, USA

167. Smith, J.V.; Fernandes, J.M.O.; Jones, S.J.; Kemp, G.D. & Tatner, M.F. (2000). Antibacterial proteins in rainbow trout, Oncorhynchus mykiss. Fish and Shellfish Immunology, Vol. 10, pp. 243-260

168. Smith, J.V. & Fernandes, J.M.O. (2009). Antimicrobial peptides of the innate immune system, Chapter 8, In: Fish Defenses, Volume 1: Immunology, G. Zaccone, J. Meseguer, A. García-Ayala, B.G. Kapoor (Eds.), 241-275, Science Publishers, ISBN 978-1-57808- 327-5, Enfield, New Hampshire, USA

169. Soltani, M.; Sheikhzadeh, N.; Ebrahimzadeh-Mousavi, H.A. & Zargar, A. (2010). Effects of Zataria multiflora essential oil on innate immune responses of common carp (Cyprinus carpio). Journal of Fisheries and Aquatic Science, Vol. 5, pp. 191-199

170. Subasinghe, R. (2009). Disease control in aquaculture and the responsible use of veterinary drugs and vaccines: The issues; prospects and challenges. In: Options Méditerranéennes, Series A, No. 86: The Use of Veterinary Drugs and Vaccines in Mediterranean Aquaculture, C. Rodgers & B. Basurco (Eds.), 5-11, CIHEAM/FAO, ISBN 2-85352-422-1, Zaragoza, Spain

171. Tafalla, C.; Aranguren, R.; Secombes, C.J.; Castrillo, J.L.; Novoa, B. & Figueras, A. (2003). Molecular characterisation of sea bream (Sparus aurata) transforming growth factor beta1. Fish and Shellfish Immunology, Vol. 14, pp. 405-421

172. Tatner, M. F. & Findlay, C. (1991). Lymphocyte migration and localization patterns in rainbow trout, Onchorhynchus mykiss, studies using the tracer sample method. Fish and Shellfish Immunology, Vol. 1, pp. 107-117

173. Torroba, M. & Zapata, A.G. (2003). Aging of the vertebrate immune system. Microscopy Research and Technique, Vol. 62, pp. 477– 481

174. Tort, L.; Balasch, J.C. & Mackenzi, S. (2003). Fish immune system. A crossroads between innate and adaptive responses. Inmunología, Vol. 22, No. 3, pp. 277-286

175. Vallejo, A. N. & Ellis, A. E. (1989). Ultrastructural study of the response of eosinophil granule cells to Aeromonas salmonicida extracellular products and histamine liberators in rainbow trout Salmo gairdneri Richardson. Developmental and Comparative Immunology, Vol. 13, pp. 133-148

176. Vallejo, A.N.; Miller, N.W. & Clem. L.W. (1992). Antigen processing and presentation in teleost immune responses. Annual Review of Fish Diseases, Vol. 2, pp. 73-89

177. Vatsos, I.; Thompson, K.D & Adams, A. (2003). Starvation of Flavobacterium psychrophilum in broth, stream water and distilled water. Disease of Aquatic Organisms, Vol. 56, pp. 115-126

178. Wang, T.; Holland, J.W.; Bols, N. & Secombes, C.J. (2005). Cloning and expression of the first non-mammalian interleukin-11 gene in rainbow trout Oncorhynchus mykiss. FEBS Journal, Vol. 272, pp. 1136-1147

179. Whyte, S.K. (2007). The innate immune response of finfish: A review of current knowledge. Fish and Shellfish Immunology, Vol. 23, No. 6, pp. 1127-1151

180. Wilson, M.; Bengten, E.; Miller, N.W.; Clem, L.W.; Du Pasquer, L. & Warr, G.W. (1997). A novel chimeric Ig heavy chain from a teleost fish shares similarities to IgD.n Proceedings of the National Academy of Sciences of the USA 94, April 1997 Immunology, Vol. 94, pp. 4593- 4597

181. Wilson, T. & Carson, J. (2003). Development of sensitive, high-throughput one–tube RTPCR-enzyme hybridisation assay to detect selected bacterial fish pathogens. Disease of Aquatic Organisms, Vol. 54, pp. 127-134

182. Yano, T. (1996). The nonspecific immune system: Humoral defense. In: The Fish Immune System: Organism, Pathogen, and Environment, G. Iwama & T. Nakanishi (Eds.), 105- 157, Academic Press, ISBN 0-12-350439-2, San Diego, California, USA

183. Yin, G.; Ardó, L.; Thompson, K.D.; Adams, A.; Jeney, Z. & Jeney, G. (2009). Chinese herbs (Astragalus radix and Ganoderma lucidum) enhance immune response of carp, Cyprinus carpio, and protection against Aeromonas hydrophila. Fish and Shellfish Immunology, Vol. 26, pp. 140-145

184. Yoshida, T.; Kruger, R. & Inglis, V. (1995). Augmentation of non-specific protection in African catfish, Clarias gariepinus (Burchell), by the long-

term oral administration of immunostimulants. Journal of Fish Disease, Vol. 18, pp.195–198

185. Yoshiura, Y.; Kiryu, I.; Fujiwara, A.; Suetake, H.; Suzuki, Y.; Nakanishi, T. & Ototake, M. (2003). Identification and characterisation of Fugu orthologues of mammalian interleukin-12 subunits. Immunogenetics, Vol. 55, pp. 296-306

186. Yousif, A.N.; Albright, L.J. & Evelyn, T.P.T. (1995). Interaction of coho salmon Oncorhynchus kisutch egg lectin with the fish pathogen Aeromonas salmonicida. Diseases of Aquatic Organisms, Vol. 21, pp. 193-199

187. Zapata, A.; Diez, B.; Cejalvo, T.; Gutierrez de Frias, C. & Cortes, A. (2006). Ontogeny of the immune system of fish. Fish and Shellfish Immunology, Vol. 20, pp. 126-136

188. Zapata, A.G.; Chibá, A. & Varas, A. (1996). Cells and tissue of the immune system of fish. In: The Fish Immune System: Organism, Pathogen, and Environment, G. Iwama & T. Nakanishi (Eds.), 1-62, Academic Press, ISBN 0-12-350439-2, San Diego, California, USA

189. Zhang, J.Y.; Wu, Y.S. & Wang, J.G. (2004). Advance of phage display antibody library and its' implication prospect in aquaculture. Journal of Fisheries of China, Vol. 28, pp. 329–333

190. Zhang, J.Y.; Wang, J.G.; Wu, Y.S.; Li, M.; Li, A.H. & Gong, X.L. (2006). A combined phage display ScFv library against Myxobolus rotundus infecting crucian carp, Carassius auratus auratus (L.), in China. Journal of Fish Diseases, Vol. 29, pp. 1-7

191. Zhao, W.; Liang, M. & Zhang P. (2010). Effect of yeast polysaccharide on the immune function of juvenile sea cucumber, Apostichopus japonicus Selenka under pH stres. Aquaculture International, Vol. 18, pp. 777-786

192. Zou, J.; Secombes, C.J.; Long, S.; Miller, N.; Clem, L.W. & Chinchar, V.G. (2003). Molecular identification and expression analysis of tumor necrosis factor in channel catfish (Ictalurus punctatus). Developmental and Comparative Immunology, Vol. 27, pp. 845-858

193. Zou, J.; Yoshiura, Y.; Dijkstra, J.M.; Sakai, M.; Ototake, M. & Secombes, C. (2004). Identification of an interferon gamma homologue in fugu; Takifigu rubripes. Fish and Shellfish Immunology, Vol. 17, pp. 403-409

# Chapter 8

## DEFINING "ADVERSE ENVIRONMENTAL IMPACT" AND MAKING § 316(B) DECISIONS: A FISHERIES MANAGEMENT APPROACH

David E. Bailey[1] and Kristy A.N. Bulleit[2]

[1]Mirant Corporation, 8711 Westphalia Road, Upper Marlboro, MD 20772, USA

[2]Hunton & Williams, 1900 K Street, N.W., Washington, D.C. 20006-1109, USA

## ABSTRACT

The electric utility industry has developed an approach for decisionmaking that includes a definition of Adverse Environmental Impact (AEI) and an implementation process. The definition of AEI is based on lessons from fishery management science and analysis of the statutory term "adverse environmental impact" and is consistent with current natural resource management policy. The industry has proposed a definition focusing on "unacceptable risk to the population's ability to sustain itself, to support reasonably anticipated commercial or recreational harvests, or to perform its normal ecological function." This definition focuses not on counting individual fish or eggs cropped by the various uses of a water body, but on preserving populations of aquatic organisms and their functions in the aquatic community. The definition recognizes that assessment of AEI should be site-specific and requires both a biological decision and a balancing of diverse societal values. The industry believes that the definition of AEI should be implemented in a process that will maximize the overall societal benefit of the § 316(b) decision by considering the facility's physical location, design, and operation, as well as the local biology. The approach considers effects on affected fish and shellfish populations and the benefits of any necessary best technology available (BTA) alternatives. This is accomplished through consideration of population impacts, which conversely allows consideration of the benefits of any necessary BTA modifications. This in turn allows selection of BTAs that will protect potentially affected populations in a cost-effective manner. The process also employs risk assessment with stakeholder participation, in accordance with EPA's Guidelines for Ecological Risk Assessment. The information and tools are now available to make informed decisions about site-specific impacts that will ensure protection of aquatic ecosystems and best serve the public interest.

## INTRODUCTION

Generating electric power requires cooling water to condense steam after it is used in steam-powered turbines. Withdrawing cooling water from surface waters for this purpose can impinge fish on screens and entrain fish and shellfish, eggs, and larvae. Impingement is the entrapment of fish or shellfish on screens that are used to prevent condenser blockage. Entrainment is the passing of organisms through the cooling water system, which may cause mortality from exposure to heat, physical stress, or chemicals.

In § 316 of the Clean Water Act, Congress included a subsection (a) to allow variances from thermal standards, if it is demonstrated that there will be "protection and propagation of a balanced, indigenous population of shellfish, fish and wildlife in and on the waterbody." Immediately following is § 316(b), which states that any standard applicable to a point source under § 301 or § 306 of the Act "shall require that the location, design, construction, and capacity of cooling water intake structures reflect the best technology available for minimizing adverse environmental impact." The U.S. Environmental Protection Agency (EPA), driven by a lawsuit in federal district court in New York State, is conducting a rulemaking to implement § 316(b)[1].

The purpose of this paper is to contribute to the development of § 316(b) regulations that will both protect living aquatic resources and reflect sound social policy. It addresses the following topics:

- The history of § 316(b) and EPA's current approach to the rulemaking
- The need for a definition of "adverse environmental impact"
- The need for a rule based on the tools and principles of fisheries management science
- The need for a rule that maximizes net social benefit
- A suggested approach that meets these needs.

## A BRIEF HISTORY OF § 316(B) AND EPA'S 316(B) RULEMAKING

Congress enacted § 316(b) of the Clean Water Act in 1972. The language of § 316(b) first appeared in the Conference Report on the 1972 Federal Water

Pollution Control Act Amendments in a section called "Thermal Discharges." There was no comparable language in earlier House or Senate bills and little testimony or debate in the record explaining its sudden appearance. It appears, in fact, to have been an afterthought[2]. In December 1973, little more than a year after the statute was enacted, EPA proposed a rule to implement § 316(b). The rule was finalized in 1976. Both the proposed and final versions referenced EPA Development Documents, which described factors and design alternatives to consider when making a § 316(b) determination. A preamble to the 1976 final rule said that "decisions relating to the best technology available are to be made on a case-by-case basis." The rule was short-lived, for the Fourth Circuit Court of Appeals set it aside on procedural grounds.1 In 1977, EPA published a draft guidance document, but this was never finalized[2,3]. For over 20 years, § 316(b) has been widely implemented on a site-specific basis, guided by the 1977 draft guidance rather than by regulations. In 1993, several environmental groups filed suit against EPA in a U.S. district court in New York, seeking to compel EPA to issue regulations to implement § 316(b).2 EPA and the environmental plaintiffs settled the case and agreed to a rulemaking schedule in a consent agreement entered by the court. EPA's final rule for new facilities was published in the Federal Register, December 18, 2001, and a new proposed rule for existing facilities was published April 9, 2002. Although new and existing facilities do deserve different treatment under § 316(b), many issues raised by the proposed new facilities rule will be the same as or similar to the issues for existing sources.

## THE NEED FOR A DEFINITION OF "ADVERSE ENVIRONMENTAL IMPACT"

In the "Phase I" rulemaking for new facilities, EPA reports that it has received numerous comments addressing how "adverse environmental impact" (AEI) should be defined[4]. A definition is important because it establishes the basis for resource protection and provides a standard for selecting best technology available (BTA), in cases where BTA is required. While a number of possible definitions of AEI have been offered, the following definition, proposed by the Utility Water Act Group (UWAG), is both scientifically sound and socially relevant for § 316(b) decisionmaking: "Adverse environmental impact is a reduction in one or more representative indicator species that (1) creates an unacceptable risk to the population's ability to sustain itself, to support reasonably anticipated commercial or recreational harvests, or to perform its normal ecological function and (2) is attributable to the operation of the cooling water intake structure"[5]. This definition focuses on protection at the population level. As stated in AFS Policy Statement #1, a goal of fisheries management is "to ensure selfsustaining populations that would support commercial and recreational fishing both now and in the future"[6]. As Suter and Barnthouse concluded, "(t)he reproducing population is the smallest ecological unit that is persistent on the human time scale, and hence the lowest level that we can meaningfully protect"[7]. Despite this emphasis on population-level effects, it is recognized that for species whose populations are at critically low levels, the population can become endangered, in which case the protection of individual organisms through the Endangered Species Act3 is appropriate. In addition to the federal statute, many states have enacted similar endangered species legislation.4 These statutes, already in place, should and will be applied no matter what § 316(b) regulatory process EPA ultimately adopts. The proposed AEI definition set out above also acknowledges that ecosystem integrity, structure, and function must be protected and, from a fisheries management perspective, that reasonably expected harvests should not be impaired. Finally, the recommended definition of AEI incorporates the idea of risk and therefore invokes risk management as part of the AEI decisionmaking process.

## THE NEED FOR A RULE BASED ON THE TOOLS AND PRINCIPLES OF FISHERIES MANAGEMENT SCIENCE

The effect of cooling water intake structures (CWIS) on fisheries is fundamentally similar to the effects of recreational and commercial harvesting of fish and associated effects of bycatch and bait collection. One primary difference is which species are affected. Fishery harvesting, of course, targets species that are desirable for human or animal food consumption and sport interest, while CWIS losses are a function of the interaction of fishery populations with the CWIS. CWIS vulnerability tends to be highly variable, depending on the CWIS location, design, and species' life history and behavior. Nevertheless, the similarities between losses from fishing and CWIS losses are such that CWIS effects on the fishery can be evaluated using the same basic approaches used by state and federal fishery managers to manage their commercial and recreational fisheries. The species and sizes of fish and shellfish impinged and entrained can be quantified and evaluated in the context of fishery management tools, including long-term populating monitoring, annual harvest levels, models, and natural resource protection regulations. As part of their management efforts, fisheries managers have learned to manage complex trade-offs. For example, increasingly they are being asked to weigh trade-offs between game, nongame, native, and nonnative species management[8].

The fisheries management approach views the fishery as a renewable resource that can be managed. It recognizes that the federal government need not protect every fish (leaving aside endangered species, which require special treatment), let alone every egg, but should instead preserve the fishery resource itself. Fisheries managers know that a certain level of cropping of fish stocks can occur without destroying a population's ability to sustain itself. How low is too low? While the fishery science literature does not provide a definitive answer to this question, NMFS believes that a prudent rule can be established as follows: Two of the best known models in the fishery science literature find that, on average, the stock size at MSY (maximum sustainable yield) is approximately 40% of the stock size that would be obtained if fishing mortality were zero (the pristine level). . . . Also, the fishery science literature contains several suggestions to the effect that any stock size below about 20% of the pristine level should be cause for serious concern. In other words, a stock's capacity to produce MSY on a continuing basis may be jeopardized if it falls below a threshold of about one-fifth the pristine level (emphasis added)[9].

## Commonly Used Fishery Reference Points

Due to similarities of CWIS impacts and commercial and recreational fishing impacts, fishery management tools have been commonly applied to evaluate these impacts[57]. Regulations issued by NMFS and the Fish and Wildlife Service (FWS) incorporate the concept of "optimum yield" of a fishery, based in turn on the concept of "maximum sustainable yield" (MSY) (50 C.F.R. 600.310(c)(1)(i) (1999)). MSY is defined as "the largest long-term average catch or yield that can be taken from a stock or stock complex under prevailing ecological and environmental conditions" (id.). Currently, tools such as Biomass per Recruit (BPR) and spawning stock measures are more in favor than MSY. NMFS recognizes that maximum productivity from a stock can be achieved by reducing the stock size by as much as 60% and that the population will be able to sustain or replace itself until the stock size is reduced by about 80%. Fishery managers consider removal of 70 to 80% of an unfished stock's biomass (Spawning Stock Biomass or SSB) and 65 to 80% of a stock's reproductive potential (Spawning Stock Biomass per Recruit or SSBPR) to be safe, given the compensatory reserve inherent in most fish stocks[10,11]. "Spawning Stock Biomass per Recruit" (SSBPR) is the total weight of a mature spawning stock that would be generated over the lifetime of an individual recruit[12]. When reliable estimates of the compensatory capacity of a population exist, spawner-recruit models can be used to develop more realistic and less conservative biological reference points[13]. As with the SSBPR approach, spawner-recruit analyses show that mortality due to entrainment and impingement is likely to have negligible effects on the abundance or yield of a fish population unless that population is already being fished at a level that greatly exceeds $F_{msy}$. Biological reference points and quantitative assessment tools used in fisheries management can also be used to evaluate the likelihood that entrainment and impingement mortality will reduce the reproductive capacity of a fish population to a level that warrants management concern. Fisheries management concepts, therefore, provide scientifically sound principles for determining whether cooling-water withdrawals can cause "adverse environmental impact" to vulnerable fish populations.

## Risk Assessment

No matter how sound the definition of AEI and the available assessment tools, a decisionmaking process that must decide "how much is too much" cannot escape uncertainty[15]. Assessing AEI inevitably calls for an assessment of risk to affected populations (or, for new facilities, potentially affected populations), to the aquatic community, and to the fishery. EPA's Ecological Risk Assessment Guidelines[16] provide a three-phase process of problem formulation, analysis, and risk characterization useful for AEI decisionmaking. The final product is a risk description that includes an interpretation of ecological adversity and descriptions of uncertainty and lines of evidence. In short, the effect of cooling water intake structures on fisheries has many similarities to the effects of commercial and recreational fishing and associated effects (bycatch and removal of bait fish). Thus, the same general field and analytical methods developed for use in fishery management can be and have been applied to assess the effects of a CWIS on fish and shellfish in waterbodies from which cooling water is withdrawn.

# THE NEED FOR A RULE THAT MAXIMIZES NET SOCIAL BENEFIT

## Balancing Fishery Protection and Other Uses

The CWA establishes the protection of fisheries as a national goal [Clean Water Act § 101(a)(2), 33 U.S.C. § 1251(a)(2)]. Many states have likewise adopted this goal.6 However, society has many goals for management and use of water resources, such as flood control, public water supply, agriculture, industrial water supply, and commercial and recreational fishing. Each of these uses results in impacts to fisheries, and it would be irrational to manage or regulate water resources solely for a single use such as maximizing fish production. While any of these uses could be eliminated, to do so would result in a significant social cost. To take just one example, hydroelectric power is one of the most significant in terms of volume withdrawn from a waterbody, but it also provides significant benefits such as (1) flood protection, (2) preservation of water during high-flow periods for use during low-flow periods, (3) recreational benefits, (4) increased fish habitat, (5) power production, and (6) economic development. To be sure, hydropower has deleterious effects, such as habitat fragmentation, blocking of the passage of fish, and effects on dissolved oxygen. But massive efforts are underway to mitigate these effects through impact assessments under the National Environmental Policy Act and relicensing proceedings by the Federal Energy Regulatory Commission. Perhaps the most significant impact on fish — particularly in estuarine and marine waterbodies — is fishery exploitation[17]. In addition to the direct harvest of fish, fishery impacts occur through bycatch and bait fish removal. Another manner in which fisheries can be affected is by the deliberate introduction of nonnative species into waterbodies to promote recreational fisheries — e.g., introduction of Pacific salmon into the Great Lakes to create a recreational trout fishery and introduction of gizzard shad into reservoirs as a food source to increase sport fish populations. In addition to water withdrawals and fishery harvests, human activities can alter fish populations in other ways. For example, land development or agricultural activities can cause sedimentation, habitat loss, and nutrient enrichment and affect dissolved oxygen levels and/or water temperature and clarity[18] and ultimately impact fisheries. Water transportation can also impact fisheries as a result of construction of navigation channels and shipping (e.g., the Welland Canal, which introduced the sea lamprey into the Great Lakes, affecting the lake trout fishery) and the associated navigational use of the waterways, which can introduce exotic species in ballast water. It is in this broader context of multiple impacts on fisheries and competing societal costs and benefits that we should approach the task of protecting fisheries from entrainment and impingement, while still providing a reliable source of electric power. Fig. 1 illustrates the three key aspects of sound § 316(b) decisionmaking. These aspects are (1) evaluation of biological conditions in the vicinity of the CWIS and assessment of the impact or potential impact to the fishery; (2) analysis of the location of the CWIS (i.e., waterbody type and local aquatic community where the facility is located); and (3) CWIS design considerations.

## Biological Conditions and CWIS Impacts

Fishery management/assessment methods and tools that are available to assess fisheries and impacts from the interaction of the CWIS and the fishery were discussed earlier in this paper. Other authors — including EPA in the Economic and Engineering Analyses Report developed for the Phase I § 316(b) rule[19] — have documented that very large numbers of organisms may become entrained or impinged at a single facility. If this is so, why haven't CWIS impacts been a more prominent national issue? There are a number of reasons:

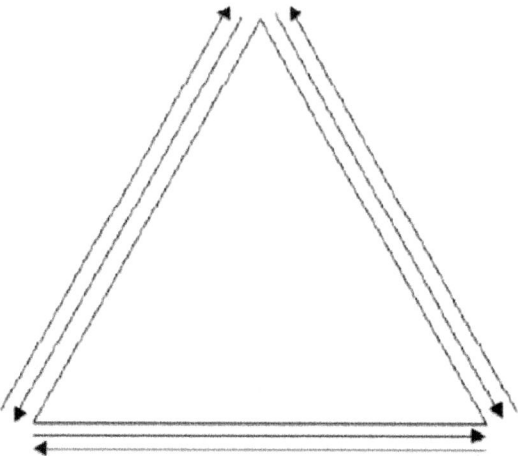

Waterbody Type &
Physical Location
  • Waterbody Sensitivity
  • Social Uses of Waterbody
  • Social Impacts of Technology

Facility Design
  • Technology
    Effectiveness
  • Technology Cost
  • Overall Technology
    Benefits and Impacts

Biological Conditions
and CWIS Impacts
  • Life History
  • Population Status
  • Fishery Management
    Objectives

**Figure 1:** Key components for effective 316(b) decisionmaking.

- § 316(a) and (b) Studies — Many states have already developed and implemented §§ 316(a) and (b) regulatory programs, including Maryland, Delaware, New York, New Jersey, Tennessee, South Carolina, California, Michigan, Ohio, Illinois, Alabama, Kentucky, Indiana, and others.7 Studies conducted by companies located in these states (and, in some instances, independent studies conducted by the states themselves), including some long-term studies, provide a good baseline for understanding power plant fishery impacts[20]. Long-term data on one reservoir, Lake Wheeler, collected by the Tennessee Valley Authority[21], shed light on the relationship between long-term once-through cooling operation and the status of the fish community in the lake. Browns Ferry Nuclear (BFN) currently operates two units supported by six intake pumps with a rated total capacity of 2,312 MGD. BFN units were placed in operation between 1974 and 1977 (originally the plant supported three units). Reservoir-wide monitoring was discontinued in

1980, but cove rotenone samples were continued to provide a minimum data base on fish community in the vicinity of BFN, particularly in support of BFN's thermal variance monitoring program for the Alabama Department of Environmental Management. Cove rotenone samples have been collected annually during August and September at three sites since 1969. The data base, therefore, includes five years of pre-operational reservoir data (1969 to 1974) against which the long-term operational impacts of the plant can be compared. Details on sampling, species examined (19 species were examined, and, for each species, data were collected for three size classes: young-ofyear, intermediate, and harvestable or adult), results, and analyses performed on the data are provided in TVA[21]. Although standing stock estimates for the reservoir exhibit extreme fluctuations, regression analysis revealed no significant increasing or decreasing trend for either total numbers (fish/hectare) or biomass (kg/ha) during the 30 years of monitoring.

- Survival — Early § 316(b) studies assumed 100% mortality to entrained organisms. Later studies, however, evaluated the survival rate of entrained organisms, many of them considering both immediate and latent mortality. EPRI recently completed a comprehensive review of entrainment mortality studies[22]. Fig. 2 presents a summary of findings demonstrating significant survival, in some cases exceeding 90%. Many of the recreationally important species had high survival rates, such as striped bass (mean survival rate 61%) and weakfish (mean survival rate 79%), while others, such as herrings and anchovies, had survival rates of approximately 25%[22,23]. Likewise, an entrainment mortality study for zooplankton at the Anclote power station in Florida demonstrated that the survival rate was quite high[24,26].

- Stakeholder and Regulator Judgment — Many biologists working for stakeholders, and regulatory and resource agencies as well, have judged that waterbodies where cooling water intakes operate are not impaired by entrainment and impingement. This view is reflected in the previous Administration's Clean Water Action Plan, which does not identify entrainment or impingement as a source of resource degradation[25].

- Empirical Information — Examples of successful fisheries in cooling ponds show that CWIS do not necessarily create adverse impact. Cooling ponds are constructed solely for the purpose of providing condenser cooling water, thereby eliminating the need for large withdrawals from a major source waterbody. Although a very high percentage of cooling pond water normally passes through the CWIS, many of these ponds support naturally reproducing fisheries[27,28,29]. While in some instances studies resulted in actions by facilities to modify their intake structures to reduce impingement or entrainment or both, or to implement offsite enhancements to avoid AEI, in most cases no significant adverse environmental impact was identified.

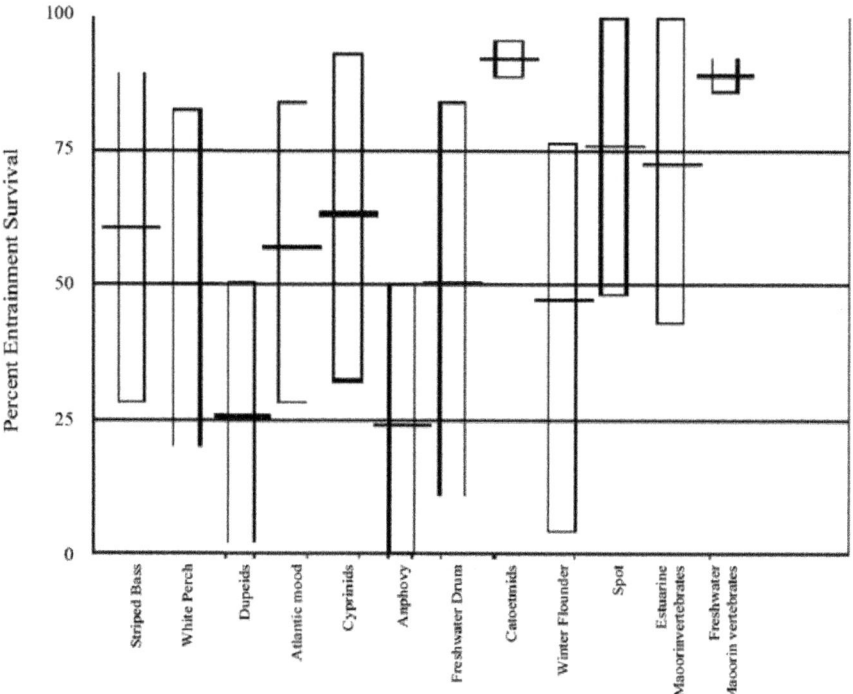

**Figure 2:** An illustration of the range in entrainment survival observed across various groups of fish (Source: EPRI 2000, Figure 3-1).

- Behavioral or Life History Factors — By virtue of their behavior or life history, many fish are able to avoid CWIS impacts[30,31,32]. For example, in freshwater many fish species lay eggs in nests or attached to substrate or vegetation, making them unavailable for entrainment. At Chalk Point, a power plant located on a tidal portion of the Patuxent River, it was initially assumed that up to 76% of each year's population in the river could be lost to entrainment. As a result of behavioral studies, however, the station determined that, due to regional movement, diurnal position in the water column, and the ability of larvae to avoid entrainment, the estimates of losses were reduced to 10 to 20%[33].

- Compensation — As noted by Myers[34], the concept of "compensation" is fundamental to understanding and managing biological resources.

For any biological population to persist, reductions in population size caused by natural environmental fluctuations must result in increased survival, growth, or fecundity of the remaining individuals[35,36,37]. Mechanisms of compensation have been well studied in both terrestrial and aquatic systems. The compensatory response to reductions in population size is the key factor that permits fish populations to sustain themselves despite enormous natural mortality for early life stages and even intensive harvesting of adults.

Long-term research surveys have demonstrated compensation in a variety of marine, estuarine, and freshwater fish species. Field experiments in which fish population sizes are artificially manipulated have also been used to demonstrate compensation[34]. UWAG has identified approximately 50 recent scientific studies (many published in the last 10 years) demonstrating specific mechanisms responsible for compensation in a variety of fish species[38]. The National Research Council (NRC) has recognized the importance of compensation for modern fisheries management:

Many species appear to have strongly compensatory (spawnerrecruit) relationships; that is, per capita recruitment increases significantly as stock size decreases. Reference levels are now more commonly based on a % (SSBPR), but the percentage is often specified by analogy with other stocks or by using the results (of comparisons among other biological reference points). A knowledge of the compensatory capacity of the stock is necessary to define the most appropriate (biological reference points) for a stock. Even without such knowledge, however, a conservative % (SSBPR) still can be selected. (Citation omitted).[13, p. 44]

Spawner-recruit relationships of the type discussed by the NRC are used to manage two estuarine-dependent fish species, striped bass, and weakfish[39,40]. Methods discussed by the NRC can be used to incorporate the concept of compensation in management strategies for species for which spawner-recruit data are not available. Fisheries scientists have demonstrated the importance of compensation for ensuring the continued persistence of fish populations, and fisheries managers routinely consider compensation when establishing harvesting regulations. While the precise quantification of compensation can be difficult, its occurrence cannot be disputed. The above factors are presented not to suggest that CWIS impacts are always insignificant, but rather to put impingement and entrainment impacts in perspective. In the vast majority of cases, CWIS impacts have not been determined to be a substantial limiting factor for fisheries; thus, in most cases the elimination of these impacts would not be expected to substantially improve fisheries.

## Facility Design

Where adverse environmental impacts are identified, a wide range of CWIS technologies designed to reduce impacts are available, as documented in a recent EPRI report[41] and summarized in Taft 2000[42]. The EPRI report identifies a wide array of technologies available for protecting impingeable organisms, including barrier nets, angled screens, and technologies designed to take advantage of fish behavior. For protecting entrainable organisms, wedgewire screens, fine mesh screens and, more recently, the Gunderboom8 have been demonstrated to be effective in certain waterbody types and for certain species, although these technologies have limitations in some waterbody types or for protection of certain species or have not yet been evaluated in a full range of waterbody conditions[43,44,45,46,47,48,49]. In addition, while not part of the CWIS, wet closed-cycle and dry cooling systems significantly reduce or eliminate the need for condenser cooling water. While some have advocated that these systems be designated as "best technology available" (BTA) for § 316(b) purposes, they can have significant negative environmental effects that would preclude their universal application. Both types of system also have significant energy requirements that reduce the efficiency and increase the fuel consumption of the generating facility. This inefficiency results in increased fuel use and air pollutant emissions, which in turn can affect water quality and fisheries by deposition of nitrogen emissions. Wet closed-cycle cooling, which can reduce cooling water requirements by up to 98%, causes consumptive water use with fishery consequences during lowflow conditions in freshwater. Wet closed-cycle cooling towers may also be unsuitable due to their noise and vapor plumes[50]. Additionally, both wet closed-cycle and dry cooling systems have significant space requirements and aesthetic impacts. The associated increase in impervious surface (especially from dry cooling systems) can impact water quality and fisheries. Wet closed-cycle cooling systems are frequently used as components of new generation construction projects, but due to their potential environmental disbenefits, they would be a poor choice for universal BTA from a net social benefit perspective[50,51].

## Waterbody Type and Physical Location

Location considerations include characteristics such as waterbody type (marine, estuarine, riverine, or lake), the aquatic physical environment (e.g., hydro- and thermodynamics, depth, and water quality conditions in the vicinity of the facility), and the local terrestrial setting (e.g., urban, rural, or industrial; topography, space constraints, and proximity to facilities such as airports, historically important sites, etc.). Such factors directly affect the feasibility of certain CWIS technologies. In particular, use of wet closed-cycle or dry cooling systems — with their associated space requirements, noise, and aesthetic issues — can have significant effects on local communities.

It is therefore important that decisions balance tradeoffs in these factors to make sound decisions. The Dickerson Station on the freshwater free-flowing Potomac and the Chalk Point Station on the tidal Patuxent, both facilities located in Maryland, provide a useful example of this importance. Maryland uses the AFS fishery replacement values to quantify the value of economic losses for BTA impingement decisionmaking. These values were $11,281/yr for Dickerson and $28,450/yr (after barrier net deployment) at Chalk Point. The Department of Natural Resources estimated that the economic value of entrainment losses was approximately $1000/yr (1981 dollars) at Dickerson and had a net present value of $1.3 million (i.e., 1989 dollar loss projected over the life of the facility) at Chalk Point. The values are low in contrast to the cost of wet closed cycle cooling, estimated to be on the order of $100 million at Chalk Point and somewhat less at Dickerson, even without considering the environmental disadvantages of this technology.

## MAKING § 316(B) DECISIONS: A PROPOSED PROCESS THAT MEETS THE NEEDS IDENTIFIED ABOVE

An approach to § 316(b) decisions that takes advantage of fishery management tools and balances multiple waterbody uses and social considerations must also be manageable and implementable from a regulatory perspective. The major components of an approach currently under development that incorporates these needs are described below. This approach establishes some distinctions between § 316(b) decisions for existing facilities and those for new facilities.

### Decision Process for Existing Sources

The proposed approach is based on the definition of AEI presented earlier, which focuses on population- and community-level impacts and fishery use protection. It includes the elements listed below.

#### *Use of Representative Indicator Species*

Previous work has demonstrated that it is not necessary to study each and every species in a waterbody. Rather, species can be selected based on recreational or commercial importance, roles in the food chain, and/or vulnerability. Previous work has identified most of the species typically vulnerable to CWIS impacts, and sitespecific screening studies can confirm the selection of species for further study as necessary

## Determination of Adverse Impact

Three alternative approaches are proposed for making § 316(b) adverse impact decisions. The first approach uses explicit criteria that are sufficiently stringent to support a decision that the facility presents no risk of adverse impact. The second approach uses a process based on the principles of EPA's risk assessment/risk management framework. The third allows decisions based on previously conducted site-specific § 316(b) studies, if it can be shown that they meet certain standards.

## Use of Screening Criteria

It is important that the decisionmaking criteria be clear and explicit to facilitate easy implementation. The criteria are not performance standards. Instead, they are designed to be well below a level that could reasonably be expected to result in AEI. This approach addresses the issue of uncertainty by setting criteria at these low thresholds. Several specific criteria being evaluated include

- Location. This criterion is based on determining if a CWIS is located in a waterbody or portion of a waterbody that cannot support aquatic life at any significant level due to poor water quality, such as anoxia, or lack of habitat. For example, if the CWIS withdraws its intake water from an anoxic zone which cannot support impingeable and entrainable organisms important to the fishery, it would be very unlikely to result in AEI.

- Facility design. If a facility employs a CWIS which is designed or has features to minimize impingement and/or entrainment, or makes use of technologies such as wet closed-cycle cooling, it would present no appreciable risk of AEI. In this situation, the technology must be demonstrated to be effective. If the technology is known to be effective only for impingement, for example, then the issue of entrainment will still need to be assessed.

- Percentage of waterbody used. Use of this criterion is suggested for entrainable organisms in smaller waterbodies such as freshwater rivers, lakes and reservoirs. A criterion value of 5% (or less) of the 90% exceedance flow of a river or of the volume of the biological zone of influence9 in a lake or reservoir, measured when entrainable life stages of RIS are present, is proposed. This approach essentially is based on a 95% protection standard, which is believed to be adequately protective for freshwater locations.

- Biological criteria. The low-risk biological criteria being evaluated again are limited to use in freshwater rivers, reservoirs, and lakes other than the

Great Lakes. Criteria of 5% (or less) loss of a non-harvested species and 1% (or less) loss of a harvested species are being considered as values that would generally pose no risk of adverse impact. These values are low compared to generally allowable fishery harvest management levels. Other biological criteria, also being considered, would take advantage of well-designed long-term monitoring programs and measures such as the multi-metric criteria developed by Duke Power and TVA, in use at many southern reservoirs.

## Use of Risk Assessment and Risk Management Principles

A second method of AEI decisionmaking involves use of EPA's ecological risk assessment/risk management guidelines. This approach includes active stakeholder participation, in which natural resource managers and interested members of the public identify populations of special interest for assessing the potential impact of the CWIS. These form the basis of the next step, which is identification of appropriate methods and analytical approaches. Finally, study endpoints are established to allow easy AEI decisionmaking after data collection. The process can address uncertainty by balancing comprehensiveness of study design, use of fishery information for species of concern, and modeling assumptions.

## Use of Previously Conducted § 316(b) Studies

This approach makes use of the extensive § 316(b) studies already conducted at many facilities. Any studies that are not reasonably current would have to be evaluated to ensure that the studies are representative of the current facility design and biological conditions and that the data collection methods and analytical tools remain valid. In particular, sufficient information must be provided to show that the populations examined continue to be appropriate in terms of fishery management objectives. The objective of this approach is to take full advantage of the previous investment in data collection and evaluation conducted by regulators. Fishery managers and other stakeholders would be able to participate in the evaluation through the NPDES process. The above three decisionmaking approaches could be used independently or in combination. For example, screening criteria could be used initially to provide focus to determine appropriate RIS, with the final decisionmaking done using the risk-based approach. Finally, the decision process for existing facilities would incorporate three additional features: using cost-benefit analysis to maximize net benefit, allowing "environmental enhancements" in appropriate circumstances, and reviewing BTA determinations if new information showed that circumstances had changed.

## *Maximum Net Benefit*

If the decisionmaking process outlined above shows that an existing CWIS is creating, or will create, an appreciable risk of AEI, then the decisionmaker must decide what is "best technology available" or BTA to "minimize" AEI. UWAG's economic consultant advises that the most rational way to make this decision is to choose the technology that maximizes net benefits (that is, benefits minus costs). To use this approach, the permit applicant would have to identify all reasonably available intake structure technologies that would reduce the impact to the aquatic community and be feasible at the site. The applicant would also estimate the costs and benefits of each such technology, including the impacts of the CWIS on aquatic biota, in addition to the monetary costs of construction and operation, energy costs, and environmental costs such as air pollution, aesthetics, and land use. Summing the costs and benefits for each "available" technology, the permittee would choose as "best" the one that had the highest net benefit. Industry believes that cost-benefit analyses suitable for BTA selection can be developed based on existing tools and methods, such as by adopting some features of EPA's BEN model for evaluating the benefits of violating environmental laws or of the methods used to evaluate natural resource damages[52].

## *Environmental Enhancements*

"Environmental enhancements" are actions taken by a facility determined to cause AEI (or a facility that wishes to settle a dispute over its permit) to compensate for the CWIS losses to the affected RIS species rather than install a CWIS technology. Environmental enhancements — such as wetlands creation or fish stocking — are one means of compensating for CWIS losses. In some cases, the most limiting factor for the aquatic populations is not the CWIS, but rather (1) low dissolved oxygen as a result of nutrient enrichment or (2) lack of habitat for spawning and nursery functions[53]. In such cases, by investing dollars addressing the most limiting environmental factors, the facility may spur a more significant recovery to the population than could be achieved through installation of a CWIS technology. Such actions, as long as they are directly related to the fishery, can result in a greater net social benefit than installation of BTA. Enhancements have been used effectively at a number of stations, including Salem, John Sevier, and Chalk Point. Florida Power Corporation's Crystal River fish hatchery is another successful enhancement program. Such environmental enhancements are not "intake technologies" (and therefore cannot be "BTA" nor be required by authority of § 316(b)), but the § 316(b) regulatory framework is flexible enough to allow them to be used, if offered by the permit applicant. They can be employed as a cost-effective means of addressing adverse environmental impact, potentially resulting in environmental benefits greater than use of BTA alone.

## Periodic BTA Review

Once an existing facility has gone through the process of determining that the CWIS is BTA, the BTA status would need to be revisited at the time of permit renewal if the regulatory agency had information showing that the previous studies were no longer valid (for example, that biological conditions had changed). Factors that might result in a change of BTA status would include modifications to the CWIS design or operation or significant changes in the waterbody.

## Decision Process for New Facilities

The process described above is for existing facilities. For new facilities, a "Two Track" approach has been proposed that would allow a company seeking to build a new facility that will require use of surface cooling water either (1) to commit to a highly protective (indeed often over-protective) technology at the outset or (2) to engage in a site-specific analysis to determine whether the intake would create an appreciable risk of AEI and, if so, what would be the BTA for the site.

## Track 1: The "Fast Track"

Track 1, the "Fast Track," would allow the applicant to commit to one of the following highly protective technologies, in return for expedited permitting without the need for pre-operational or operational studies in the source waterbody by using one of two options:

- Option 1: Employ a technology that limits intake flow to the flow that would be required by wet closed-cycle cooling for a given amount of generation at that site and design the average approach velocity (measured in front of the intake screens or the opening to the cooling water intake structure) to be no more than 0.5 ft/s; or

- Option 2: Employ a technology that will achieve a level of protection from impingement and/or entrainment that is reasonably consistent with Option 1. This option is intended to permit facilities to use either standard technologies, or new ones, that have been demonstrated to be effective for the species, type of waterbody, and flow volume proposed for their use. Examples of candidate technologies include:

a.   Wedgewire screens, where there is constant flow, as in rivers;

b.   Traveling fine mesh screens with a fish return system designed to minimize entrainment and impingment mortality; and

c.   Gunderbooms, at sites where they would not be rendered ineffective by high flows or fouling

"Reasonably consistent with" means that an acceptable alternative technology should provide a level of protection within the range expected under Option 1 achieved by flow reductions associated with wet closed-cycle cooling and a 0.5 ft/s approach velocity for the type of waterbody on which the facility is to be sited. Use of highly protective technologies should eliminate the need for periodic BTA review. The effectiveness of wet closed-cycle cooling is well documented. The other technologies listed above promise a level of protection reasonably consistent with that of wet closed-cycle cooling. To prevent impingement, the Gunderboom is designed to have a low approach velocity (almost unmeasurable) and uses a very fine mesh to provide entrainment protection[42]. Wedgewire screens are designed to minimize entrainment and impingement through a combination of small slot width (0.5 to 2 mm) and an approach velocity of less than 0.5 ft/s[42,54].

For fine mesh screens, the survival of fish collected on the screens is speciesand life- stage-specific[41,42]. Survival of many species can be very high, exceeding 90% even at velocities above 0.5 ft/s. As for entrainment, again the effect of fine mesh screens varies by species, but the data indicate that, if control mortality is taken into account, fine mesh screens can reduce entrainment mortality by 90% or more for some species. Other species, such as bay anchovy, have a high mortality both naturally and after encountering fine mesh screens. Nevertheless, given the present state of knowledge, it is reasonable to include fine mesh screens (with a properly designed fish return system) as a candidate technology for some sites that can reduce overall losses to a level (i.e., 90% or better) reasonably consistent with wet closed-cycle cooling. Option 2 of Track 1 encourages alternative or innovative intake structure technologies. A proponent of a new alternative technology would conduct a laboratory or site-specific study appropriate for the waterbody type and species of concern prior to employment of the technology. If the demonstration was successful, after the facility deployed the new technology, monitoring would be conducted as appropriate to validate performance. At a few sites, there could potentially be unusual species-specific circumstances in which Fast Track technologies meeting the above criteria would not be sufficient to avoid AEI. While the number of such sites is likely to be very small, the evaluation process should give permit writers the authority to require additional protective measures if the permitting agency has information to support a finding that exceptional conditions exist such that the proposed facility could affect one or more populations in a way that would not be prevented by other federal or state requirements (such as the Endangered Species Act) and thus has the potential to cause AEI.

### Track 2: A More Tailored Approach

Track 2 of the proposed Two Track approach is similar to the decisionmaking process for existing facilities summarized above. It differs in that Track 2 for new facilities can make use only of predictive fishery management tools, rather than retrospective ones. Track 2 would be for facilities that wished to pursue use of a less stringent BTA. In these cases the applicant could evaluate AEI using the risk screening criteria or the risk assessment/risk management AEI evaluation methods for existing sources. For the population percent reduction criteria, source waterbody type, data availability and assessment, and analytical tool availability will determine the difficulty of predicting impingement rates in a sufficiently quantitative manner. Where this cannot be done, new facilities will need to plan for some kind of technology to protect fish from impingement. The Two Track decision process, then, is both efficient and flexible, and it has one very important advantage: it avoids worsening the already-present "energy crisis" now affecting California and possibly soon other states[55,56,57]. Track 1, the Fast Track, is available for speeding new generating facilities online in parts of the country where they are needed most, in return for a commitment to highly (often overly) protective intake structure technology, and also encourages innovative technologies. Track Two allows a close look at the features of any proposed site and avoids arbitrary, less efficient, restrictions.

## CONCLUSION

The Clean Water Act requires that cooling water intake structures minimize, where it exists, "adverse environment impact." In order to be able to determine whether AEI exists or is threatened, and if it exists to decide how to minimize it, one must first have a definition. The definition needs to ensure the protection of living resources. And the process for "minimizing" AEI needs to strike a balance among competing social needs. Tools are available today to accomplish both these goals. The science of fisheries management provides concepts (like maximum sustainable yield), tools (like biological modeling), and knowledge (such as knowledge of how fish populations compensate for losses) that will allow cooling water users and regulatory agencies to make sound § 316(b) decisions that will protect the living fishery resources. Cost-benefit analysis, drawing on experience of calculating the benefits of environmental violations and natural resource damages, provides a tool for choosing an intake technology that maximizes the net benefits to society. Given a workable definition of AEI and the tools to assess and minimize it, one needs, finally, a decisionmaking process that allows the tools to be used appropriately. The electric utility industry has proposed such a process, one that provides both the opportunity to bring new generating plants online quickly, in return for installing highly (often overly) protective intake technology, and the flexibility to look closely at site characteristics when assessing the risk of AEI, and taking advantage of site characteristics as appropriate to concurrently protect the environment and produce energy efficiency.

## REFERENCES

1.  U.S. Environmental Protection Agency (2000) National pollutant discharge elimination system--regulations addressing cooling water intake structures for new facilities; proposed rule. 65 Fed. Reg. 49,060 (August 10, 2000).

2.  Anderson II, W.A. and Gotting, E.P. (2001) Taken in over intake structures? Section 316(b) of the Clean Water Act. 26 Colum. J. Envtl. L. 1, 1–79

3.  U.S. Environmental Protection Agency (1977) Guidance for evaluating the adverse impact of cooling water intake structures on the aquatic environment: Section 316(b) P. L. 92-500 (Draft).

4.  U.S. Environmental Protection Agency (2001) Notice of data availability; national pollutant discharge elimination system—regulations addressing cooling water intake structures for new facilities. 66 Fed. Reg. 28,853 (May 25, 2001).

5.  Utility Water Act Group (UWAG) (2000) Comments of the Utility Water Act Group on EPA's proposed § 316(b) rule for new facilities and ICR No. 1973. 01.

6.  American Fisheries Society (2000) AFS Policy Statement #1: North American Fisheries Policy (Revised). http://www.fisheries.org/Public_Affairs/Policy_Statements/ps_1.shtml

7.  Barnthouse, L. (1993) Ecological Risk Assessment 26. (in the section called "Assessment Concepts") Suter, G., Ed. cited in Anderson, II, W.A. and Gotting, E.P. (2001) Taken in over intake structures? Section 316(b) of the Clean Water Act. 26 Colum. J. Envtl. L. 1, 3, 19–21.

8.  Beamesderfer, R. (2000) Deciding when intervention is effective and appropriate. Fisheries 25(6), 18–23.

9.   National Marine Fisheries Service (1998) Magnuson-Stevens Act Provisions; National Standard Guidelines; Final Rule. 63 Fed. Reg. 24,212, 24,219 (May 1, 1998).

10.  Mace, P. and Sissenwine, M. (1993) How much spawning per recruit is enough? In Smith, S.J., Hunt, J.J., and Rivard, D., Eds. Risk Evaluation and Biological Reference Points for Fisheries Management. Can. Spec. Publ. Fish. Aquat. Sci. 120, 101–118.

11.  Smith, S.J., Hunt, J.J., and Rivard, D. (Eds.). (1993) Risk Evaluation and Biological Reference Points for Fisheries Management. Can. Spec. Publ. Fish. Aquat. Sci. 120,

12.  Goodyear, C.P. (1993) Spawning stock biomass per recruit in fisheries management: foundation and current use. In Smith, S.J., Hunt, J.J., and Rivard, D., Eds. Risk Evaluation and Biological Reference Points for Fisheries Management. Can. Spec. Publ. Fish. Aquat. Sci. 120, 67–81.

13.  National Research Council (1998) Improving Fish Stock Assessments. National Academy Press, Washington, D.C. 177 pp.

14.  Electric Power Research Institute (1999) Catalog of Assessment Methods for Evaluating the Effects of Power Plant Operations on Aquatic Communities. Report No. TR-112013.

15.  Weeks, H., and Berkeley, S. (2000) Uncertainty and precautionary management of marine fisheries: Can the old methods fit the new mandates? Fisheries 25(12), 6–14,15.

16.  U.S. Environmental Protection Agency. (1998) Guidelines for Ecological Risk Assessment. 63 Fed. Reg. 26,845–26,924 (May 14, 1998).

17.  Auster, P.J. (2001) Defining thresholds for precautionary habitat management actions in a fisheries context. North Am. J. Fisheries Mgmt. 21, 1–9.

18.  Virginia Secretary of Natural Resources (2000) Annual Report on Status of Tributary Strategies, Chesapeake Bay Agreement and Water Quality for Virginia's Chesapeake Bay and Tributaries, p. 5.

19.  U.S. Environmental Protection Agency (2000) Economic and Engineering Analyses of the Proposed § 316(b) New Facility Rule. EPA-821-R-00-019.

20.  Richkus, W.A. and McLean, R. (2000) Historical overview of the efficacy of two decades of two decades of power plant fisheries impact assessment activities in Chesapeake Bay. In Wisniewski, J., Ed. Power Plants & Aquatic Resources: Issues and Assessments. Env. Sci.Policy 3, S283–S293.

21. Tennessee Valley Authority (1998) Browns Ferry Nuclear Plant Thermal Variance Monitoring Program Final Report. TVA Water Management Environmental Compliance. Norris, TN, 54 pp; including supplemental statistical analyses, 10 pp.

22. Electric Power Research Institute (2000) Review of Entrainment Survival Studies: 1970- 2000. Report No. 1000757.

23. Cannon, T.C., Jinks, S.M., King, L.R., and Lauer, G.J. (1978) Survival of entrained ichthyoplankton and macroinvertebrates at Hudson River power plants. In Jensen, L.D., Ed. Proceedings of the Fourth National Workshop on Entrainment and Impingement. EA Communications, Melville, NY. pp. 71–89.

24. Melton, B.R. and Serviss, G.M. (2000) Florida Power Corporation – Anclote Power Plant entrainment survival of zooplankton. In Wisniewski, J., Ed. Power Plants & Aquatic Resources: Issues and Assessments. Env. Sci. Policy 3, S233–S248.

25. U.S. Environmental Protection Agency and U.S. Department of Agriculture. (1998) Clean Water Action Plan: Restoring and Protecting America's Waters, pp. 7–9. http://cleanwater.gov/action/toc.html

26. Ronafalvy, J., Cheesman, R.R., and Matejek, W.M. (2000) Circulating water traveling screen modifications to improve impinged fish survival and debris handling at Salem Generating Station. In Wisniewski, J., Ed. Power Plants & Aquatic Resources: Issues and Assessments. Env. Sci. Policy 3, S377–S382.

27. Electric Power Research Institute (1979) Synthesis and Analysis of Ecological Information from Cooling Impoundments. Report No. EA-1054, Vol. 1.

28. Electric Power Research Institute (1980) Evaluation of a Cooling-Lake Fishery. Volume 1: Introduction, Water Quality & Summary. Prepared by Illinois Natural History Survey. Report No. EA-1148.

29. Electric Power Research Institute (1986) Sport Fishery Potential of Power Plant Cooling Ponds. Prepared by Southern Illinois University at Carbondale. Report No. EA-4838.

30. Taber, C.I. (1969) The Distribution and Identification of Larval Fishes in the Buncombe Creek Arm of Lake Texoma with Observations of Spawning Habits and Relative Abundance. [PhD dissertation] University of Oklahoma. 120 pp.

31. Balon, E.K. (1975) Reproductive guilds of fishes: A proposal and definition. J. Fish. Res. Bd. Can. 32, 821–864.

32. Jones, P.W., Martin, F.D., and Hardy, Jr., J.D. (1978) Development of fishes of the MidAtlantic Bight. Prepared by Chesapeake Biological Laboratory of the University of Maryland for U.S. Fish and Wildlife Service. FWS/OBS-78/12.

33. Bailey, D.E., Loos, J.J., and Perry, E.S. (1998) Studies of cooling water intake structure effects at Potomac Electric Power Company Generating Stations. EPRI Clean Water Act Section 316(b) Technical Workshop. Coolfont Conference Center, Berkeley Springs, WV.

34. Myers, R. (2000) Appendix B to Comments of the Utility Water Act Group on EPA's Proposed § 316(b) Rule for New Facilities and ICR No. 1973.01. 35. Rose, K.A., Cowan, Jr., J.H., Winemiller, K.O., Myers, R.A., and Hilborn, R. (2001) Compensatory density-dependence in fish populations: Importance, controversy, understanding, and prognosis. In press.

35. Van Winkle, W. (2000) A perspective on power generation impacts and compensation in fish populations. In Wisniewski, J., Ed. Power Plants & Aquatic Resources: Issues and Assessments. Env. Sci. Policy 3, S425–S431.

36. Boreman, J. (2000) Surplus production, compensation, and impact assessments of power plants. In Wisniewski, J., Ed. Power Plants & Aquatic Resources: Issues and Assessments. Env. Sci. Policy 3, S445–S449.

37. Meronek, T.G., et al. (1996) A review of fish control projects. North Am. J. Fisheries Mgmt. 16, 63–74.

38. National Marine Fisheries Service (1998) Report of the 26th Northeast Regional Stock Assessment Workshop. Stock Assessment Review Committee Consensus Summary of Assessments. Northeast Fisheries Science Center Reference Document 98-03. Woods Hole, Massachusetts.

39. National Marine Fisheries Service (2000) Report of the 30th Northeast Regional Stock Assessment Workshop. Stock Assessment Review Committee Consensus Summary of Assessments. Northeast Fisheries Science Center Reference Document 00-03. Woods Hole, Massachusetts.

40. Electric Power Research Institute (1999) Fish Protection at Cooling Water Intakes: Status Report. Report No. TR-114013.

41. Taft, E.P. (2000) Fish protection technologies: A status report. In Wisniewski, J., Ed. Power Plants & Aquatic Resources: Issues and Assessments. Env. Sci. Policy 3, S349–S359.

42. Taft, E.P., Horst, T.J., and Downing, J.K. (1981) Biological evaluation of a fine-mesh traveling screen for protecting organisms. Presented at

the Workshop on Advanced Intake Technology, San Diego, CA, April 22–24, 1981.

43. Brueggemeyer, B., Cowdrick, D., and Durrell, K. (1988) Full-scale operational demonstration of fine mesh screens at power plants. In Proceedings of the Conference on Fish Protection at Steam and Hydro Plants, San Francisco, CA, October 28–30, 1987. Electric Power Research Institute CS/EA/AP-5663-SR.

44. Veneziale, E.J. (1991) Fish protection with wedge wire screens at Eddystone Station. In Proceedings of the American Power Conference.

45. Hanson, B.N., Bason, W.H., Beitz, B.E., and Charles, K.E. (1978) A practical intake screen which substantially reduces entrainment. In Fourth National Workshop on Entrainment and Impingement, Chicago, IL, December 5, 1977.

46. Lifton, W. (1979) Biological aspects of screen testing on the St. Johns River, Palatka, Florida. Prepared for Passive Intake Screen Workshop, Chicago, IL, December, 1979.

47. Lawler, Matusky & Skelly Engineers (1997) Lovett Generating Station Gunderboom Evaluation Program 1996. Prepared for Orange and Rockland Utilities, Inc.

48. Lawler, Matusky & Skelly Engineers (1998) Lovett Generating Station Gunderboom Evaluation Program 1998. Prepared for Orange and Rockland Utilities, Inc.

49. U.S. Nuclear Regulatory Commission (1996) Generic Environmental Impact Statement for License Renewal of Nuclear Plants. Main Report. Final Report. NUREG-1437, Vol. 1. pp. 163–183.

50. Argonne National Laboratory (1992) Impact on the Steam Electric Power Industry of Deletion of § 316(a) of the Clean Water Act: Phase 2, Energy and Environmental Impacts.

51. U.S. Environmental Protection Agency (1999) BEN Users Manual. Office of Enforcement and Compliance Assurance.

52. U.S. Environmental Protection Agency. (2000) National Water Quality Inventory: 1998 Report to Congress. EPA 841-R-00-001.

53. Ehrler, C. and Raifsnider, C. (2000) Evaluation of the effectiveness of intake wedgewire screens. In Wisniewski, J., Ed. Power Plants & Aquatic Resources: Issues and Assessments. Env. Sci. Policy 3, S361–S368.

54. Smith, R. and Emshwiller, J.R. (2001) Why California isn't the only place bracing for electrical shocks. Wall St. J. Apr. 26, 2001. Page A1 col. 6.

55. National Energy Policy Development Group (2001) National Energy

Policy. U.S. Government Printing Office, Washington, D.C. ISBN 0-16-050814-2.

56.  North American Electric Reliability Council (2001) 2001 Summer Assessment: Reliability of the Bulk Electricity Supply in North America. 64 pp.

# Chapter 9

## DEMOGRAPHIC DIVERSITY AND SUSTAINABLE FISHERIES

Masami Fujiwara
Department of Wildlife and Fisheries Sciences, Texas A&M University, College Station, Texas, United States of America

## ABSTRACT

Fish species are diverse. For example, some exhibit early maturation while others delay maturation, some adopt semelparous reproductive strategies while others are iteroparous, and some are long-lived and others short-lived. The diversity is likely to have profound effects on fish population dynamics, which in turn has implications for fisheries management. In this study, a simple density-dependent stage-structured population model was used to investigate the effect of life history traits on sustainable yield, population resilience, and the coefficient of variation (CV) of the adult abundance. The study showed that semelparous fish can produce very high sustainable yields, near or above 50% of the carrying capacity, whereas long-lived iteroparous fish can produce very low sustainable yields, which are often much less than 10% of the carrying capacity. The difference is not because of different levels of sustainable fishing mortality rate, but because of difference in the sensitivity of the equilibrium abundance to fishing mortality. On the other hand, the resilience of fish stocks increases from delayed maturation to early maturation strategies but remains almost unchanged from semelparous to long-lived iteroparous. The CV of the adult abundance increases with increased fishing mortality, not because more individuals are recruited into the adult stage (as previous speculated), but because the mean abundance is more sensitive to fishing mortality than its standard deviation. The magnitudes of these effects vary depending on the life history strategies of the fish species involved. It is evident that any past high yield of long-lived iteroparous fish is a transient yield level, and future commercial fisheries should focus more on fish that are short-lived (including semelparous species) with high compensatory capacity.

# INTRODUCTION

Organisms exhibit a wide range of life history strategies [1]. For example, some have early maturation while others delay maturation, some adopt semelparous reproductive strategies while others are iteroparous, and some are long-lived and others short-lived. Such demographic diversity is likely to have profound effects on population dynamics. As fisheries management worldwide faces the challenge of managing fish stocks that encompass broad demographic diversity, there is great interest in investigating the relationship between life history traits and population dynamics [2], [3], [4], [5], [6], [7], and the need to adjust fish stock management based on fish life history strategies, e.g. [8], [9], [10], [11], [12]. In this study, a simple population model was used to assess the effects of demographic diversity on population dynamics under fishing mortality.

One of the most important concepts in fishery management is sustainable yield, see [13], [14],[15]. In a general sense, sustainable yield is a consistent catch over a long (often infinite) period of time, e.g. [16] and often equated to the catch level that results in a stable equilibrium abundance under a deterministic fishery model. When it is at the maximum level, the sustainable yield is called the maximum sustainable yield. When a model-based MSY is available, the desirable yield or fishing mortality can be determined after incorporating precautionary measures that reflect uncertainties, including fluctuations in the environment, errors in parameter estimates, and deficiencies in model formulations [17]. As sustainable yield plays an important role in fishery management, I investigate how sustainable yield and MSY varies with different life history traits of target fish and the fishing mortality rate.

Another objective of this study was to investigate how the sensitivity of transient population dynamics is affected by the life history traits of fish and the fishing mortality rate. Understanding transient dynamics is important in the management of natural resources [18],[19], [20] because a large part of what we actually observe are transient dynamics. In this study, transient population dynamics are measured in two ways. The first involves estimating the coefficient of variation (CV) of adult fish abundance [21] under stochastically fluctuating juvenile survival, which is commonly thought to be the fish population parameter most sensitive to fluctuating environmental conditions. A recent study demonstrated that increased fishing mortality also increases the CV of adult fish abundance [21]. There is great interest in understanding how this measure is affected by various factors, e.g. [22] because unpredictability associated with a large fluctuation in fish abundance will reduce the optimal fishing quota [23]. The second measure of transient dynamics involves calculation of the resilience of the population abundance near a stable equilibrium point.

Resilience is a measure of the time that takes for a population to return to asymptotic dynamics after a perturbation, e.g. [24], [25], [26]. Resilience is a measure of intermediate-term transient dynamics, whereas the CV of adult fish abundance is a measure of short-term transient dynamics. In this study, the CV of adult fish abundance, resilience, and MSY were used to characterize short-, intermediate-, and long-term dynamics, respectively, of fishery models incorporating demographic diversity.

In addition to life history traits, a major factor affecting population dynamics is density- dependent regulation, which makes model equations non-linear. Density dependence is necessary for fishery sustainability, and the processes involved have been the subject of much research since Verhulst [27] developed the logistic model. Theoretical understanding of the potential dynamics that can arise from deterministic density-dependent population models is well established. The dynamics converge asymptotically to a stable equilibrium, cycle, aperiodic loop, or chaotic attractor, e.g. [28], [29], [30]. Two types of density- dependent processes are commonly used in fisheries population models. The first is an over-compensatory density-dependent process represented by the Ricker model [31], and the second is compensatory process represented by the Beverton-Holt model [32]. The focus of this study was on density-dependent regulation that results from resource limitation (e.g. competition for available food). This type of regulation is likely to be more common and tends to lead to the Beverton-Holt density-dependent process. Furthermore, the Beverton-Holt density dependence is discrete-time equivalent to the classic logistic model.

# METHODS

## Model

The aim of the study was to investigate the effect of demographic diversity on transient and asymptotic population dynamics under various fishing mortality rates, and involved the use of a density-dependent two-stage matrix population model. Although the model is simple, it can incorporate a wide range of life history traits by varying parameter values. In particular, when the life history is semelparous, the model is discrete-time equivalent to a simple logistic (Shaffer) model, which is still widely used in fishery modeling. Therefore, the model presented herein is more general and applicable to real fish populations than the majority of existing fishery population models. Finally, because of the simplicity, asymptotic abundance, sustainable yield, MSY, and resilience can be calculated analytically. This allowed exploration of the model under a wide range of parameter values.

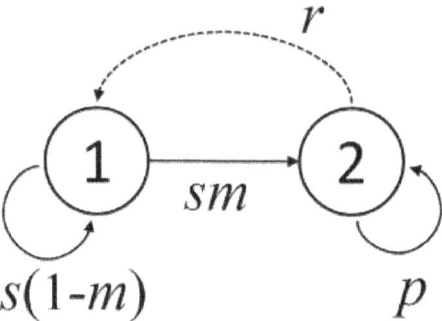

**Figure 1:** Lifecycle graph of a two-stage model with stage 1 for juveniles and stage 2 for adults. Arrows indicate possible contributions of one stage to the other. $s$: annual survival rate of juveniles, $m$: annual maturation rate, $p$: annual survival rate of adults, and $r$: annual fertility rate. doi:10.1371/journal.pone.0034556.g001

**Table 1:** Juvenile and adult stage duration after the first reproduction under the parameter values considered in the analysis. doi:10.1371/journal.pone.0034556.t001

| | | $m=0.1$ | | | $m=0.5$ | | | $m=0.9$ | | |
| | | Figure Panel | Juvenile Duration (years) | Adult Duration (years) | Figure Panel | Juvenile Duration (years) | Adult Duration (years) | Figure Panel | Juvenile Duration (years) | Adult Duration (years) |
|---|---|---|---|---|---|---|---|---|---|---|
| $s=0.1$ | $p=0.0$ | 4a | 1.10 | 1 | 4b | 1.05 | 1 | 4c | 1.01 | 1 |
| | $p=0.5$ | 4d | 1.10 | 2 | 4e | 1.05 | 2 | 4f | 1.01 | 2 |
| | $p=0.9$ | 4g | 1.10 | 10 | 4h | 1.05 | 10 | 4i | 1.01 | 10 |
| $s=0.5$ | $p=0.0$ | 3a | 1.82 | 1 | 3b | 1.33 | 1 | 3c | 1.05 | 1 |
| | $p=0.5$ | 3d | 1.82 | 2 | 3e | 1.33 | 2 | 3f | 1.05 | 2 |
| | $p=0.9$ | 3g | 1.82 | 10 | 3h | 1.33 | 10 | 3i | 1.05 | 10 |
| $s=0.9$ | $p=0.0$ | 2a | 5.26 | 1 | 2b | 1.82 | 1 | 2c | 1.10 | 1 |
| | $p=0.5$ | 2d | 5.26 | 2 | 2e | 1.82 | 2 | 2f | 1.10 | 2 |
| | $p=0.9$ | 2g | 5.26 | 10 | 2h | 1.82 | 10 | 2i | 1.10 | 10 |

The model consists of two stages (Fig. 1). The first (the juvenile stage) is for reproductively immature individuals, and the second (the adult stage) is for mature individuals. The model includes four population parameters: juvenile survival rate ($s$), adult survival rate ($p$), maturation rate ($m$), and fertility ($r$). It was assumed that the time unit was one year, and therefore the rates were annual rates. Fertility ($r$) is the product of the survival of adults until the reproductive season and the annual per capita fecundity (the number of eggs). This type of matrix population model is called a post-breeding model [33], in which the population abundance is determined immediately after spawning event. Because the survival rate of eggs and juveniles are often substantially different for fish, it is assumed that the multiplicative difference between the egg and juvenile survival is also implicitly included in $r$. When the four parameters in the model are density independent, the densities of juveniles ($n_1$) and adults ($n_2$) are "projected" from year $t$ to the next year, as follows:

$$\begin{bmatrix} n_1 \\ n_2 \end{bmatrix}_{t+1} = \begin{bmatrix} s(1-m) & r \\ sm & p \end{bmatrix}\begin{bmatrix} n_1 \\ n_2 \end{bmatrix}_t,$$  (1)

where the subscripts of the vectors denote year. The matrix is in general termed a population matrix. The dominant eigenvalue of the matrix gives the annual asymptotic population growth rate, and an associated right eigenvector gives the relative asymptotic densities between the two stages, see [31].

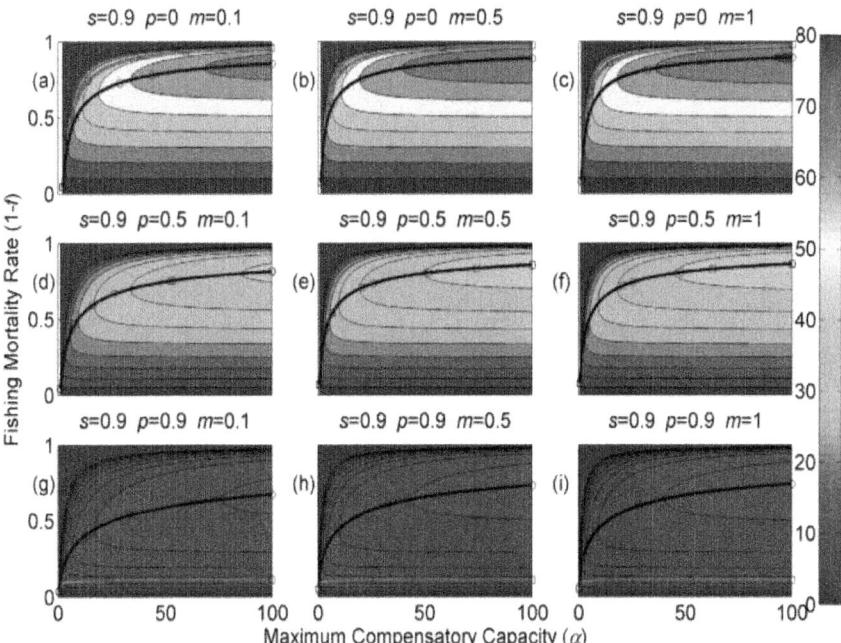

**Figure 2:** Sustainable yield (shown in contours) as a function of the annual fishing mortality rate and the maximum compensatory capacity when $s=0.9$. The black curves with circle markers indicate the annual fishing mortality rate at the maximum sustainable yield as a function of the maximum compensatory capacity. The red curves with square markers indicate the annual fishing mortality rate when the equilibrium adult abundance is 50% of the carrying capacity (i.e. asymptotic adult abundance when there is no fishing mortality) as a function of the maximum compensatory capacity. Each panel represents a life history strategy defined by the parameters shown above.
doi:10.1371/journal.pone.0034556.g002

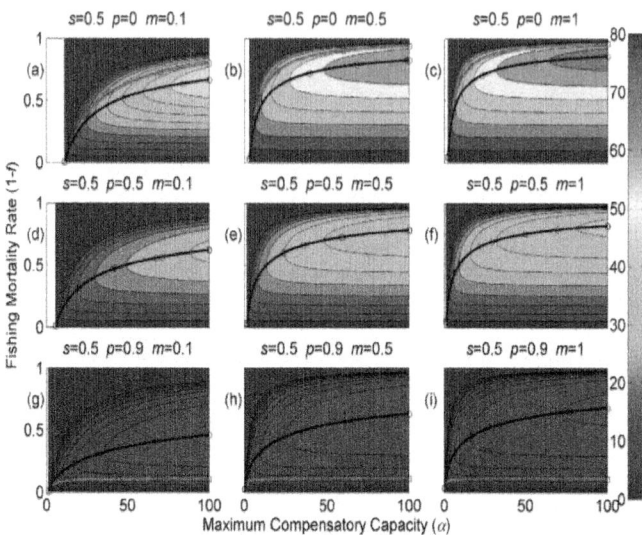

**Figure 3:** Sustainable yield (shown in contours) as a function of the annual fishing mortality rate and the maximum compensatory capacity when $s=0.5$. The black curves with circle markers indicate the annual fishing mortality rate at the maximum sustainable yield as a function of the maximum compensatory capacity. The red curves with square markers indicate the annual fishing mortality rate when the equilibrium adult abundance is 50% of the carrying capacity (i.e. asymptotic adult abundance when there is no fishing mortality) as a function of the maximum compensatory capacity. Each panel represents a life history strategy defined by the parameters shown above. doi:10.1371/journal.pone.0034556.g003

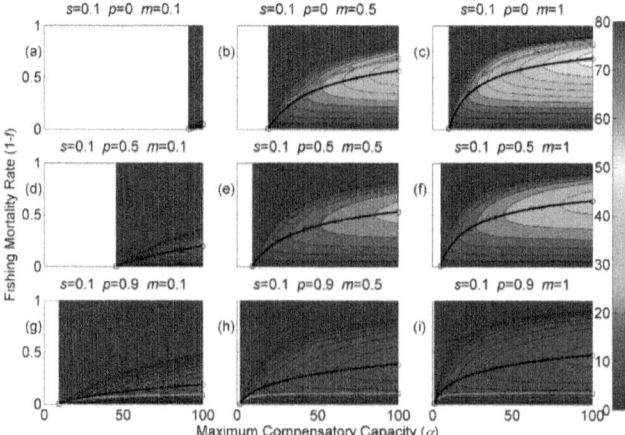

**Figure 4:** Sustainable yield (shown in contours) as a function of the annual fishing mortality rate and the maximum compensatory capacity when $s=0.1$. The black curves

with circle markers indicate the annual fishing mortality rate at the maximum sustainable yield as a function of the maximum compensatory capacity. The red curves with square markers indicate the annual fishing mortality rate when the equilibrium adult abundance is 50% of the carrying capacity (i.e. asymptotic adult abundance when there is no fishing mortality) as a function of the maximum compensatory capacity. Each panel represents a life history strategy defined by the parameters shown above. doi:10.1371/journal.pone.0034556.g004

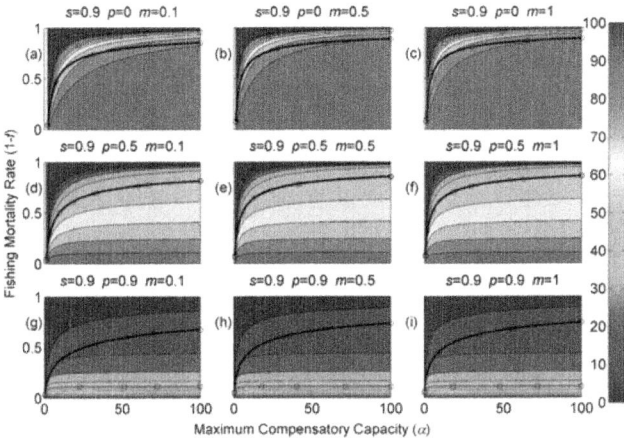

**Figure 5:** Equilibrium adult abundance (shown in contours) as a function of the annual fishing mortality rate and the maximum compensatory capacity when $s=0.9$. The black and red curves are the same as shown in Figure 2. Each panel represents a life history strategy defined by the parameters shown above. doi:10.1371/journal.pone.0034556. g005

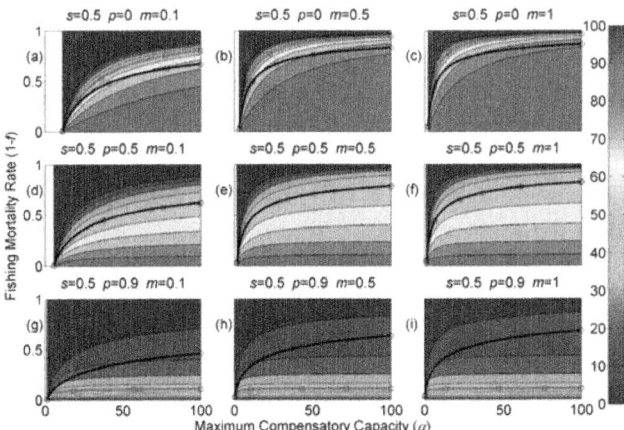

**Figure 6:** Equilibrium adult abundance (shown in contours) as a function of the annual fishing mortality rate and the maximum compensatory capacity when $s=0.5$. The black

and red curves are the same as shown in Figure 3. Each panel represents a life history strategy defined by the parameters shown above. doi:10.1371/journal.pone.0034556. g006

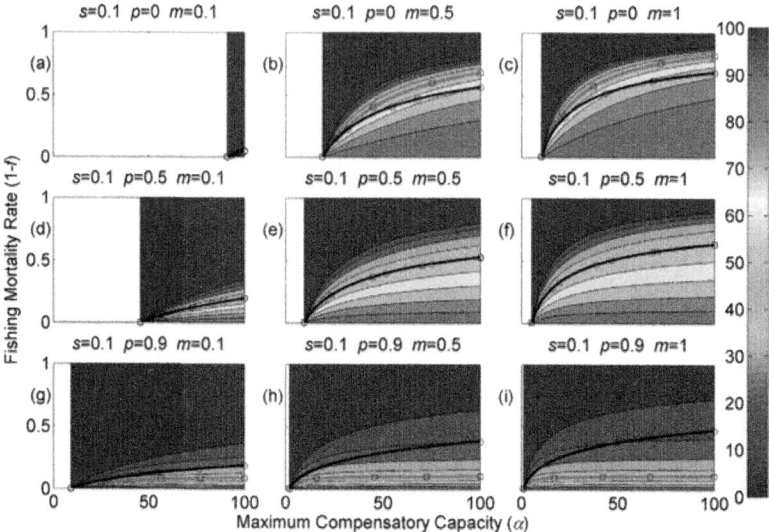

**Figure 7:** Equilibrium adult abundance (shown in contours) as a function of the annual fishing mortality rate and the maximum compensatory capacity when $s$=0.1. The black and red curves are the same as shown in Figure 4. Each panel represents a life history strategy defined by the parameters shown above. doi:10.1371/journal.pone.0034556. g007

By varying parameter values, various life history strategies can be incorporated into the model. For example, by increasing $m$ from a low to a high value, the life history strategy is modified from maturing early (precocious) to maturing late (delayed maturation). Together $m$ and $s$ determine the mean age of maturation. By increasing $p$ from 0 toward 1 the life history strategy changes from semelparous to iteroparous. It is noted that even if $p$=0, a positive value of $r$ ensures that some individuals will reproduce once before their death. The generation time of the organisms is determined by $m$, $s$ and $p$. Under this model, the average time individuals spend in the juvenile stage before maturing is given by $(1-s(1-m))^{-1}$, and the average time individuals spend in the adult stage is given by $(1-p)^{-1}$. These average stage durations can be used to approximate actual species of fisheries interest using modeled life history strategies.

Fishing mortality affects adult survival. It is incorporated into the model by multiplying the rate of surviving from fishing mortality ($f$)

with parameters in the projection matrix. Thus, $1-f$ is the annual fishing mortality rate. With fishing mortality included, the equation becomes:

$$\begin{bmatrix} n_1 \\ n_2 \end{bmatrix}_{t+1} = \begin{bmatrix} s(1-m) & rf \\ sm & pf \end{bmatrix} \begin{bmatrix} n_1 \\ n_2 \end{bmatrix}_t. \qquad \text{(2)}$$

The fertility term is also multiplied by $f$ because $r$ includes the adult survival rate as described previously. The sequence of event is that (1) fishing mortality and natural adult mortality and then (2) spawning. The use of a post-breeding matrix population model avoids spurious results in which a population is persistent with the adult fishing mortality rate of 1 because newly recruited adults always have at least one chance of reproduction. An alternative way of modeling is to have a pre-breeding model, and the <2,1> entry of the population matrix is multiplied by $f$ instead of the <1,2> entry.

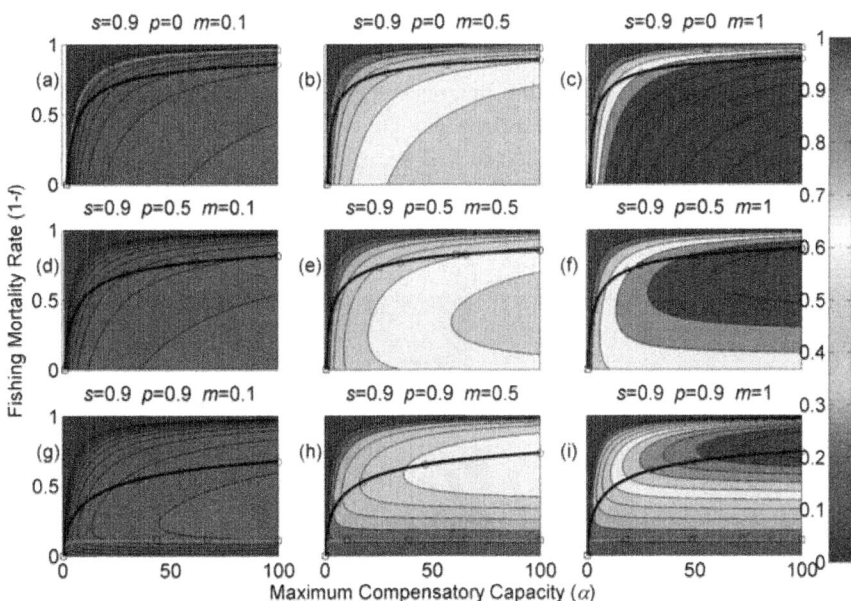

**Figure 8:** Resilience of the equilibrium densities (shown in contours) as a function of the annual fishing mortality rate and the maximum compensatory capacity when $s=0.9$. The black and red curves are the same as shown in Figure 2. Each panel represents a life history strategy defined by the parameters shown above. doi:10.1371/journal. pone.0034556.g008

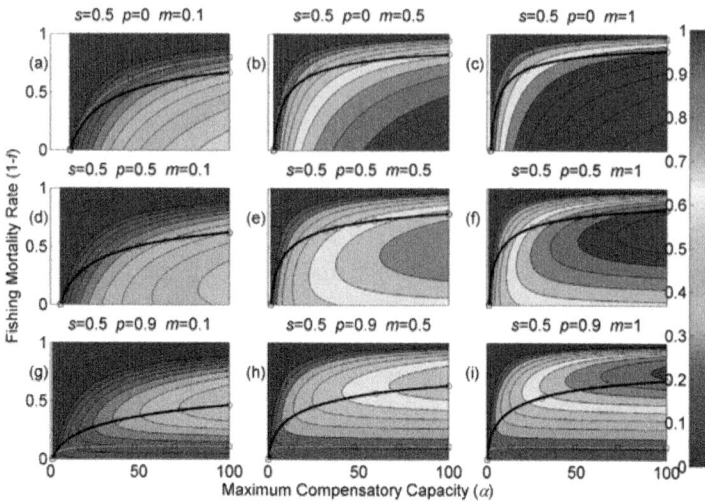

**Figure 9:** Resilience of the equilibrium densities (shown in contours) as a function of the annual fishing mortality rate and the maximum compensatory capacity when $s=0.5$. The black and red curves are the same as shown in Figure 3. Each panel represents a life history strategy defined by the parameters shown above. doi:10.1371/journal.pone.0034556.g009

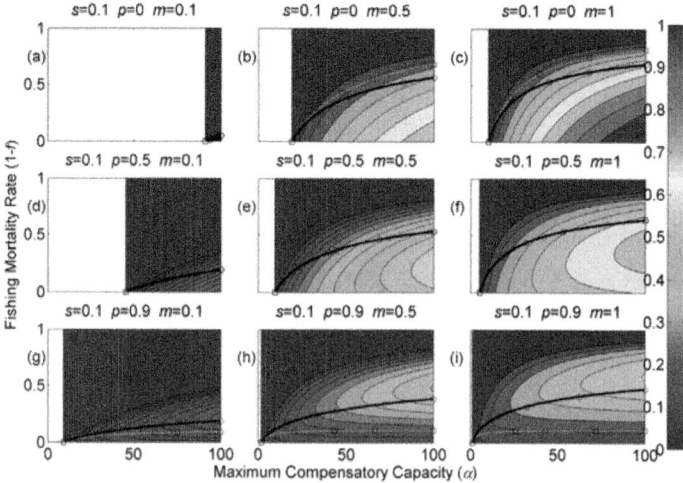

**Figure 10:** Resilience of the equilibrium densities (shown in contours) as a function of the annual fishing mortality rate and the maximum compensatory capacity when $s=0.1$. The black and red curves are the same as shown in Figure 4. Each panel represents a life history strategy defined by the parameters shown above. doi:10.1371/journal.pone.0034556.g010

The Beverton-Holt density-dependent process is also incorporated into the model. Here, it was assumed that the effect of the density-dependent process is on fecundity and/or the survival during the first year of life. Therefore, it was incorporated into the fertility term, which includes both fecundity and the difference in survival rate between egg and juvenile stage, as:

$$r_t = \frac{\alpha}{1 + \beta f n_2^{(t)}}, \qquad (3)$$

which replaces $r$ in equation (2). Parameter $\alpha$ is the maximum per capita fertility rate at a low adult abundance, and it is referred to here as the maximum compensatory capacity ($\alpha$). For a given value of $\alpha$, parameter $\beta$ determines the equilibrium adult abundance. It can be shown that under this model if there is an equilibrium point at which both juvenile and adult densities are positive, the equilibrium point is stable.

In equation (1), there are four population parameters. However, under the stable equilibrium point the dominant eigenvalue of the population matrix is 1. Therefore, the number of free population parameters is three under the equilibrium point without fishing mortality, and consequently the fertility rate at the equilibrium point was chosen to be varied as a function of $m$, $s$ and $p$. Thus, by setting the eigenvalue to 1, the fertility term at the equilibrium under no fishing mortality ($r^*$) can be solved as follows.

$$r^* = \frac{(1-p)(ms-s+1)}{ms} \qquad (4)$$

There are two density-dependent parameters ($\alpha$ and $\beta$) in equation (3). However, for given values of $\alpha$, $r^*$ and an equilibrium adult abundance $n_2^*$, parameter $\beta$ is expressed as:

$$\beta = \frac{1}{n_2^*}\left(\frac{\alpha}{r^*} - 1\right), \qquad (5)$$

when fishing mortality rate is 0 ($f=1$). In the subsequent analyses, the equilibrium adult abundance without any fishing mortality was set at 100 under all life history traits. Hereafter, this level is referred to as the carrying capacity. Setting the carrying capacity at 100 allows any reduction in the equilibrium abundance because of fishing mortality to be interpreted as a percent decline from the carrying capacity. Similarly, sustainable yield can also be related to the percent of the carrying capacity.

The use of a simple model means various simplifying assumptions were made. For example, instead of modeling both biomass and fish abundance (common practices in fishery modeling), the focus was placed solely on the latter. Similarly, instead of categorizing fish into different sizes as well as maturation status, the model only incorporates two stages; juveniles and adults. Consequently, although the model can incorporate a wide range of fish life history strategies, it does not incorporate all strategies. Instead, the model was designed to present a broad picture of the population dynamics of fish over a wide range of life history strategies, thus complementing existing efforts to build and analyze more complex fishery models, which are often stock specific.

The model in this study is closely related to stock recruitment/replacement-line models, which are frequently used in fisheries management, e.g. [14], [15]. In the current model, the stock recruitment relationship, e.g. [34] is given by the relationship between $n_{2,t}$ (stock) and $n_{1,t+1}sm$ (new recruitment). The slope (often the inverse of the slope) of the replacement line, which is stock abundance per new recruit, is given by the ratio between $n_{2,t+1}$ and $(n_{1,t}sm)$. However, the model includes population parameters more explicitly, enabling the investigation of life history variations.

The model is also closely related to the surplus production model, which is often expressed in terms of ordinary differential equations. For example, the most basic surplus production model uses a logistic growth model with added instantaneous fishing mortality, see [14], [35]. The current model is equivalent to the surplus production model except it includes life history strategy of fish and it is modeled with life history strategy events including reproduction, compensatory density dependence, survival, maturation and fishing mortality occurring in a discrete sequence rather than simultaneously. The Beverton-Holt density dependence is also a discrete-time equivalent to the density dependence under the logistic equation.

## Analyses

The ways in which equilibrium fishery yield, resilience, and the CV of adult abundance responded to changes in the parameters in the model (i.e. life history traits) were investigated. In particular, $s$, $m$ and $p$ were varied among low, intermediate and high values ($s$: 0.1, 0.5, 0.9; $m$: 0.1, 0.5, 1.0; $p$: 0, 0.5, 0.9). Average stage durations were calculated for each combination of parameters in the absence of fishing mortality. The equilibrium fishery yield, resilience, and CV of adult abundance were then calculated over ranges of $\alpha$ and fishing mortality $(1-f)$, as described below.

Fishery yield at equilibrium $(Y^*)$ is given by

$$Y^* = (1-f)n_2^*,\tag{6}$$

where $n_2^*$ is the equilibrium adult abundance under a fishing morality rate of $1-f$. The expression for the equilibrium adult abundance is obtained from equations (2) and (3) as

$$n_2^* = \frac{1}{\beta}\left(\frac{\alpha ms}{(1-pf)(ms-s+1)} - \frac{1}{f}\right),$$

(7)

by setting the dominant eigenvalue of the population matrix (2) to be 1 and solving the equation for the adult abundance.

The fishing mortality rate at MSY satisfies the equation that is derived by taking the derivative of $Y^*$ with respect to $f$, and setting the derivative to 0. In equation (6), $n_2^*$ is also a function of $f$, and consequently the expression for the fishing mortality rate at MSY is not simple. However, the procedure for analytically taking the derivative is straightforward. The fishing mortality rate at MSY can be substituted into equations (6) and (7) to obtain the MSY.

The resilience (R) of a density-dependent matrix population model around the equilibrium point measures how quickly a population returns to an equilibrium point after a perturbation, and it is given by the negative slope of a log transformed difference between population abundance and the equilibrium point approaching 1. It is given by:

$$R = -\log(\lambda_J),$$

(8)

where $\lambda_J$ is the absolute value of the dominant eigenvalue of a linearized population projection matrix around the equilibrium point [33]. The linearized population projection matrix in this study was:

$$A^* + \left[\frac{\partial A}{\partial n_1}\mathbf{n}^* \quad \frac{\partial A}{\partial n_2}\mathbf{n}^*\right] = \begin{bmatrix} s(1-m) & \dfrac{\alpha f}{\left(1+\beta f n_2^*\right)^2} \\ sm & pf \end{bmatrix},$$

where * indicates that the population is at the equilibrium point and all of the derivatives are also evaluated at the equilibrium point. The linearized matrix is similar to the original population matrix because only one of the four parameters experience density dependence in the model. As it is a 2×2 matrix, obtaining the dominant eigenvalue of this matrix was also straightforward.

The CV of adult abundance [21] was estimated by incorporating stochastic perturbations into $s$, and simulating the population over 1000 time steps starting from the equilibrium abundance of a deterministic model. In this simulation, $s$ was allowed to fluctuate according to the Beta distribution with means of 0.1, 0.5, and 0.9, and a constant variance of 0.01, without any serial autocorrelation. After each 1000 time step simulation, the CV of adult abundance was estimated.

# RESULTS

The average stage durations of various life history traits considered in this study are shown inTable 1 for reference. Hereafter, species with $p$=0.0, 0.5, and 0.9 are referred to as semelparous, short-lived iteroparous, and long-lived iteroparous, respectively. Similarly, species with a juvenile stage longer than 1.5 years are referred to as having delayed maturation, while others are considered to be precocious.

Asymptotic fishery yields as a function of fishing mortality and the maximum compensatory capacity are shown in the contours in Figures 2, 3, 4 (for $s$=0.9, 0.5, and 0.1, respectively). For example, Figure 2a shows asymptotic fishery yield for one of the semelparous precocious life history strategies. For a given value of maximum compensatory capacity, the sustainable yield initially increases as fishing mortality is increased from 0. This is because fishers catch more with increased effort. However, it peaks at some intermediate value of fishing mortality and declines thereafter because the equilibrium abundance of adults declines rapidly with increased mortality when the mortality rate is high. The grey area in the left side of some panels indicates that the population is not sustainable even without fishing mortality because of a low maximum compensatory capacity. As we go down the panels in each figure, life history is semelparous ($p$=0) short-lived interoparous ($p$=0.5), and long-lived iteroparous ($p$=0.9). As we move from the left to right panels, maturation rate increases.

The black curves with circles show the level of fishing mortality at the maximum sustainable yield for a given value of maximum compensatory capacity. As maximum compensatory capacity is increased, the population can be fished at a higher rate because the losses from the fishery are better compensated for, and the MSY also increases.

The red curves with square markers show the annual fishing mortality rate that reduces the adult abundance to 50% of the carrying capacity (i.e. the equilibrium abundance under no fishing mortality). This mortality level is referred to as the $0.5\,K$ mortality. It should be noted, in the model, the equilibrium abundance is measured prior to fishing mortality but after recruitment. Under a traditional surplus production model with a logistic equation and instantaneous yield, which may be constant or proportional to fish abundance, the MSY is achieved when the population abundance is reduced to 50% of the carrying capacity. Figures 2, 3, 4 shows that the MSY fishing mortality is higher than the $0.5\,K$ mortality when the survival rate of adults is high ($p$=0.9). Conversely, the MSY fishing mortality is lower than the $0.5\,K$ mortality when the survival rate of adults is low ($p$=0). They are equal when the survival rate of adults is at the intermediate value ($p$=0.5) under all juvenile survival and maturation rates.

It is evident that the sustainable yield at a given value of fishing mortality and MSY declines from the semelparous ($p=0$) to the iteroparous ($p=0.5$) strategies, and also from short-lived iteroparous ($p=0.5$) to long-lived iteroparous ($p=0.9$) strategies. Therefore, the sustainable yield is the property associated with the adult parameter. This is despite the fact that the fishing mortality rate at MSY only slightly decreases from the semelparous to the long-lived iteroparous. For long-lived iteroparous species the maximum sustainable yield is always less than 10% of the carrying capacity, whereas for semelparous species the MSY can be greater than 50% of the carrying capacity under some parameter values. This results from differences in equilibrium abundance under fishing mortality among the various life history strategies.

The dark blue region in the upper left corner in each panel of Figures 2, 3, 4 indicates the region of unsustainable fisheries. As $s$ is reduced (i.e. from Figure 2 to 3 and Figure 3 to 4), the region of unsustainable fisheries increases, and this is more pronounced when $m$ is low. This occurs because, at a reduced $s$, few individuals are recruited into the adult stage. Consequently, only a slight increase in the adult mortality as a consequence of fishing will cause fishing to become unsustainable.

Equilibrium adult abundance as a function of fishing mortality and maximum compensatory capacity are shown in the contours in Figures 5, 6, 7 (for $s=0.9$, 0.5, and 0.1, respectively).Figure 5a shows the equilibrium adult abundance for one of the semelparous precocious life history strategies. For a given value of maximum compensatory capacity, as fishing mortality rate is increased, equilibrium adult abundance declines. When maximum compensatory capacity is low, the decline starts early and quickly reaches very low abundance whereas, when maximum compensatory capacity is high, the population can maintain high abundance with higher fishing mortality rate. The black curve with circles and the red curves with squares are the same as before: fishing mortality rate at MSY and $0.5\,K$ abundance, respectively. As adult survival rate (i.e. as we go down the panels) is increased, the equilibrium adult abundance declines substantially. On the other hand, maturation rate has almost no effect on the equilibrium adult abundance. These trends remain the same with lower juvenile survival rate (Fig. 6 and 7). However, as juvenile survival rate declines, the equilibrium abundance declines faster with increasing fishing mortality rate.

The resilience of a population around the equilibrium point is shown in Figures 8, 9, 10 (for $s=0.9$, 0.5, and 0.1, respectively). With low maximum compensatory capacity, the population resilience always declines with increasing fishing mortality. This means that if there is a perturbation to the population, such as a natural or anthropogenic disaster, it will take longer to return to the equilibrium point as fishing mortality increases. Conversely, when maximum compensatory capacity is high, the resilience initially increases with increased fishing mortality, peaks, and then declines. However, irrespective of whether maximum compensatory capacity is high or low, when the fishing mortality exceeds the level for MSY, in most cases, the resilience declines with increasing fishing mortality.

Resilience is also affected by life history traits (Fig. 8, 9, 10). It increases with increasing maturation rate $m$ and juvenile survival rate $s$. Although resilience decreases with increasing adult survival rate $p$, the change is by small amount. Therefore, the resilience is the property associated with parameters of juvenile stage. Finally, increased maximum compensatory capacity also increases resilience because of high compensatory capability.

The CV of the adult abundance increased with increased fishing mortality in all cases (Fig. 11,12, 13). This implies that the population fluctuates more with higher fishing mortality rate. This occurs regardless of whether the fish have iteroparous, semelparous, precocious, or delayed maturation life history strategies.

## DISCUSSION

A simple stage-structured model was used to investigate how fishing mortality affects the population dynamics of fish with different life history strategies. Although simple, the model can encompass a wide variety of life history strategies. For example, Winemiller and Rose [5]investigated 10 life history traits of 216 North American fish species. The study revealed that the species can be categorized by three attributes: juvenile survivorship, generation time, and fecundity. The model in the present study also included three life history parameters that were varied to represent demographic diversity. Juvenile survival was explicitly included in the model, and generation time and fertility (instead of fecundity) were also functions of the three parameters in the model. Therefore, similar life history variation was investigated in the current study, but was parameterized in different ways. Fecundity (number of eggs produced) was not explicitly included in the model because the number of eggs that a fish produces is often a local/regional adaptation to the environment [36]. Of more importance to the overall population dynamics is the product of the survival rate of adults over the time-scale of a population model (1 year in the model in this study) and their fecundity (or fecundity and the survival of offspring over the time-scale of a population model in pre-breeding model).

The results suggest that the sustainable yield is reduced from long-lived iteroparous species (panels g, h, i in Fig. 2, 3, 4) to short-lived iteroparous species (panels d, e, f in Fig. 2, 3, 4) and also from short-lived iteroparous to semelparous (panels g, h, i in Fig. 2, 3, 4) under the same fishing mortality rate. This results from reduced adult equilibrium abundance, and suggests that the sensitivity of the equilibrium abundance increases as the duration of an adult stage increases. This means that, although we can rebuild over-exploited longer-lived fish by stopping its exploitation, after resuming exploitation, population abundance will decline again even with a low level of fishing mortality rate.

The MSY also exhibits large differences among different life history strategies. For example, there is a large difference in the MSY between long-lived iteroparous and semelparous fish (Fig. 2, 3, 4). For long-lived iteroparous species, the MSY was always less than 8% (often much less than that) of the carrying capacity. This result suggests that the large yields that fishers may have obtained in the past with some long-lived fish are transient yield and cannot be sustained at the same level. The only way to achieve those yields (although still transiently) is to rebuilt the stocks to historical levels. It is possible that the low MSY for long-lived iteroparous species will not be economically viable for many fishery stocks. However, the MSY of semelparous fish can be 10 fold greater than that of long-lived iteroparous species under the same fishing mortality.

In contrast to MSY and sustainable yield, fishing mortality rate at MSY was similar among different life history strategies (Fig. 2, 3, 4). Semelparous fish have slightly higher fishing mortality at MSY than short-lived iteroparous species, and short-lived iteroparous fish have slightly higher fishing mortality at MSY than long-lived iteroparous species. However, the differences are small, and the factor that is differentially affected by fishing mortality among different life history strategies is the equilibrium abundance of adults. If we only want to know the model-based prediction of fishing mortality rate at MSY, it is not necessary to incorporate life history of organisms into the analysis. The information on the compensatory capacity of a fish population is sufficient. However, this does not mean that we should fish at the level of model-based MSY because, when population abundance is suppressed to a very low level, other factors such as environmental fluctuation and depensatory processes may affect the population.

In general, fishing mortality rate at MSY is the additional mortality rate that achieves the maximum production of the population. Under the logistic equation (Schaffer) model, it happens to be the level that suppresses the equilibrium abundance to a half of the carrying capacity (the $0.5\ K$ level). However, under the two-stage model with the Beverton-Holt density dependence affecting a fertility term, mortality rates at $0.5\ K$ and MSY levels are different. The exception was when the adult natural mortality rate was 0.5 (Fig. 2, 3, 4), but it should be noted that this exception is specific to the model used in the study. Therefore, in general, it cannot be assumed that MSY is achieved when the population abundance is at the 0.5 $K$ level.

The maximum compensatory capacity ($\alpha$) of a population affects the sustainable yield of fish; the greater $\alpha$ is, the greater the maximum sustainable yield is (Fig. 2, 3, 4). This makes intuitive sense, because, if $\alpha$ is high, a reduction in abundance because of fishing mortality can be better compensated for, and in turn allows higher fishing mortality rate. The mortality rate at MSY is also affected more by $\alpha$ than other parameters. Furthermore, a model for competition between stage-structured populations suggests the importance of this parameter[37]. The current result re-emphasizes the importance of accurately estimating density-dependent parameters, e.g. [38]. Unfortunately, $\alpha$ is probably the parameter in the model that is most difficult to estimate from field observations.

The effects of juvenile survival and maturation rate appear to have synergetic effects. For example, when the juvenile survival or maturation rate is high, reducing the other population parameter has only a small effect on the sustainable yield (Fig. 2, 3, 4). When the maturation rate is low, individuals tend to remain immature for longer. However, the low maturation rate is compensated for by increased reproduction. Consequently, individuals will accumulate in the juvenile stage as long as juvenile survival is high, and greater abundance in the juvenile stage will result in enough number of individuals maturing each year to maintain a population level. Similarly, when juvenile survival is low, a smaller proportion of individuals will survive to maturity, but reduced survival is compensated for by increased reproduction. As long as the maturation rate is sufficiently high, the compensation is sufficient for enough number of individuals to mature each year to maintain equilibrium abundance. However, when juvenile survival or the maturation rate is low, reducing the other parameter has a pronounced effect on the sustainable yield. When both rates are low the population tends to be nonviable even though adults had high fertility rate, which is evidenced by the large grey areas in the figures.

The resilience of a population is an important population dynamics characteristic in fishery management (Fig. 8, 9, 10). The model suggests that fish with relatively low maximum compensatory capacity or those that are caught at levels above the MSY will have reduced resilience as fishing mortality is increased. This means that if there are additional mortalities caused by natural or anthropogenic events, these effects will last longer when the species are subject to high fishing mortality. This can have severe effects on economic sustainability of fisheries. Resilience is not a factor that is commonly incorporated into fishery management. I suggest it should be considered in future management decisions.

Resilience is also affected by life history traits, and fish stocks that exhibit the most resilience are those with an early maturing semelparous life history strategy and a high maximum compensatory capacity. Fish with such life history strategies will return to the equilibrium abundance more rapidly following a perturbation than stocks with a different life history strategy. Long-lived iteroparous species have the lowest resilience, but a change from short-lived iteroparous to long-lived iteroparous life history does not appear to change the resilience as much as the reduction that results from the change from delayed maturation to precocious reproductive strategies. This suggests that resilience is a quality of strategies associated with the juvenile stage. Thus, long-lived fish can have high resilience if they mature early and produce a large number of offspring. In order to determine the speed of recovery of over-exploited fish populations, we should focus on examining the early life-stage of the fish rather than how long adults can live.

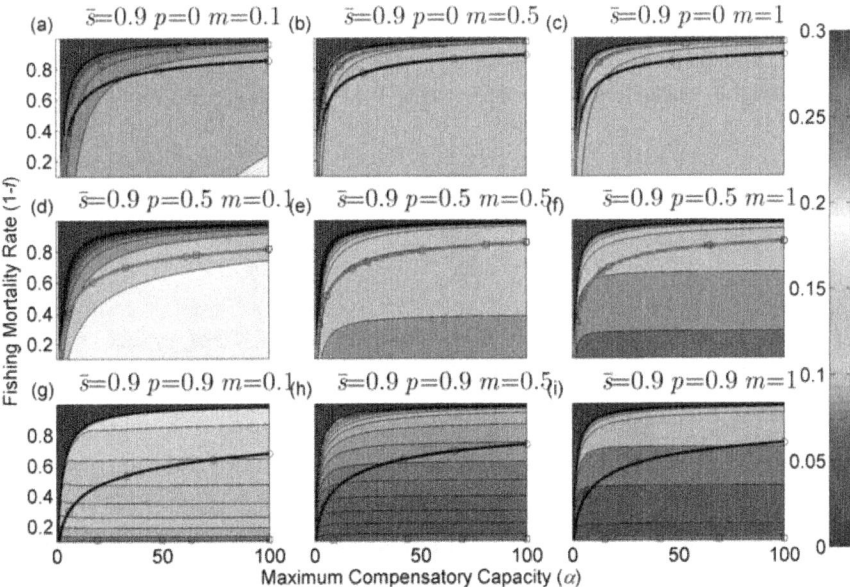

**Figure 11:** Coefficient of variation (CV) of adult abundance (shown in contours) as a function of the annual fishing mortality rate and the maximum compensatory capacity when juvenile survival fluctuates stochastically when $\bar{s} = 0.9$. The black and red curves are the same as shown in Figure 2. Each panel represents a life history strategy defined by the parameters shown above. doi:10.1371/journal.pone.0034556.g011

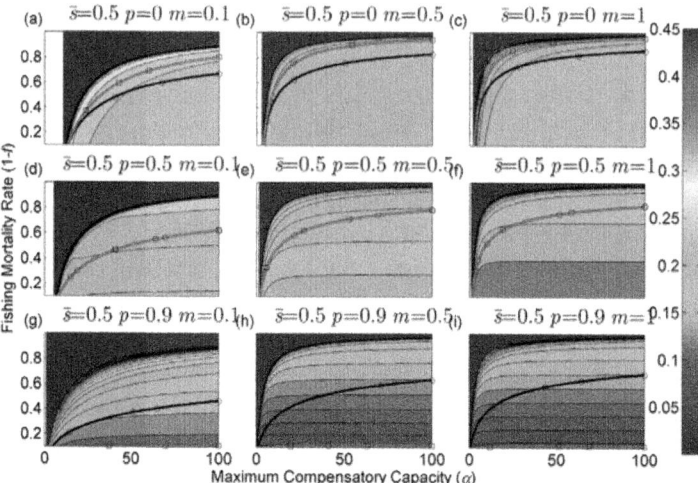

**Figure 12:** Coefficient of variation (CV) of adult abundance (shown in contours) as a function of the annual fishing mortality rate and the maximum compensatory capacity when juvenile survival fluctuates stochastically when $\bar{s} = 0.5$. The black and red curves are the same as shown in Figure 3. Each panel represents a life history strategy defined by the parameters shown above. doi:10.1371/journal.pone.0034556.g012

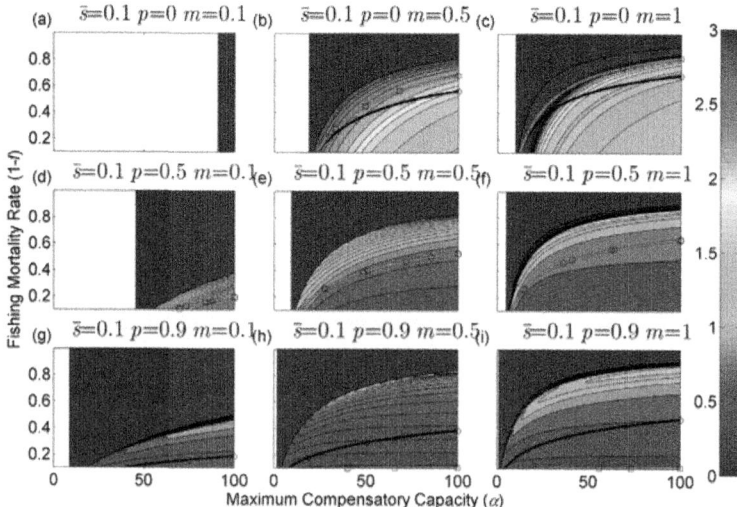

**Figure 13:** Coefficient of variation (CV) of adult abundance (shown in contours) as a function of the annual fishing mortality rate and the maximum compensatory capacity when juvenile survival fluctuates stochastically when $\bar{s} = 0.1$. The black and red curves are the same as shown in Figure 4. Each panel represents a life history strategy defined by the parameters shown above. doi:10.1371/journal.pone.0034556.g013

The CV of adult abundance always increased with increasing fishing mortality when a stochastic perturbation was introduced into the juvenile survival rate (Fig. 11, 12, 13). The results were somewhat surprising because I hypothesized that a stock with low resilience would have a higher autocorrelation in adult abundance, which would in turn increase the variance of the abundance. Contrary to this hypothesis, the CV of adult abundance always increased with increasing fishing mortality. Examination of the mean and standard deviation of the adult abundance showed that fishing mortality reduces the mean abundance, which reduces its standard deviation. If both were reduced at the same rate, the CV of adult abundance would remain the same, but the mean was reduced faster than the standard deviation. Consequently, the CV of adult abundance increased with increasing fishing mortality. Resilience is associated with the dominant eigenvalue of the linearized projection matrix, which measures a longer-term transient dynamics. Consequently, resilience appears to be a measure of intermediate time-scale dynamics, whereas the CV of adult abundance may be viewed as a measure of short time-scale dynamics. However, the CV is not necessarily a measure of how much a population fluctuates because it appears to be affected more by the equilibrium abundance.

An increase in the CV of adult abundance with increasing fishing mortality has been observed, and its potential cause was attributed to an increase in newly recruited individuals in the adult stage [21], which causes increased variance in the adult abundance. However, the results presented here show that an increase in the CV of adult abundance also occurs with semelparous fish, which comprise only newly recruited individuals in the adult stage. The cause of the increased CV of adult abundance is that the mean adult abundance is more sensitive than its standard deviation, and the sensitivity increases with increasing adult duration.

The analyses in this study were based on a model containing various assumptions (see *Model* section). A simple model was intentionally used to provide general insights into how demographic diversity affects the response of fish stocks to fishing mortality. The model did not include factors that many fishery biologists may consider important in understanding fish population dynamics. Amongst these are the effect of age and/or size on population parameters, autocorrelation in environmental fluctuations, depensatory processes under low population density (the Allee effect), other types of compensatory density-dependent processes, differential effects of fisheries on different size classes, changes in parameters caused by interactions among populations (e.g. predation and competition), and changes in population parameters resulting from rapid evolution as a response to fishing pressures. The results presented here should form the basis for further investigations on how these other factors might affect the optimal management of fish stocks.

## ACKNOWLEDGMENTS

I thank Mark Boyce and two anonymous reviewers for constructive comments on the previous version of the manuscript.

## AUTHOR CONTRIBUTIONS

Conceived and designed the experiments: MF. Performed the experiments: MF. Analyzed the data: MF. Contributed reagents/materials/analysis tools: MF. Wrote the paper: MF.

## REFERENCES

1.  Stearns SC (1992) The evolution of Life Histories: Oxford University Press.

2.  Adams PB (1980) Life-history patterns in marine fisheries and their consequences for fisheries management. Fishery Bulletin 78: 1–12.

**3.**  .Balon EK (1975) Reproductive guilds of fishes - proposal and definition. Journal of the Fisheries Research Board of Canada 32: 821–864.

4.  Winemiller KO (2005) Life history strategies, population regulation, and implications for fisheries management. Canadian Journal of Fisheries and Aquatic Sciences 62: 872–885.

5.  Winemiller KO, Rose KA (1992) patterns of life-history diversification in North-American fishes - implications for population regulation. Canadian Journal of Fisheries and Aquatic Sciences 49: 2196–2218.

6.  Goodwin NB, Grant A, Perry AL, Dulvy NK, Reynolds JD (2006) Life history correlates of density-dependent recruitment in marine fishes. Canadian Journal of Fisheries and Aquatic Sciences 63: 494–509.

7.  Pinsky ML, Jensen OP, Ricard D, Palumbi SR (2011) Unexpected patterns of fisheries collapse in the world's oceans. Proceedings of the National Academy of Sciences of the United States of America 208: 8317–8322.

8.  Williams EH, Shertzer KW (2003) Implications of life-history invariants for biological reference points used in fishery management. Canadian Journal of Fisheries and Aquatic Sciences 60: 1037–1037.

9.  Fromentin JM, Fonteneau A (2001) Fishing effects and life history traits: a case study comparing tropical versus temperate tunas. Fisheries Research 53: 133–150.

10. Shuter BJ, Abrams PA (2005) Introducing the symposium "Building on Beverton's legacy: life history variation and fisheries management". Canadian Journal of Fisheries and Aquatic Sciences 62: 725–729.

11.   Schindler DE, Hilborn R, Chasco B, Boatright CP, Quinn TP, et al. (2010) Population diversity and the portfolio effect in an exploited species. Nature 465: 609-U102.

12.   Brooks EN, Powers JE, Cortes E (2010) Analytical reference points for age-structured models: application to data-poor fisheries. Ices Journal of Marine Science 67: 165–175.

**13.**   .Hilborn R, Walters CJ (1992) Quantitative Fisheries Stock Assessment: Choice, Dynamics & Uncertainty. Boston, MA: Kluwer Academic Publishers.

14.   Quinn TJ, Deriso RB (1999) Quantitative Fish Dynamics. New York: Oxford University Press, Inc..

15.   Haddon M (2001) Modelling and Quantitative Methods in Fisheries. Boca Raton, FL: Chapman & Hall/CRC.

16.   USDOC (2007) Magnuson-Stevens Fishery Conservation and Management Act: As amended through January 12, 2007. In: Commerce USDO. PL 94–265 and 109–479:

17.   Cadrin SX, Pastoors MA (2008) Precautionary harvest policies and the uncertainty paradox. Fisheries Research 94: 367–372.

18.   Hastings A (2004) Transients: the key to long-term ecological understanding? Trends in Ecology & Evolution 19: 39–45.

19.   Wiedenmann J, Fujiwara M, Mangel M (2009) Transient population dynamics and viable stage or age distributions for effective conservation and recovery. Biological Conservation 142: 2990–2996.

20.   Worden L, Botsford LW, Hastings A, Holland MD (2010) Frequency responses of age-structured populations Pacific salmon as an example. Theoretical Population Biology 78: 239–249.

21.   Hsieh CH, Reiss CS, Hunter JR, Beddington JR, May RM, et al. (2006) Fishing elevates variability in the abundance of exploited species. Nature 443: 859–862.

22.   Anderson CNK, Hsieh CH, Sandin SA, Hewitt R, Hollowed A, et al. (2008) Why fishing magnifies fluctuations in fish abundance. Nature 452: 835–839.

23.   Hannesson R (1994) Bioeconomic analysis of fisheries. FAO Fisheries Report DN 0429–9337, no. 499. FAO, Rome, Italy.

24.   Beddington JR, Free CA, Lawton JH (1976) Concepts of stability and resilience in predator-prey models. Journal of Animal Ecology 45: 791–816.

25. Harrison GW (1979) Stability under environmental-stress - resistance, resilience, persistence, and variability. American Naturalist 113: 659–669.

26. DeAngelis DL (1980) Energy-flow, nutrient cycling, and ecosystem resilience. Ecology 61: 764–771.

27. Verhulst PF (1838) Notice sur la loi que la population poursuit dans son accroissement. Correspondance Mathématique et Physique 10: 113–121.

28. May RM (1976) Simple mathematical-models with very complicated dynamics. Nature 261: 459–467.

**29.** .Costantino RF, Cushing JM, Dennis B, Desharnais RA (1995) Experimentally-induced transitions in the dynamic behavior of insect populations. Nature 375: 227–230.

**30.** .Costantino RF, Desharnais RA, Cushing JM, Dennis B (1997) Chaotic dynamics in an insect population. Science 275: 389–391.

31. Ricker WE (1954) Stock and rescruitment. Journal of the Fisheries Research Board of Canada 11: 559–623.

**32.** .Beverton RJH, Holt SJ (1957) On the Dynamics of Exploited Fish Populations. : Ministry of Agriculture, Fisheries and Food, London (republished by Chapman & Hall in 1993):

33. Caswell H (2001) Matrix Population Models: Construction, Analysis, and Interpretation. Sunderland: Sinauer Associates, Inc..

34. Myers RA, Barrowman NJ (1996) Is fish recruitment related to spawner abundance? Fishery Bulletin 94: 707–724.

35. Jennings S, Kaiser MJ, Reynolds JD (2001) Marine Fisheries Ecology. Malden, MA: Blackwell Publishing.

36. Winemiller KO, Rose KA (1993) Why do most fish produce so many tiny offspring. American Naturalist 142: 585–603.

37. Fujiwara M, Pfeiffer G, Boggess M, Day S, Walton J (2011) Coexistence of competing stage-structured populations. Scientific Reports 1: DOI:10.1038/srep00107.

38. Rose KA, Cowan JH (2003) Data, models, and decisions in US Marine Fisheries management: Lessons for ecologists. Annual Review of Ecology Evolution and Systematics 34: 127–151.

# CITATION

## CHAPTER 1

Carlos Edwar de Carvalho Freitas, Alexandre A. F. Rivas, Caroline Pereira Campos, Igor Sant'Ana, James Randall Kahn, Maria Angélica de Almeida Correa and Michel Fabiano Catarino (2012). The Potential Impacts of Global Climatic Changes and Dams on Amazonian Fish and Their Fisheries, New Advances and Contributions to Fish Biology, Prof. Hakan Turker (Ed.), ISBN: 978-953-51-0909-9, InTech, DOI: 10.5772/54549.

## CHAPTER 2

Bimal Mohanty, Sasmita Mohanty, Jnanendra Sahoo and Anil Sharma (2010). Climate Change: Impacts on Fisheries and Aquaculture, Climate Change and Variability, Suzanne Simard (Ed.), ISBN: 978-953-307-144-2, InTech,

## CHAPTER 3

Selim Sekkin and Cavit Kum (2011). Antibacterial Drugs in Fish Farms: Application and Its Effects, Recent Advances in Fish Farms, Dr. Faruk Aral (Ed.), ISBN: 978-953-307-759-8, InTech, DOI: 10.5772/26919.

## CHAPTER 4

Tomislav Vladić (2011). Ejaculate Allocation and Sperm Competition in Alternative Reproductive Tactics of Salmon and Trout: Implications for Aquaculture, Recent Advances in Fish Farms, Dr. Faruk Aral (Ed.), ISBN: 978-953-307-759-8, InTech, DOI: 10.5772/26270.

# CHAPTER 5

Y. Hamed, Sh. Salem, A. Ali and A. Sheshtawi (2011). Environmental Effect of Using Polluted Water in New/Old Fish Farms, Recent Advances in Fish Farms, Dr. Faruk Aral (Ed.), ISBN: 978-953-307-759-8, InTech, DOI: 10.5772/27591.

# CHAPTER 6

Polovina JJ, Woodworth-Jefcoats PA (2013) Fishery-Induced Changes in the Subtropical Pacific Pelagic Ecosystem Size Structure: Observations and Theory. PLoS ONE 8(4): e62341. doi:10.1371/journal.pone.0062341

# CHAPTER 7

Cavit Kum and Selim Sekkin (2011). The Immune System Drugs in Fish: Immune Function, Immunoassay, Drugs, Recent Advances in Fish Farms, Dr. Faruk Aral (Ed.), ISBN: 978-953-307-759-8, InTech

# CHAPTER 8

David E. Bailey and Kristy A.N. Bulleit, "Defining "Adverse Environmental Impact" and Making § 316(b) Decisions: A Fisheries Management Approach," TheScientificWorldJOURNAL, vol. 2, pp. 147-168, 2002. doi:10.1100/tsw.2002.191

# CHAPTER 9

Fujiwara M (2012) Demographic Diversity and Sustainable Fisheries. PLoS ONE 7(5): e34556. doi:10.1371/journal.pone.0034556

# INDEX